Lecture Notes
in Physics

20

Statistical Mechanics and Mathematical Problems

[redacted]

Battelle Seattle 1971 Rencontres

Edited by A. Lenard, Indiana University

Springer-Verlag
Berlin · Heidelberg · New York 1973

In accord with that part of the charge of its founder, Gordon Battelle, to assist in the education of men, it is the commitment of Battelle to encourage the distribution of information. This is done in part by supporting conferences and meetings and by encouraging the publication of reports and proceedings. Towards that objective, this publication, while protected by copyright from plagiarism or unfair infringement, is available for the making of single copies for scholarship and research, or to promote the progress of science and the useful arts.

ISBN 3-540-06194-0 Springer-Verlag Berlin · Heidelberg · New York
ISBN 0-387-06194-0 Springer-Verlag New York · Heidelberg · Berlin

Offsetprinting and bookbinding: Julius Beltz, Hemsbach/Bergstr.

FORWORD

This volume is the record of the lectures delivered at the Rencontres in Mathematics and Physics during the summer of 1971.

The Rencontres, organized and supported by the Battelle Memorial Institute, have as their aim the encouragement of fruitful and active collaboration by Mathematicians and Physicists in selected areas of common concern.

The maturing and growth of modern mathematical physics is one of the striking intellectual developments of the last two decades. Many beautiful results were produced, and many new areas for the application of modern mathematical methods were found. Even more importantly, mathematical physics fosters a cooperative and unifying spirit between practitioners of different areas of expertise. Within mathematical physics statistical mechanics was - and remains - one of the areas most profiting from this development. Statistical mechanics offers many problems for mathematical analysis, with an impressive range in the nature of the questions asked, in the level of generality, and in the degree of difficulty. It has many ties to other subjects besides physics and mathematics.

It was natural, then, to select statistical mechanics as the central theme for the 1971 Rencontres.

The formal program consisted of a number of lecture series, each an exploration in depth of a major topic. The participants met twice a day to hear the lectures. Besides these, there were informal seminars, meetings devoted to the discussion of new developments, active personal collaboration in research, and last but not least, that elusive (but most important) activity of exposing oneself to the ideas of others in an unhurried and creative atmosphere.

Neither could there be, nor would it have been desirable to have, a narrowly conceived topical unity in the main lectures. Each contributes to the clarification of some significant problem, or range of problems. All of them are mathematical in nature (what in older days used to be called, somewhat redundantly, "rigorous results").

The lectures of Oscar Lanford deal with an explanation scheme for the classical statistical mechanics of thermal equilibrium. It has always been "known," but rarely discussed precisely, that "macrosopic quantities do not fluctuate." The recognition of this, crucial to the whole rationale for statistical mechanics, goes back to the founders of the subject, notably Gibbs, but it was only in recent years that precise formulation of these facts were given. Lanford's lectures contain a systematic discussion of this issue, together with its connection with the classical ways of describing thermal equilibrium from a microscopic point of view. The main technical tool is the concept of entropy which is used here in a sense more general, as well as more precise, than is customary. The novelty of the treatment is the idea of attaching an entropy notion not to a state of the system but rather to any macroscopic quantity ("finite range observable"). This entropy concept is related, in turn, to the more usual concepts, and its role in the explanation scheme is explored.

The Lectures of Joel Lebowitz and Elliott Lieb deal with the problem of proving the limiting behaviour (thermodynamic limit) of quantum systems of many particles interacting only through a Coulomb force. The fascination in this problem consists in the fact that in nature we see, almost exclusively, the manifestations of the Coulomb force, yet mathematically this force law is particularly inconvenient in trying to prove theorems. The goal is achieved by modifying in an ingenious manner some of the conventional methods necessary to deal with the problem of thermodynamic limits. In all problems of this kind there are two basic ingredients, a "monotonicity" problem and a "lower boundedness problem". The monotonicity problem appears in a particularly sophisticated garb in the work of Lebowitz and Lieb. The boundedness problem ("stability") is taken care of by the theorem of Dyson and Lenard. The lectures of Andrew Lenard are devoted to a somewhat stream-lined version proof of this theorem.

Although not unique anymore as a non-trivial soluble problem, the Ising model still attracts continuing interest as a testing ground for general ideas. The classic exact solution of Onsager refers only to the macroscopic equation of state, but there are many interesting questions about the model which cannot be answered in equally explicit terms. The lectures of Giovanni Gallavotti, Anders Martin-Löf and Salvador Miracle-Solé deal with one of these: the statistical structure of the state in the two-phase coexistence region. The analysis of this problem, initiated by Sinai and Minlos, is pursued here in great detail. One outcome of the analysis is that - as expected on intuitive grounds - there are exactly two pure phases for this model. Furthermore, it appears that the two phases are (with probability 1) well separated by a single line whose statistical properties admit a detailed analysis.

The last contribution to this volume is the set of lectures by Masamichi Takesaki. These are concerned with some purely mathematical problems in the theory of operator algebras. It is a truly remarkable fact that the central concept (the modular operator and associated automorphism group) which is so crucial in bringing out the symmetry in the theory, has a natural interpretation, and in fact has been discovered independently, in the physical context of quantum statistical mechanics. The ideas of Tomita, following a long line of work by mathematicians, fit in perfectly with the analysis of the so-called KMS-condition by Haag, Hugenholtz and Weenink. States satisfying this condition, as well as the associated representations of the operator algebras, show a certain symmetry whose existence has far reaching consequences. The Takesaki lectures are devoted to the exposition of this theory to which the lecturer himself contributed significantly.

It is hoped that the publication of this volume will contribute constructively to the enterprise of mathematical physics by making the lectures available to a wider circle of readers. The lectures were delivered in an enthusiastic spirit; if the care with which the manuscripts were prepared communicates itself to the readers of this volume, then its purpose will be well served.

Speaking for the participants at the Rencontres, the editor wishes to thank the lecturers for their contribution, and Professor David Ruelle for organizing a number of stimulating seminars. He wishes to acknowledge with gratitude the generous support of the Battelle Memorial Institute, and the fine facilities of their Seattle Research Center. Thanks are due to Dr. F. J. Milford for his cooperation in the organization of the Rencontres; to the staff of the Center for general helpfulness; and to the Springer-Verlag Publishing Company for accepting responsibility for publication. The editor is also indebted for the secretarial help received in the Mathematics Department of Indiana University, and in particular for the competent typing of Miss Betty Gehrke. Final typing in preparation of the manuscripts for publication was done at Battelle Seattle Research Center by Mrs.Vera Swile, Mrs. Lorraine Pritchett and Mrs.Alta Zapf.

1972
Bloomington, Indiana

A. Lenard

TABLE OF CONTENTS

OSCAR E. LANFORD III - Entropy and Equilibrium States in Classical Statistical Mechanics

A. Thermodynamic Limits
1. Introduction and Preliminaries 1
2. Thermodynamic Limits along Special Sequence of Boxes 14
3. Thermodynamic Limits along More General Sequences of Regions 28
4. A Digression: Sums of Independent Random Variables 35
5. The Canonical Ensemble 49
6. The Grand Canonical Ensemble 69

B. Invariant Equilibrium States
1. Preliminaries 77
2. States of Infinite Systems 82
3. Maximization of the Entropy 87
4. Invariant Equilibrium States 91
5. Tangents to the Graph of the Pressure 95

C. Gibbs States
1. Definitions and General Properties 100
2. Dobrushin's Uniqueness Theorem 107

A. LENARD - Lectures on the Coulomb Stability Theorem
1. Introduction 114
2. The Problem 114
3. NTC-Inequalities 116
4. More Inequalities 120
5. The Electrostatic Inequality 123
6. Cubical Partition of Space 124
7. Lower Bound for Energy in Terms of K_1 126
8. An Inequality for K_1, K_2, and T 128
9. Antisymmetric Functions 130
10. The Role of Antisymmetry: An Inequality for T and K_2 132
11. Completion of the Proof 134

E.H. LIEB and J.L. LEBOWITZ - Lectures on the Thermodynamic Limit for Coulomb Systems
1. Introduction 136
2. Packing a Domain with Balls 145
3. Thermodynamic Limit for Spherical and General Domains 147
4. Systems with Net Charge 151
5. Grand Canonical Ensemble 154
6. The Microcanonical Ensemble for Neutral Systems 158

G. GALLAVOTTI, A. MARTIN-LÖF, and S. MIRACLE-SOLÉ. - Some Problems Connected with the Description of Coexisting Phases at Low Temperatures in the Ising Model
Introduction 162
1. Notations and Definitions 164
2. The Translationally Invariant Equilibrium States 166
2A. Appendix 171
3. Description of the Phase Separation and Definition of the Surface Tension 174

4. Cluster Theory for a Pure Phase 176
5. The Phase Separation 188
5A. Appendix 192
6. The Surface Tension 198
7. Concluding Remarks 202

M. TAKESAKI - States and Automorphisms of Operator Algebras. Standard Representations and the Kubo-Martin-Schwinger Boundary Condition
Introduction 205
1. A von Neumann Algebra with a Cyclic and Separating Vector 208
2. The Polar Decomposition of the Involution 210
3. Bounded Elements 214
4. The Resolvent of the Modular Operator 215
5. The One-Parameter Automorphism Group Defined by the Modular Operator 221
6. The Kubo-Martin-Schwinger Boundary Condition 228
7. The Conditional Expectations 229
8. The Radon-Nikodym Theorems 232
9. Notes 243

List of Attendees 247

ENTROPY AND EQUILIBRIUM STATES IN CLASSICAL STATISTICAL MECHANICS

Oscar E. Lanford III *

A. THERMODYNAMIC LIMITS

A1. INTRODUCTION AND PRELIMINARIES

The objective of statistical mechanics is to explain the macroscopic properties of matter on the basis of the behavior of the atom and molecules of which it is composed. One of the most striking facts about macroscopic matter is that in spite of being fantastically complicated on the atomic level — to specify the positions and velocities of all molecules in a glass of water would mean specifying something of the order of 10^{25} parameters — its macroscopic behavior is describable in terms of a very small number of parameters, e.g., the temperature and density for a system containing only one kind of molecule. We will begin these lectures by describing an explanation-scheme to account for this fact. The term explanation-scheme is intended to convey the fact that what is outlined here is not a finished piece of work — there are many difficult theorems to be proved before we can be sure that the proposed explanation is correct, and it will surely be necessary to modify the technical details of the formulation — but rather a set of ideas which seem to me to offer the most promising avenue for understanding the effectiveness of statistical mechanics for computing the properties of matter. My discussion will draw heavily on ideas developed by Ruelle in an important but little known paper [10].

The explanation I want to consider is based on two main ideas:

(a) We look only at observables of a special kind; roughly speaking, those which test correlations between particles which are not too far apart. This description is a little obscure; we will give a more precise one in a moment.

(b) We consider the behavior of the system as the number of particles becomes very large.

The spirit of our approach is to try to argue that, for a very large system, all observables are "essentially" determined by the energy and the density. By this we mean that, if we consider systems of fixed density but of increasing size, and hence of increasing number of particles, the probability distribution of each observable (approximately normalized) with respect to Lebesgue measure on each energy surface,

---

* Department of Mathematics, University of California, Berkeley, California. Alfred P. Sloan Foundation Fellow. Preparation for this article began while the author was exchange professor at the Centre Universitaire de Marseille-Luminy, and completed at the University of California with partial financial support from NSF grant GP-15735.

approaches a delta function, i.e., the value of the observable is very near some equilibrium value for all but a very small fraction of the points on the energy surface, this fraction becoming arbitrarily small as the system becomes very large.

Once the somewhat fuzzy description given in the preceding paragraph is made precise, and adjusted to take into account the existence of phase transitions, it becomes a purely mathematical problem (and perhaps not a totally hopeless one) to prove that with very high probability any observable is near its equilibrium value, probability always being computed with respect to Lebesgue measure in configuration (or phase) space. It is a much more profound problem to understand why events which are very improbable with respect to Lebesgue measure do not occur in nature. I, unfortunately, will have nothing to say about this latter problem.

We now specify more precisely the sort of observables we will consider. To start with, we will restrict ourselves to observables depending only on the positions of the particles we are considering. Removing this restriction would not create any serious difficulties at all, but it would complicate the notation slightly. The problem of eliminating it is left as a (straightforward) exercise. To motivate the definition of observable, we think of the potential energy $U(q_1,\ldots,q_n)$. This is a sequence of functions of an indefinite number of variables, or, to be slightly more elegant, a function on

$$\bigcup_n (\mathbb{R}^\nu)^n$$

with some special properties:

a. For each n, $U(q_1,\ldots,q_n)$ is a symmetric function of $q_1,\ldots,q_n$.
b. For each n, $U(q_1+a,\ldots,q_n+a) = U(q_1,\ldots,q_n)$ for all $a \in \mathbb{R}^\nu$.

(Here ν denotes the number of dimensions of the space in which the particles move; we are of course most interested in $\nu = 3$.) There is another essential property of physically reasonable interactions: particles which are far apart do not interact very strongly. In other words, if we have two clusters of particles, say $q_1, \ldots, q_n$ and $q_1', \ldots, q_m'$, and if all particles in the first cluster are far from all particles in the second cluster, then

$$U(q_1,\ldots,q_n,q_1',\ldots,q_m') \approx U(q_1,\ldots,q_n) + U(q_1',\ldots,q_m') \quad .$$

We will make the simplest (and most restrictive) version of this assumption and assume that we have exact equality.

Thus, we define a *finite range observable** to be a function f on $\bigcup_{n=0}^{\infty} (\mathbb{R}^\nu)^n$

---

* The use of the word "observable" in this restricted sense (note that the number of particles in a given region of space is not an observable) is not intended to have any profound significance; it is merely a convenient term for a class of functions we want to consider. The reader who prefers a more neutral terminology is invited to substitute the expression "correlation quantity" for "observable."

(i.e., a function $f(q_1,\ldots,q_n)$ on $(\mathbb{R}^\nu)^n$ for each n) satisfying:

(a) (Continuity): For each n, $f(q_1,\ldots,q_n)$ is a continuous function on $(\mathbb{R}^\nu)^n$,

(b) (Symmetry): For each n, $f(q_1,\ldots,q_n)$ is a symmetric function,

(c) (Translation invariance): For each n, and each $a \in \mathbb{R}^\nu$, $f(q_1 + a,\ldots,q_n + a) = f(q_1,\ldots,q_n)$,

(d) (Normalization): $f(q_1) = 0$

(e) (Finite range): There exists a real number $R < \infty$ such that, if

$$|q_i - q_j'| > R \quad \text{for all } i,j \quad ,$$

then

$$f(q_1,\ldots,q_n,q_1',\ldots,q_m') = f(q_1,\ldots,q_n) + f(q_1',\ldots,q_m') \quad .$$

(The smallest such R will be called the *range* of the observable.) Let us mention some examples. If we consider an interaction $U(q_1,\ldots,q_n)$ specified, say, by a finite-range continuous two-body potential Φ (i.e., Φ is a continuous function of compact support on $\mathbb{R}^\nu$ and $U(q_1,\ldots,q_n) = \frac{1}{2}\sum_{i\neq j}\Phi(q_i - q_j)$ for all n), then $U(q_1,\ldots,q_n)$ is a finite-range observable. A second example is the number of pairs of particles with separation less than some preassigned distance R_0. (This function is not quite a finite-range observable since it is not continuous). A more general class of examples can be constructed in the following way: Start with a continuous function $h(q_2,\ldots,q_j)$ of compact support in $(\mathbb{R}^\nu)^{j-1}$. If we define

$$(\Sigma h)(q_1,\ldots,q_n) = \frac{1}{j!}\sum_{i_1,\ldots,i_j}{}' h(q_{i_2} - q_{i_1}, q_{i_3} - q_{i_1},\ldots,q_{i_j} - q_{i_1}) \quad .$$

where $\sum'_{i_1,\ldots,i_j}$ means the sum over all j-tuples of distinct indices, then Σh is a finite-range observable. (We will refer to an observable of this special form as a *j-particle observable* or a *j-particle correlation quantity*.)

The definition of observable we have given is not only philosophically restrictive but also technically restrictive. In particular, the potential energy defined by a two-body potential Φ is not an observable unless Φ is continuous and of compact support. Since we will want to treat the potential energy as an observable, this means that our definition rules out most interesting interactions. These restrictions are, in fact, not really necessary. The assumption that observables are continuous functions rather than general Borel functions is used at only one place in the argument we are going to give, and we have made it only to avoid cluttering the reasoning at that point. The assumption that observables (and, in particular, the interaction) have strictly finite range is more serious; to replace it by the assumption that the interaction drops off reasonably quickly at infinity would involve complicating many of the proofs enormously but without introducing many new ideas. In the interest of trying to make the essential ideas as clear as possible, we have chosen to consider only finite-range interactions; the reader who wants to see what is involved in weakening this restriction is referred to Chapter 3 of Ruelle [9].

We would now like to do the following: we fix an interaction and choose another observable f. We choose also two numbers ε (the energy per particle) and ρ (the density). We consider a "large" region Λ in $\mathbb{R}^\nu$ and an integer N such that $\frac{N}{V(\Lambda)} \approx \rho$. In Λ^N, the configuration space for N particles in Λ, we consider the "surface" where $\frac{U(q_1,\dots,q_N)}{N} = \varepsilon$. Ignoring technicalities about possible singularities of this surface, we consider "normalized surface measure" and compute the distribution of the observable f/N with respect to this surface measure. We then let Λ become bigger and bigger, allowing N to increase also so that $\frac{N}{V(\Lambda)}$ approaches ρ, but holding ε fixed. If, as we do this, we find that the distribution for f/N approaches a δ-function, then we can say that, for large systems, the value of the observable f is "essentially" determined by the energy per particle ε and the density ρ, and we have obtained a reasonably satisfactory justification for the empirical fact that the observed quantity f/N always has the same value for a system with a given density and energy per particle.

The program described schematically above is a little beyond what present techniques allow us to handle. In particular, it is much harder to prove results about a single value of the energy per particle than for a range of values. Since the range of values can be as narrow as we like, this is not completely unsatisfactory from a conceptual point of view. We will choose, therefore, an open interval I_ε of values for the energy per particle, and replace the energy surface

$$\{(q_1,\dots,q_N)\varepsilon\Lambda^N: \frac{U(q_1,\dots,q_N)}{N} = \varepsilon\}$$

in the above discussion with the thickened energy surface

$$\{(q_1,\dots,q_N)\varepsilon\Lambda^N: \frac{U(q_1,\dots,q_N)}{N} \in I_\varepsilon\} \quad ,$$

equipped with normalized Lebesgue measure. To study the distribution of f with respect to this measure, it is convenient to define:

$$\mathcal{V}(\Lambda,N,f,J) = \frac{1}{N!} \times \text{Lebesgue measure of}$$

$$\{(q_1,\dots,q_N)\varepsilon\Lambda^N: \frac{U(q_1,\dots,q_N)}{N} \in I_\varepsilon$$

$$\text{and } \frac{f(q_1,\dots,q_N)}{N} \in J\} \quad .$$

For the purposes of this definition, Λ is any bounded open (or even Borel) set in $\mathbb{R}^\nu$, N is any integer, and J is any open interval. The probability, with respect to normalized Lebesgue measure on the thickened energy surface, of finding f/N in the interval J is then given by:

$$\frac{V(\Lambda,N,f,J)}{V(\Lambda,N,f,(-\infty,\infty))}$$

(the factors of $1/N!$, introduced for convenience later on, cancel). We will investigate the distribution of f/N as Λ and N become large by studying the asymptotic behavior of $V(\Lambda,N,f,J)$. This will be a long and involved process, and it seems useful, before starting on the details of the argument, to sketch the broad outline. We will do this in a sloppy way in that some of the assertions we make will be subject later on to technical modifications; thus, the statements in the next few paragraphs should not be taken completely literally.

In outline, then, the asymptotic behavior of $V(\Lambda,N,f,J)$ as Λ and N become large in such a way that $\frac{N}{V(\Lambda)}$ approaches ρ is either

(a) $V(\Lambda,N,f,J)$ goes to zero faster than exponentially in N

or

(b) $V(\Lambda,N,f,J) \sim e^{Ns(\rho,f,J)}$, where $s(\rho,f,J)$ does not depend on Λ and N except through the ratio $\frac{N}{V(\Lambda)} = \rho$. To be more precise, we are asserting the existence of a limit of

$$\frac{1}{N} \log V(\Lambda,N,f,J)$$

as Λ and N become large in such a say that $\frac{N}{V(\Lambda)}$ approaches ρ; we will denote the limit of $s(\rho,f,J)$. It is $-\infty$ if (a) holds and otherwise has some finite value. The sense in which Λ becomes large will be made precise later; for the moment, it suffices to think of the limit as being taken along any sequence of bounded open sets Λ_n with, say, piecewise smooth surfaces, such that the surface area of Λ_n does not grow faster than $[V(\Lambda_n)]^{\frac{\nu-1}{\nu}}$. In particular, the existence of the limit implies that, to first approximation, $V(\Lambda,N,f,J)$ does not depend on the shape of Λ.

Note that the limit $s(\rho,f,J)$ gives only very crude information about $V(\Lambda,N,f,J)$; this quantity can change by any multiplicative factor varying slower than exponentially with N without changing the limit. Conversely, if we have two intervals J_1 and J_2 such that

$$s(\rho,f,J_1) > s(\rho,f,J_2)$$

then the ratio of the probabilities of finding f/N in J_1 and in J_2 is

$$\frac{V(\Lambda,N,f,J_2)}{V(\Lambda,N,f,J_1)}$$

which behaves asymptotically like

$$\exp\{N[s(\rho,f,J_2) - s(\rho,f,J_1)]\}$$

and therefore goes to zero exponentially as $N,\Lambda \to \infty$. In particular, if $s(\rho,f,J) < s(\rho,f,(-\infty,\infty))$, then the probability that f/N is in J is extremely small for large systems. These remarks might give rise to the suspicion that $s(\rho,f,J)$ actually does not depend on J; this, unfortunately, does sometimes happen, but it will become clear in the course of the development of the theory that in most interesting situations there is enough dependence on J to allow us to draw physically interesting conclusions. The choice of the letter s for the limit of $\frac{1}{N} \log V$ is supposed to suggest a connection between this limit and the thermodynamic entropy. There is indeed such a connection; $s(\rho,f,(-\infty,\infty))$ is the configurational entropy per particle for a system with the specified interaction, density ρ, and potential energy per particle in I_ε. The reasons for this interpretation will be explained later.

Bypassing for the moment the problem of proving the existence of the limit defining s, we want now to analyze the dependence of s on J. To simplify the writing of the formulas, we will suppress ρ and f from the notation (as we have already suppressed the interaction and the interval I_ε of allowed energies per particle.) Thus, we consider $s(J)$ as a set function defined on the open intervals. It is nearly evident that this set function is increasing. If $J \subset J'$, then

$$V(\Lambda,N,J) \le V(\Lambda,N,J') \quad ,$$

so

$$s(J) \le s(J') \quad .$$

In fact, we have a stronger property. Let $J = J' \cup J''$ (where J, J' and J'' are all open intervals). Then

$$V(\Lambda,N,J') \vee V(\Lambda,N,J'') \le V(\Lambda,N,J)$$

$$\le V(\Lambda,N,J') + V(\Lambda,N,J'') \le 2[V(\Lambda,N,J') \vee V(\Lambda,N,J'')] \quad .$$

(The middle inequality comes from the fact that, except for a factor $\frac{1}{N!}$, $V(\Lambda,N,J)$ is the Lebesgue measure of the set of configurations with f/N in J which is the union of the sets of configurations with f/N in J' and in J''). Taking the logarithm and dividing by N gives:

$$\frac{1}{N} \log[V(\Lambda,N,J') \vee V(\Lambda,N,J'')] \le \frac{1}{N} \log[V(\Lambda,N,J)]$$

$$\le \frac{1}{N} \log 2 + \frac{1}{N} \log[V(\Lambda,N,J') \vee V(\Lambda,N,J'')] \quad .$$

When we pass to the limit $N,\Lambda \to \infty$, the $\frac{1}{N} \log 2$ drops out and we get:

$$s(J') \vee s(J'') \le s(J' \cup J'') \le s(J') \vee s(J''), \text{ i.e.,}$$

$$s(J' \cup J'') = s(J') \vee s(J'') \quad . \tag{*}$$

Thus, the set function $s(J)$, although constructed from the additive set function

$V(J)$, is the opposite of additive. One way of constructing a set function $s(J)$ satisfying (*) is to start from a point function $s(x)$ and define

$$s(J) = \sup_{x \in J} s(x) \quad .$$

The property (*) alone does not suffice to prove that $s(J)$ is of this form. We will need a continuity property of $s(J)$, which we state here without proof, and which we will have to verify when we prove the existence of the limit defining s:

For any open interval J, $s(J) = \sup\{s(\hat{J})$: $\hat{J}$ a bounded open interval with closure contained in $J\}$. (+)

We now need:

Lemma A1.1.

Let $s(J)$ *be a set function defined on open intervals in* $\mathbb{R}$ *satisfying* (*) *and* (+). *Define, for each* $x \in \mathbb{R}$,

$$s(x) = \inf\{s(J): J \text{ an open interval containing } x\}$$

Then $s(x)$ *is upper semi-continuous and*

$$s(J) = \sup\{s(x): x \in J\} \quad .$$

Proof: The semi-continuity is a complete triviality: If J is any open interval containing x, and if (x_n) is a sequence converging to x then, for large n, $x_n \in J$ so $s(x_n) \le s(J)$. Thus,

$$\limsup_n s(x_n) \le s(J) \quad .$$

This is true for all open J containing x, so

$$s(x) \ge \limsup_n s(x_n) \quad ,$$

which implies that $s(x)$ is upper semi-continuous. To prove that

$$s(J) = \sup\{s(x): x \in J\} \quad ,$$

we first remark that, from the definition of $s(x)$,

$$s(J) \ge \sup\{s(x): x \in J\} \quad .$$

Assume, therefore, that

$$s(J) > \sup\{s(x):\ x \in J\} + \varepsilon \quad \text{for some} \quad \varepsilon > 0 \ .$$

Then, by (+), we can find a bounded interval $\hat{J}$, whose closure is contained in J, such that

$$s(\hat{J}) > \sup\{s(x):\ x \in J\} + \varepsilon$$

Now, for each $x \in J$, we can choose an interval J_x, containing x and contained in J, such that

$$s(J_x) < s(x) + \varepsilon \leq \sup\{s(x):\ x \in J\} + \varepsilon$$

Cover $\hat{J}$ with a finite set of such J_x's, i.e.,

$$\hat{J} \subset J_{x_1} \cup \ldots \cup J_{x_n} \ .$$

Then, by (*),

$$s(\hat{J}) \leq \sup_i\{s(J_{x_i})\} < \sup\{s(x):\ x \in J\} + \varepsilon \ ,$$

and this contradicts the choice of $\hat{J}$.

The interpretation of the function $s(x)$ should now be fairly clear. If J is some very small ("infinitesimal") interval containing x, then, as $\Lambda, N \to \infty$

$$V(\Lambda, N, J) \approx e^{Ns(x)}$$

Of course, if we fix Λ, N, and let the interval J shrink to the point x, we will usually have

$$V(\Lambda, N, J) \to 0$$

so it is essential to take the limit $\Lambda, N \to \infty$ before shrinking the interval J to x. Note also that, if

$$s(x) < s((-\infty,\infty)) = \sup_x\{s(x)\} \ ,$$

then the probability of finding f/N near x is very small when the system is large.

So far, we do not really know very much about the function $s(x)$. The key to analyzing its behavior is the following:

<u>Theorem A1.2.</u>

$s(x)$ *is a concave function of* x, *i.e.*,

$$s(\alpha x + (1 - \alpha)x_2) \geq \alpha s(x_1) + (1 - \alpha)s(x_2)$$

for all

$$x_1, x_2 \in \mathbb{R},\ 0 \le \alpha \le 1 \quad .$$

Remark: Although we have not been emphasizing the fact, it should be kept in mind that $s(x)$ can take on the value $-\infty$. (We will see shortly, however, that it is bounded above.) If $s(x_1) = -\infty$, the concavity inequality is vacuously satisfied.

Proof: The heart of the matter is to prove that

$$s(\tfrac{1}{2} x_1 + \tfrac{1}{2} x_2) \ge \tfrac{1}{2} s(x_1) + \tfrac{1}{2} s(x_2) \tag{1}$$

Then the concavity inequality is proved for dyadic rational α by induction, and for real α by the upper semicontinuity of $s(x)$.

To prove (1), let Λ_n be a sequence of cubes with sides tending to infinity, and let N_n be a sequence of integers such that $\frac{N_n}{V(\Lambda_n)} \to \rho$. Let R be the larger of range of the interaction and the range of the observable f being considered, and Λ'_n be a sequence of rectangular boxes made by placing two translates of Λ_n beside each other with a separation of at least R.

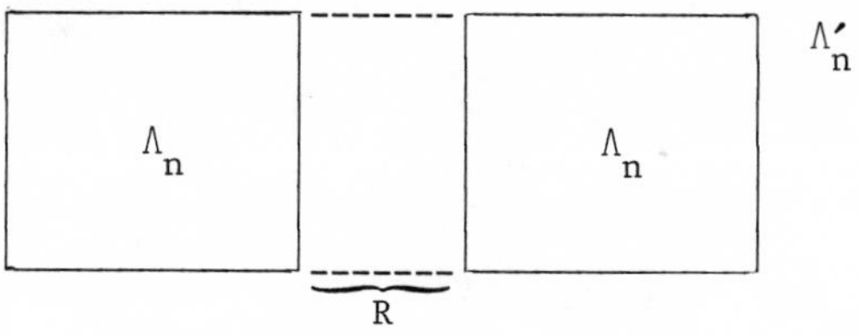

Note that

$$\lim_{n\to\infty} \frac{2N_n}{V(\Lambda'_n)} = \rho$$

We will show that, for any two intervals J_1 and J_2,

$$V(\Lambda'_n, 2N_n, \tfrac{1}{2} J_1 + \tfrac{1}{2} J_2) \ge V(\Lambda_n, N_n, J_1) V(\Lambda_n, N_n, J_2) \quad .$$

Once we have done this we are essentially through; taking the logarithm of this inequality and dividing by $2N_n$ gives

$$\frac{1}{2N_n} \log V(\Lambda'_n, 2N_n, \tfrac{1}{2} J_1 + \tfrac{1}{2} J_2) \ge \frac{1}{2} \frac{1}{N_n} \log V(\Lambda_n, N_n, J_1) + \frac{1}{2} \frac{1}{N_n} \log V(\Lambda_n, N_n, J_2) \quad .$$

Passing to the limit $n \to \infty$ gives

$$s(\tfrac{1}{2} J_1 + \tfrac{1}{2} J_2) \ge \tfrac{1}{2} s(J_1) + \tfrac{1}{2} s(J_2) \quad .$$

Letting J_1 and J_2 shrink to the points x_1 and x_2 proves

$$s(\frac{1}{2} x_1 + \frac{1}{2} x_2) \geq \frac{1}{2} s(x_1) + \frac{1}{2} s(x_2)$$

and thus proves the theorem. To prove

$$V(\Lambda'_n, 2N_n, \frac{1}{2} J_1 + \frac{1}{2} J_2) \geq V(\Lambda_n, N_n, J_1) V(\Lambda_n, N_n, J_2)$$

we note the following: Since the two translates of Λ_n contained in Λ'_n are separated by a distance R, one way of making a configuration of $2N_n$ particles in Λ'_n with $\frac{U}{2N_n} \in I_\varepsilon$ and $\frac{f}{2N_n} \in \frac{1}{2} J_1 + \frac{1}{2} J_2$ is to put N_n of the particles in each of the translates of Λ_n, with those in one of the translates arranged with $\frac{U}{N_n} \in I_\varepsilon$ and $\frac{f}{N_n} \in J_1$ and those in the other translate arranged with $\frac{U}{N_n} \in I_\varepsilon$ and $\frac{f}{N_n} \in J_2$. For each choice of which N_n particles to put in which translate we get a contribution to the configuration space volume equal to

$$[(N_n)!]^2 [V(\Lambda_n, N_n, J_1) V(\Lambda_n, N_n, J_2)]$$

There are $\frac{(2N_n)!}{(N_n!)^2}$ ways of choosing which particles to put in which translate, and different choices give disjoint regions of $(\Lambda'_n)^{2N_n}$, so we get

$$V(\Lambda'_n, 2N_n, \frac{1}{2} J_1 + \frac{1}{2} J_2)$$

$$\geq \frac{1}{(2N_n)!} \frac{(2N_n)!}{(N_n!)^2} (N_n!)^2 \; V(\Lambda_n, N_n, J_1) V(\Lambda_n, N_n, J_2)$$

$$= V(\Lambda_n, N_n, J_1) V(\Lambda_n, N_n, J_2) \quad ,$$

as desired.

From the fact that $s(x)$ is concave, it follows that $\{x: \; s(x) > -\infty\}$ is an interval (which may be open, half-open, or closed and which may be finite, semi-finite or all of $\mathbb{R}$.) If we let J denote the interior of this interval, then the following two lemmas imply that $s(x)$ is continuous on J:

Lemma A1.3.

$$s(x) \leq 1 - \log \rho \quad \textit{for all} \quad x$$

Lemma A1.4.

A function which is concave, finite, and bounded above on a convex open set in Euclidean space is continuous on that set.

<u>Proof of Lemma A1.3</u>. Trivially, $\mathcal{V}(\Lambda,N,f,J) \leq \frac{V(\Lambda)^N}{N!}$ for all Λ,N,f,J (since the right-hand side is $\frac{1}{N!}$ times the total configuration space volume). By Stirling's formula

$$\frac{1}{N} \log \mathcal{V}(\Lambda,N,f,J) \leq \log[e \frac{V(\Lambda)}{N}] + (1) \quad ;$$

passing to the limit $N,\Lambda \to \infty$, then to the limit $J \downarrow x$, gives the lemma.

<u>Proof of Lemma A1.4</u>. A concave function on an open set in Euclidean space is automatically locally bounded below (i.e., bounded below on a neighborhood of each point). This is true since, if $\xi_1, \ldots, \xi_j$ are points of the domain whose convex hull has ξ_0 in its interior, then $h(\xi) \geq \inf_i\{h(\xi_i)\}$ on the convex hull, which is by assumption a neighborhood of ξ_0. Thus, proving the lemma reduces to showing that a function which is bounded and concave on an open set containing 0 is continuous at 0. This is true even in Banach spaces; the proof is as follows: Suppose $h(x)$ is concave and $|h(x)| \leq M$ on $|x| < r_0$. Then, for $|x| < \frac{r_0}{3n}$; $|y| < \frac{r_0}{3n}$, we have:

$$h(y) \geq (1 - \frac{1}{n})h(x) + \frac{1}{n} h(x + n(y-x)), \text{ i.e.,}$$

$$h(y) - h(x) \geq \frac{1}{n} [h(x + n(y-x)) - h(x)] \geq -\frac{2M}{n} \quad .$$

(Since $|x + n(y - x)| \leq |x| + n|y - x| \leq \frac{r_0}{3n} + \frac{2r_0 n}{3n} < r_0$.)

Interchanging the roles of x and y gives

$$|h(x) - h(y)| \leq \frac{2M}{n} \quad \text{if} \quad |x| < \frac{r_0}{3n} \, , \; |y| < \frac{r_0}{3n} \quad ,$$

This proves continuity at 0 and hence proves the lemma.

The possibilities for the qualitative behavior of $s(x)$ are now rather restricted. In particular, it takes on its supremum either:

a) never

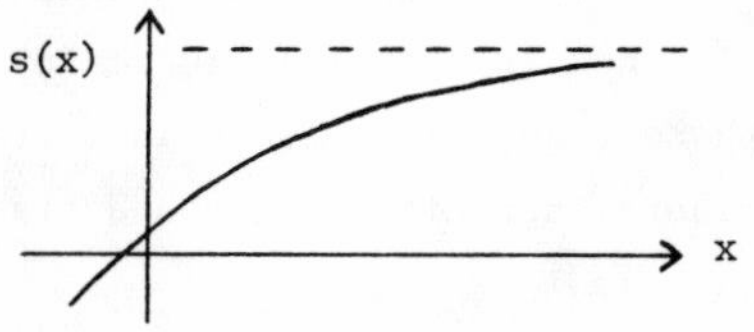

b) exactly once:

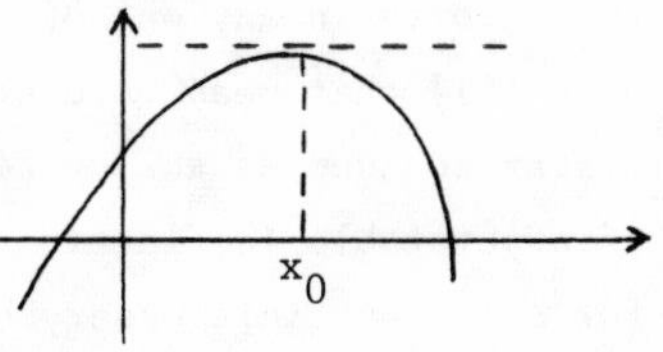

c) or on an interval

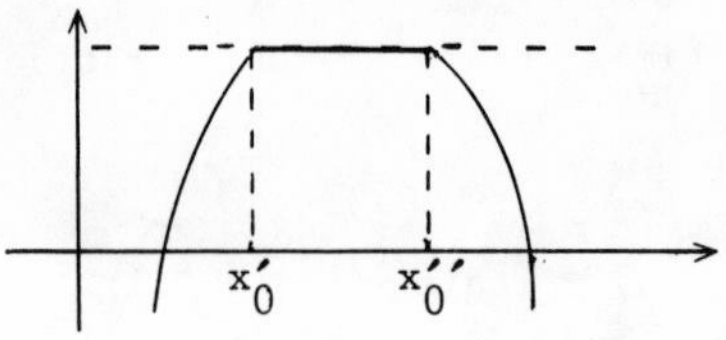

Let us analyze the significance of the second kind of behavior. Let $s = \sup\{s(x): x \in \mathbb{R}\} = s((-\infty,\infty))$. For any $\varepsilon > 0$, the probability of having $\frac{f}{N}$ larger than $x_0 + \varepsilon$ behaves asymptotically like

$$\frac{e^{Ns((x_0 + \varepsilon,\infty))}}{e^{Ns}} = \exp[-N(s - \sup\{s(x): x_0 + \varepsilon < x < \infty\})] \quad .$$

But

$$\sup\{s(x): x_0 + \varepsilon < x < \infty\} = s(x_0 + \varepsilon)$$

is strictly smaller than s, so this probability goes to zero exponentially with the size of the system. Similarly, the probability of having f/N smaller than $x_0 - \varepsilon$ goes to zero exponentially with the size of the system. Thus: For any $\varepsilon > 0$, the probability of having $\frac{f}{N}$ outside $[x_0 - \varepsilon, x_0 + \varepsilon]$ goes to zero exponentially with the size of the system. This, then, is a strong form of the assertion that the distribution of f/N approaches a delta function as the system becomes large.

One can make a similar analysis of case (a). Here, for any constant c, the probability of finding f/N smaller than c goes to zero exponentially as the system becomes large, i.e. f/N converges to ∞. In case (c), the statements one can make are somewhat less satisfactory; certainly for any $\varepsilon > 0$ the probability of having f/N larger than $x_0'' + \varepsilon$ or smaller than $x_0' - \varepsilon$ goes to zero as the system becomes large, so the distribution for f/N becomes concentrated in the interval $[x_0', x_0'']$. The behavior inside this interval is, however, entirely undetermined by the above considerations; for example, it is perfectly possible that the distribution of f/N still approaches a δ-function, but more slowly than in case (b). Case (c) is sometimes said to correspond to the occurrence of a phase transition. This is certainly true in some cases, but it would require a rather liberal definition of phase transition to make it true in general.

Anticipating a bit, we can make a few remarks about the occurrence of these three behaviors. In the first place, behavior of the first kind will be shown to imply that the mean value of f/N goes to infinity in any one of the standard ensembles, but we know from recent work of Ruelle [11] that mean values of n-particle observables in the grand canonical ensemble stay bounded as the system becomes infinitely large, provided that the interaction is superstable (definition later). Thus, behavior of the first type is ruled out for all n-particle observables f if the interaction we are considering is superstable. In the second place, by using special techniques (the Kirkwood-Salsburg equations) one can show that, again if the interaction is superstable, there is a range of energies and densities ("low activity") such that behavior of the second kind occurs for all finite range *two-particle* observables. Even in this case, behavior like

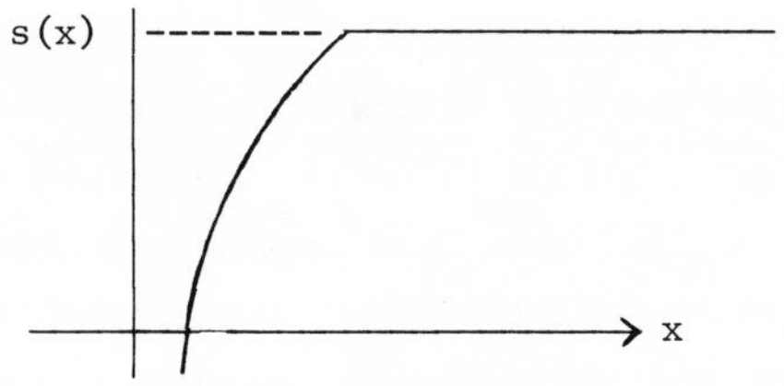

(the flat part of the graph extending all the way to infinity) seems to be typical of n-particle observables for $n > 2$. The restriction to two-particle observables is not as limiting as it might appear at first glance; there seem to be no quantities which are actually measured which depend on correlations between more than two particles. The behavior of the higher correlations is a quirk of continuous systems with "soft" interactions; it does not occur for lattice systems or for continuous systems with hard cores.

Before we begin to investigate the existence of the limit of $\frac{1}{N} \log V$, let us make a few remarks which organize things a bit:

a) We introduced the range I_ε of allowed energies as a technically necessary substitute for the uniquely defined energy value we really wanted to consider. We can get a little closer to considering a single value for the energy by letting I_ε shrink to a single point after we have passed to the limit $\Lambda, N \to \infty$. The techniques needed for doing this are exactly those used in replacing the set function $s(J)$ by the point function $s(x)$. In particular, there is a function $s(\varepsilon,x)$ defined on $\mathbb{R}^2$ such that (with a self-explanatory notation)

$$s_{I_\varepsilon}(J) = \sup\{s(\varepsilon,x): \ \varepsilon \in I_\varepsilon, \ x \in J\}$$

for all I_ε, J. The function $s(\varepsilon,x)$ is concave in both variables jointly, by the same argument used to prove the concavity of $s(x)$ and is therefore continuous except possibly on the boundary of the set where it is finite.

b) We have been considering a single observable f. We can just as well consider a finite family of observables $f_1, \ldots, f_t$ by treating them as components of a single vector-valued observable also denoted f. The range of a vector-valued observable means the supremum of the ranges of its components. If we replace intervals J by open convex sets in $\mathbb{R}^t$, all the above analysis goes through with no significant changes. Indeed, for purposes of investigating the limit, the interaction plays the same role as any other observable so we can simplify the notation by treating it as one of the components of the vector-valued observable. Thus, what we want to investigate is the asymptotic behavior of

$$V(\Lambda,N,f,J) = \frac{1}{N!} \times \text{Lebesgue measure of } \{(q_1,\ldots,q_N) \in \Lambda^H : \frac{f(q_1,\ldots,q_N)}{N} \in J\} .$$

A2. THERMODYNAMIC LIMITS ALONG A SPECIAL SEQUENCE OF BOXES

We now set out to prove the existence of $\lim \frac{1}{N} \log V$. The proof we will give has features typical of all such proofs. One first considers a special sequence of regions for which the existence of the limit can be proved by a monotonicity argument. The properties of the limit are then analyzed as above; when we have proved enough properties of the limit (in particular, the continuity of $s(x)$), we finish the proof by showing that the limit does not depend on the particular sequence of regions chosen.

It is convenient to isolate at the outset two simple arguments which will be used over and over again in the course of the proof.

Lemma A2.1.

$V(\Lambda,N,f,J)$ is increasing in Λ (with N,f,J held fixed).

Lemma A2.2.

Let f be an observable with range R; let $\Lambda_1, \ldots, \Lambda_k$ be bounded open sets in $\mathbb{R}^\nu$ such that the distance from Λ_i to Λ_j is at least R for each $i \neq j$. Let $N_1, \ldots, N_k$ be integers, and let $J_1, \ldots, J_k$ be open convex sets in $\mathbb{R}^t$. Denote

$$N = N_1 + \ldots + N_k ; \quad J = (\frac{N_1}{N})J_1 + \ldots + (\frac{N_k}{N})J_k .$$

Then:

$$V(\Lambda_1 \cup \ldots \cup \Lambda_k, N, f, J) \geq \prod_{i=1}^{k} V(\Lambda_i, N_i, f, J_i) .$$

Lemma A2.1 is sufficiently obvious not to require a proof; the proof of Lemma A2.2 has essentially been given in the proof of Theorem A2.2; we will simply recapitulate the argument. Suppose we consider

$$\begin{aligned} &q_1, \ldots, q_{N_1} \in \Lambda_1; \quad f(q_1,\ldots,q_{N_1}) \in N_1 J_1 , \\ &q_{N_1+1}, \ldots, q_{N_1+N_2} \in \Lambda_2, \; f(q_{N_1+1},\ldots,q_{N_1+N_2}) \in N_2 J_2 , \ldots \qquad (1) \\ &\ldots, q_{N-N_k+1}, \ldots, q_N \in \Lambda_k; \quad f(q_{N-N_k+1},\ldots,q_N) \in N_k J_k . \end{aligned}$$

Then, because the different Λ_j's are separated by at least the range of the observable,

$$f(q_1,\ldots,q_N) = f(q_1,\ldots,q_{N_1}) + \ldots + f(q_{N-N_k+1},\ldots,q_N)$$
$$\in N_1J_1 + \ldots + N_kJ_k$$

i.e.,

$$\frac{f(q_1,\ldots,q_N)}{N} \in J \quad .$$

Thus, configurations of the form (1) contribute to $V(\Lambda_1 \cup \ldots \cup \Lambda_k, N, f, J)$, and their contribution is

$$\frac{1}{N!} \prod_{i=1}^{k} [(N_i)!\, V(\Lambda_i, N_i, f, J_i)] \quad .$$

There are $\frac{N!}{\prod_{i=1}^{k} N_i!}$ different contributions of a similar nature, corresponding to the possible ways of choosing which particles to put in which Λ_i; the sum of all these contributions gives exactly the desired lower bound for

$$V(\Lambda_1 \cup \ldots \cup \Lambda_k, N, f, J) \quad .$$

We will work, for the moment, at a fixed value of the density. At certain points, the specific volume $v = 1/\rho$ is a more convenient variable than the density (s turns out to be concave in v, not ρ), so we will use it. Let $\ell = \ell_v = v^{1/\nu}$, i.e., the length of the edge of a cube of volume v. Let R denote the range of the observable f we are considering. For each ν-tuple $\underset{\sim}{r} = (r_1,\ldots,r_\nu)$ of positive integers, we let $\Lambda_v(\underset{\sim}{r})$ denote the rectangular solid

$$\{q = (q_1,\ldots,q_\nu) \in \mathbb{R}^\nu : \; 0 < q_i < r_i\ell - R$$

(if $r_i^{\ell} - R \le 0$ for some i, then $\Lambda_v(\underset{\sim}{r}) = \emptyset$). Let $N(\underset{\sim}{r}) = \prod_{i=1}^{\nu} r_i$, and note that

$$\frac{N(\underset{\sim}{r})}{V(\Lambda_v(\underset{\sim}{r}))} = \frac{1}{(\ell - R/r_1)\ldots(\ell - R/r_\nu)} \to \frac{1}{\ell^\nu} = \frac{1}{v} \text{ as } r_1,\ldots,r_\nu \to \infty \quad .$$

We will study the behavior of $V(\Lambda_v(\underset{\sim}{r}), N(\underset{\sim}{r}), f, J)$ as $r_1,\ldots,r_\nu \to \infty$; to simplify the formulas, we will write $V(\underset{\sim}{r}, J)$, or $V_v(\underset{\sim}{r}, J)$, for $V(\Lambda_v(\underset{\sim}{r}), N(\underset{\sim}{r}), f, J)$, and we will write $\underset{\sim}{r} \to \infty$ for $r_1, \ldots, r_\nu \to \infty$.

The reason for taking the sides of $\Lambda_v(\underset{\sim}{r})$ to have length $r_1\ell - R,\ldots,r_\nu\ell - R$, rather than $r_1\ell,\ldots,r_\nu\ell$, is to make $\log V(\underset{\sim}{r}, J)$ superadditive in each component of $\underset{\sim}{r}$, i.e., to make the following lemma valid :

Lemma A2.3.

Let $\underset{\sim}{r}' = (r_1,\ldots,r_i',\ldots,r_\nu)$, $\underset{\sim}{r}'' = (r_1,\ldots,r_i'',\ldots,r_\nu)$,

$$\underset{\sim}{r} = (r_1,\ldots,r_i' + r_i'',\ldots,r_\nu) \quad .$$

Then

$$V(\underset{\sim}{r},J) \geq V(\underset{\sim}{r}',J)V(\underset{\sim}{r}'',J) \quad .$$

Proof. The lemma follows from Lemma A2.2 and the fact that we can place in $\Lambda(\underset{\sim}{r})$ translates of $\Lambda(\underset{\sim}{r}')$ and $\Lambda(\underset{\sim}{r}'')$ with separation R. (For example, for $\nu = 2$, $i = 1$, we have:

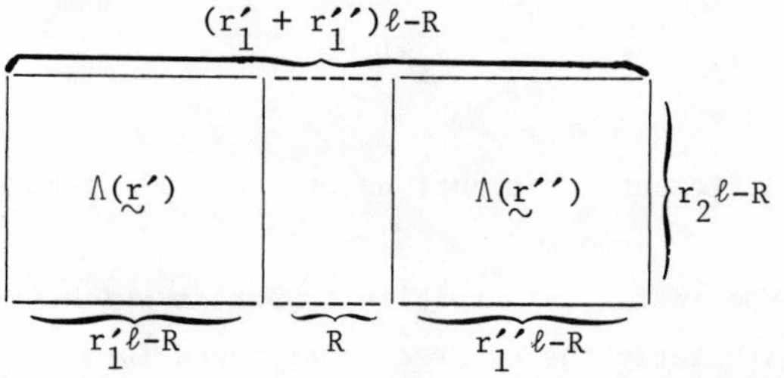

We would now like to invoke the following lemma:

Lemma A2.4.

Let $h(\underset{\sim}{r})$ *be a superadditive real-valued function defined on* $(N_+)^\nu$ *(i.e.,* $h(\underset{\sim}{r}) \geq h(\underset{\sim}{r}') + h(\underset{\sim}{r}'')$ *when* $\underset{\sim}{r}$, $\underset{\sim}{r}'$, $\underset{\sim}{r}''$ *are as in the preceding lemma.) Then*

$$\lim_{\underset{\sim}{r}\to\infty} \frac{h(\underset{\sim}{r})}{N(\underset{\sim}{r})}$$

exists and is equal to

$$\sup_{\underset{\sim}{r}} \frac{h(\underset{\sim}{r})}{N(\underset{\sim}{r})}$$

(The value $+\infty$ *is allowed for the limit).*

Proof. To simplify the notation, we consider $\nu = 1$. For any $r_0 \in N_+$, we can write, for each r,

$$r = jr_0 + k; \quad 0 \leq k < r_0 \quad .$$

Then, by superadditivity,

$$h(r) \geq jh(r_0) + kh(1) \quad ,$$

Letting $r \to \infty$, we get $j/r \to \frac{1}{r_0}, \frac{k}{r} \to 0$

$$\liminf_r \frac{h(r)}{r} \geq \frac{h(r_0)}{r_0} \quad .$$

This is true for all r_0, so

$$\liminf_r \frac{h(r)}{r} \geq \sup_{r_0} \frac{h(r_0)}{r_0} \geq \limsup_r \frac{h(r)}{r}$$, proving the lemma.

We would now like to apply the lemma to $h(\underset{\sim}{r}) = \log V(\underset{\sim}{r},J)$. Unfortunately, the lemma *is not valid* if $h(r)$ is allowed to take on the value $-\infty$, as the example $h(r) \begin{Bmatrix} = -\infty & r \text{ odd} \\ = 0 & r \text{ even} \end{Bmatrix}$ $(\nu = 1)$ shows, and $V(\underset{\sim}{r},J)$ may evidently be zero for some values of $\underset{\sim}{r}$. If, however, $h(\underset{\sim}{r}_0) > -\infty$ for one $\underset{\sim}{r}_0$, then, (denoting $(j_1r_1,\ldots,j_\nu r_\nu)$ by $\underset{\sim}{j}\underset{\sim}{r}$,) superadditivity implies that $h(\underset{\sim}{j}\underset{\sim}{r}_0) > -\infty$ for all $j \in (N_+)^\nu$ and the lemma implies that

$$\lim_{\underset{\sim}{j}\to\infty} \frac{h(\underset{\sim}{j}\underset{\sim}{r}_0)}{N(\underset{\sim}{j}\underset{\sim}{r}_0)} = \sup_{\underset{\sim}{j}} \frac{h(\underset{\sim}{j}\underset{\sim}{r}_0)}{N(\underset{\sim}{j}\underset{\sim}{r}_0)}$$

The limit, moreover, does not depend on the choice of $\underset{\sim}{r}_0$; if $h(\underset{\sim}{r}_0) > -\infty$ and $h(\underset{\sim}{r}'_0) > -\infty$, then $h(\underset{\sim}{r}_0\underset{\sim}{r}'_0) > -\infty$, and $(\underset{\sim}{j}\underset{\sim}{r}_0\underset{\sim}{r}'_0)$ is a subsequence of both $(\underset{\sim}{j}\underset{\sim}{r}_0)$ and $(\underset{\sim}{j}\underset{\sim}{r}'_0)$. Since

$$\lim_{\underset{\sim}{j}\to\infty} \frac{h(\underset{\sim}{j}\underset{\sim}{r}_0)}{N(\underset{\sim}{j}\underset{\sim}{r}_0)}, \quad \lim_{\underset{\sim}{j}\to\infty} \frac{h(\underset{\sim}{j}\underset{\sim}{r}'_0)}{N(\underset{\sim}{j}\underset{\sim}{r}'_0)}, \quad \text{and}$$

$$\lim_{\underset{\sim}{j}\to\infty} \frac{h(\underset{\sim}{j}\underset{\sim}{r}_0\underset{\sim}{r}'_0)}{N(\underset{\sim}{j}\underset{\sim}{r}_0\underset{\sim}{r}'_0)}$$

all exist, the three limits (and in particular, the first two) must be equal. Note also that, since

$$\lim_{\underset{\sim}{j}} \frac{h(\underset{\sim}{j}\underset{\sim}{r}_0)}{N(\underset{\sim}{j}\underset{\sim}{r}_0)} \geq \frac{h(\underset{\sim}{r}'_0)}{N(\underset{\sim}{r}'_0)} \quad ,$$

we can compute the limit as

$$\sup_{\underset{\sim}{r}} \frac{h(\underset{\sim}{r})}{N(\underset{\sim}{r})} \quad .$$

We are thus led to define

$$s(v,f,J) = \sup_{\underset{\sim}{r}} \frac{\log V_v(\underset{\sim}{r},J)}{N(\underset{\sim}{r})}$$

and to add the remark that, if

$$s(v,f,J) > -\infty \quad ,$$

then,

$$s(v,f,J) = \lim_{\underset{\sim}{j}\to\infty} \frac{\log V_v(\underset{\sim}{r}_0\underset{\sim}{j},J)}{N(\underset{\sim}{r}_0\underset{\sim}{j})}$$

for any $\underset{\sim}{r}_0$ such that $V_v(\underset{\sim}{r}_0,J) > 0$. The difficulty involved in passing to the sub-sequence $(j\underset{\sim}{r}_0)$, while annoying, does not affect in any essential way the arguments we will give, and, we will from now on proceed if

$$s(v,f,J) = \lim_{\underset{\sim}{r}\to\infty} \frac{1}{N(\underset{\sim}{r})} \log V_v(\underset{\sim}{r},f,J) \quad .$$

The sceptical reader is encouraged to verify that this does not affect the validity of any of our conclusions.

Thus far, $s(v,f,J)$ has been obtained as a limit of $\frac{1}{N} \log V$ along a very special sequence of regions $\Lambda(\underset{\sim}{r})$. Eventually, we will show that the same limit is obtained for much more general sequences of regions, but before doing this we need to know quite a lot about the properties of $s(v,f,J)$.

Lemma A2.5.

If $J = J_1 \cup \ldots \cup J_k$, *then*

$$s(v,f,J) = \sup\{s(v,f,J_i):\ 1 \le i \le k\} \quad .$$

Proof. For each $\underset{\sim}{r}$,

$$\sup_i\{V(\underset{\sim}{r},f,J_i) \le V(\underset{\sim}{r},f,J) < k\cdot\sup_i\{V(\underset{\sim}{r},f,J_i)$$

Taking the logarithm, dividing by $N(\underset{\sim}{r})$, and passing to the limit $\underset{\sim}{r} \to \infty$ gives the result.

Lemma A2.6.

For any open convex J, $s(v,f,J) = \sup\{s(v,f,\hat{J}):\ \hat{J}$ bounded open and convex, with closure contained in $J\}$.

Proof. By the elementary properties of Lebesgue measure,

$$V(\underset{\sim}{r},J) = \sup\{V(\underset{\sim}{r},\hat{J}):\ \hat{J} \text{ as above}\} \quad .$$

Since

$$s(v,f,J) = \sup_{\underset{\sim}{r}} \frac{1}{N(\underset{\sim}{r})} \log V(\underset{\sim}{r},J) \quad ,$$

the lemma follows at once. In more detail: We know already that

$$s(v,f,J) \geq \sup\{s(v,f,\hat{J}): \hat{J} \text{ as above}\}$$

and there is nothing to prove if $s(v,f,J) = -\infty$. Thus we assume $s(v,f,J) > -\infty$, and we let $\varepsilon > 0$. Choose $\underset{\sim}{r}_0$ so that

$$\frac{1}{N(\underset{\sim}{r}_0)} \log V(\underset{\sim}{r},f,J) > s(v,f,J) - \varepsilon \quad ;$$

then choose a bounded open convex set $\hat{J}$ with closure contained in J, such that

$$\frac{1}{N(\underset{\sim}{r})} \log V(\underset{\sim}{r},f,\hat{J}) > \frac{1}{N(r)} \log V(\underset{\sim}{r},f,J) - \varepsilon$$

Then

$$s(v,f,\hat{J}) \geq \frac{1}{N(\underset{\sim}{r})} \log V(\underset{\sim}{r},f,\hat{J}) > \frac{1}{N(\underset{\sim}{r})} \log V(\underset{\sim}{r},f,J) - \varepsilon$$

$$> s(v,f,J) - 2\varepsilon \quad .$$

Since $\varepsilon > 0$ is arbitrary

$$s(v,f,J) \leq \sup\{s(v,f,\hat{J})\} \quad ,$$

so the lemma is proved.

Now, for each $x \in \mathbb{R}^\nu$, we define

$$s(v,f,x) = \inf\{s(v,f,J): J \text{ open, convex, and containing } x\} \quad .$$

Proposition A2.7.

$s(v,f,x)$ *is upper semi-continuous in* x, *and*

$$s(v,f,J) = \sup\{s(v,f,x): x \in J\}$$

for all convex open J.

Proof. The proposition follows at once from Lemma A1.1 (generalized to more than one dimension); the hypotheses of this lemma are precisely what is proved in Lemmas A2.5 and A2.6.

Theorem A2.8.

$s(v,f,x)$ *is a concave function of* x, *and*

$$s(v,f,x) \leq 1 + \log v \quad \textit{for all} \quad v,f,x \quad .$$

Proof. The upper bound is proved exactly as before. To prove concavity, we argue just as in the proof of Theorem A1.1, i.e., we note that we can place in $\Lambda(2r_1, r_2, \ldots, r_\nu)$ two translates of $\Lambda(r_1, \ldots, r_\nu)$ with a separation of R; hence, by Lemma A2.2,

$$V((2r_1, \ldots, r_\nu), \tfrac{1}{2} J_1 + \tfrac{1}{2} J_2) \geq V((r_1, \ldots, r_\nu), J_1) V((r_1, \ldots, r_\nu), J_2) \quad .$$

Taking logarithms, dividing by $2 \cdot r_1 \cdot \ldots \cdot r_\nu$, and passing to the limit $\underset{\sim}{r} \to \infty$ gives

$$s(v,f, \tfrac{1}{2} J_1 + \tfrac{1}{2} J_2) \geq \tfrac{1}{2} s(v,f,J_1) + \tfrac{1}{2} s(v,f,J_2) \quad .$$

Letting J_1 and J_2 shrink to points, gives

$$s(v,f, \tfrac{1}{2} x_1 + \tfrac{1}{2} x_2) \geq s(v,f,x_1) + \tfrac{1}{2} s(v,f,x_2) \quad .$$

By induction, then,

$$s(v,f, \alpha x_1 + (1-\alpha) x_2) \geq \alpha s(v,f,x_1) + (1-\alpha) s(v,f,x_2)$$

for all dyadic rationals α between 0 and 1. The same inequality for general α between 0 and 1 follows from the upper semi-continuity of s in x.

Corollary A2.8.

$\{x: s(v,f,x) > -\infty\}$ *is convex and* $s(v,f,x)$ *is continuous* in x *on the interior of this set.*

Proof. The first statement is immediate; the second follows from Lemma A1.4 (the continuity of bounded concave functions) and the bound

$$s(v,f,x) \leq 1 + \log v \quad .$$

The next problem is to determine where $s(v,f,x) > -\infty$. We cannot actually find this set, but we can find its closure.

Proposition A2.9.

The closure of $\{x: s(v,f,x) > -\infty\}$ *is the closure of*

$$\bigcup_{\underset{\sim}{r}} \frac{f}{N(\underset{\sim}{r})} \left(\Lambda_v(\underset{\sim}{r})^{N(\underset{\sim}{r})}\right) \quad .$$

(In other words: for each $\underset{\sim}{r}$ we take the function $f(q_1, \ldots, q_{N(\underset{\sim}{r})})$, divide it by $N(\underset{\sim}{r})$, and take the image under the resulting function of $\Lambda_v(\underset{\sim}{r})^{N(\underset{\sim}{r})}$. This gives a set in $\mathbb{R}^\tau$ (depending on $\underset{\sim}{r}$). We take the union of these sets over $\underset{\sim}{r}$, and take the closure. The resulting set is the closure of $\{x: s(v,f,x) > -\infty\}$. Thus, except for boundary points, $\{x: s(v,f,x) > -\infty\}$ is just the set of values of $\frac{f}{N}$ consistent with the density.

Proof. If x_0 does not belong to the closure of

$$\bigcup_{\underset{\sim}{r}} \frac{f}{N(\underset{\sim}{r})} (\Lambda_v(\underset{\sim}{r})^{N(\underset{\sim}{r})}) \quad .$$

then there is an open convex set J containing x_0 which does not intersect

$$\frac{f}{N(\underset{\sim}{r})} (\Lambda_v(\underset{\sim}{r})^{N(\underset{\sim}{r})}) \quad \text{for any } \underset{\sim}{r}, \text{ i.e., which is such that}$$

$$V(\Lambda_v(\underset{\sim}{r}), N(\underset{\sim}{r}), f, J) = 0 \quad \text{for all } \underset{\sim}{r} \quad .$$

Hence, $s(v,f,J) = -\infty$, so $s(v,f,x) = -\infty$ for all $x \in J$, so x_0 is outside the closure of $\{x: s(v,f,x) > -\infty\}$.

Conversely, suppose

$$x_0 \in \frac{f}{N(\underset{\sim}{r})} (\Lambda_v(\underset{\sim}{r})^{N(\underset{\sim}{r})}) \quad .$$

Since f is continuous, this implies that, for any convex open J containing x_0, there is a non-empty open subset of $\Lambda_v(\underset{\sim}{r})^{N(\underset{\sim}{r})}$ on which $\frac{f}{N(\underset{\sim}{r})} \in J$. Thus,

$$V(\Lambda_v(\underset{\sim}{r}), N(\underset{\sim}{r}), f, J) \neq 0 \quad ,$$

so

$$s(v,f,J) \geq \frac{1}{N(\underset{\sim}{r})} \log[V(\Lambda_v(\underset{\sim}{r}), N(\underset{\sim}{r}), f, J)] > -\infty$$

Thus, every convex open J containing x_0 also contains at least one x such that $s(v,f,x) > -\infty$, so x_0 is in the closure of $\{x: s(v,f,x) > -\infty\}$. Thus,

$$\overline{\bigcup_{\underset{\sim}{r}} \frac{f}{N(\underset{\sim}{r})} (\Lambda_v(\underset{\sim}{r})^{N(\underset{\sim}{r})})} \subset \overline{\{x: s(v,f,x) > -\infty\}} \quad ,$$

and we have already proved the opposite inclusion.

Remark: In the above, we used for the first and only time the assumption that f is continuous. If we assume only that f is Borel, we can derive an analogue of the above proposition but we must replace the range of $f/N(\underset{\sim}{r})$ on $\Lambda_v(\underset{\sim}{r})^{N(\underset{\sim}{r})}$ by the essential range, i.e., the support of the image under $f/N(\underset{\sim}{r})$ of Lebesgue measure on $\Lambda_v(\underset{\sim}{r})^{N(\underset{\sim}{r})}$.

Since we know that $s(v,f,x)$ is continuous in x on the interior of $\{x: s(v,f,x) > -\infty\}$, it is useful to know that this interior is generally non-empty.

Proposition A2.10.

If the components $f_1, \ldots, f_t$ *of* f *are linearly independent functions (i.e., if for some* N, $f_1(q_1,\ldots,q_N), \ldots, f_t(q_1,\ldots,q_N)$ *are linearly independent on* $(\mathbb{R}^\nu)^N$*) then*

$$\{x: s(v,f,x) > -\infty\}$$

has non-empty interior.

Proof. Since

$$\{x: s(v,f,x) > -\infty\}$$

is convex and its closure contains $\frac{f}{N(\underset{\sim}{r})}(\Lambda_v(\underset{\sim}{r})^{N(\underset{\sim}{r})})$, it suffices to show that for some $\underset{\sim}{r}$, the convex hull of

$$f(\Lambda_v(\underset{\sim}{r})^{(N(\underset{\sim}{r})})$$

has non-empty interior. (We are using here the geometrical fact that the interior of the closure of a convex set in Euclidean space is the same as the interior of the set itself.) Since the interaction is of finite range, and since $f(q) = 0$, $f(q_1,\ldots,q_N)$ takes on the value zero for each N (take the q_i such that any two of them are separated by at least R). Hence, if $f_1(q_1,\ldots,q_N), \ldots, f_t(q_1,\ldots,q_N)$ are linearly independent, the convex hull of $f((\mathbb{R}^\nu)^N)$ has non-empty interior, and in fact, for some bounded $\Lambda \subset \mathbb{R}^\nu$, the convex hull of $f(\Lambda^N)$ has non-empty interior. Choose $\underset{\sim}{r}$ so that $N(\underset{\sim}{r}) \geq N$, and such that $\Lambda_v(\underset{\sim}{r})$ contains Λ as well as some points at a distance at least R from Λ. Then, by considering configurations of N particles in Λ and $N(\underset{\sim}{r}) - N$ particles in $\Lambda_v(\underset{\sim}{r})$ at a distance at least R from Λ we see, by letting the particles inside Λ move while holding the others fixed, that the convex hull of $f(\Lambda_v(\underset{\sim}{r})^{N(\underset{\sim}{r})})$ has a non-empty interior; hence, that

$$\{x: s(v,f,x) > -\infty\}$$

has non-empty interior.

Remarks. a) There is no real generality to be gained by considering linearly dependent set of observables; if $(f_1,\ldots,f_t)$ is a linearly dependent set and if $(f'_1,\ldots,f'_{t'})$ is a maximal linearly independent subset, then for any J, $V(\Lambda,N,f,J)$ is equal to $V(\Lambda,N,f',J')$ for an appropriately chosen J'. We will assume, for the remainder of this chapter, that *the set of observables we are considering is linearly independent*.

b) If we allow Borel functions, rather than only continuous ones, as observables, we replace the condition of linear independence by linear independence modulo

Lebesgue null functions; the argument then goes essentially as before.

c) If we want to allow our observables to take on the value $+\infty$ (to allow for interactions with hard cores), the proof of Proposition A2.10 ceases to hold. The argument given can easily be adapted to show that, if v is sufficiently large and if the components of f are linearly independent, then

$$\{x: s(v,f,x) > -\infty\}$$

has non-empty interior. A slightly more detailed argument shows that, for any specific volume strictly smaller than the close-packing volume v_0

$$v_0 = \inf\{v: s(v,f,x) \neq -\infty \text{ for at least one } x\} ,$$

the set $\{x: s(v,f,x) > -\infty\}$ has non-empty interior. So far, we have been holding v fixed; we have next to investigate the dependence of $s(v,f,x)$ on v.

Proposition A2.11.

For fixed f,x, s(v,f,x) *is an increasing* function of* v.

Proof. $\Lambda_v(\underset{\sim}{r})$ is increasing in v for each $\underset{\sim}{r}$; hence, by Lemma A2.1, $V_v(\underset{\sim}{r},f,J)$ is increasing in v for all $\underset{\sim}{r}$,f,J, so s(v,f,J) is increasing in v, so

$$s(v,f,x) > \inf\{s(v,f,J): J \text{ open, convex, containing } x\}$$

is increasing in v.

Next we will show that, with some annoying but unimportant technical restrictions, s(v,f,x) is jointly concave in (v,x). The proof will be rather roundabout, but the key step is the following lemma.

Lemma A2.12.

If $v' > \frac{1}{2} v_1 + \frac{1}{2} v_2$, *then*

$$s(v', f, \tfrac{1}{2} x_1 + \tfrac{1}{2} x_2) \geq \tfrac{1}{2} s(v_1,f,x_1) + \tfrac{1}{2} s(v_2,f,x_2) \quad .$$

(Note that the conclusion of the lemma is slightly weaker than the assertion that

$$s(\tfrac{1}{2} v_1 + \tfrac{1}{2} v_2, \tfrac{1}{2} x_1 + \tfrac{1}{2} x_2) \geq \tfrac{1}{2} s(v_1,f,x_1) + \tfrac{1}{2} s(v_2,f,x_2) \quad ,$$

but that this condition would follow at once if we knew that s was continuous from

---

* We will use "increasing" as a synonym for "non-decreasing", etc.; when we mean "strictly increasing" we will say so.

the right in v. We will, in fact, use the lemma first to prove continuity in v and then deduce concavity.)

Proof. We fix a value of $\underset{\sim}{r}$, and denote temporarily

$$\Lambda_1 = \{(q_1,\ldots,q_\nu) \in \mathbb{R}^\nu : 0 < q_i < r_i \ell_{v_1},\ 1 \le i \le \nu\}$$

$$\Lambda_2 = \{(q_1,\ldots,q_\nu) \in \mathbb{R}^\nu : 0 < q_i < r_i \ell_{v_2},\ 1 \le i \le \nu\}$$

(In other words, Λ_1 is a rectangular parallelepiped each side of which is R longer than the corresponding side of $\Lambda_{v_1}(\underset{\sim}{r})$, and Λ_2 is related to $\Lambda_{v_2}(\underset{\sim}{r})$ in the same way.) Now for large $\underset{\sim}{j}$, the volume of

$$\Lambda'_v((2r_1j_1, r_2j_2,\ldots,r_\nu j_\nu)) \approx 2v'N(\underset{\sim}{r})N(\underset{\sim}{j})$$

is strictly larger than $N(\underset{\sim}{j})$ times the volume $v_1N(\underset{\sim}{r})$ of Λ_1 plus $N(\underset{\sim}{j})$ times the volume $v_2N(\underset{\sim}{r})$ of Λ_2. It is therefore at least plausible (the reader should convince himself that it is actually true) that, for sufficiently large j, we can place in $\Lambda'_v((2r_1j_1,\ldots,r_\nu j_\nu))$ $N(\underset{\sim}{j})$ translates of Λ_1 and $N(\underset{\sim}{j})$ translates of Λ_2 in such a way that no pair of the small rectangular parallelepipeds overlaps. By centering in each translate of Λ_1 a translate of $\Lambda_{v_1}(\underset{\sim}{r})$ and in each translate of Λ_2 a translate of $\Lambda_{v_2}(\underset{\sim}{r})$, we can place in $\Lambda'_v((2j_1r_1, j_2r_2,\ldots,j_\nu r_\nu))$ $N(\underset{\sim}{j})$ translates of $\Lambda_{v_1}(\underset{\sim}{r})$ and $N(\underset{\sim}{j})$ translates of $\Lambda_{v_2}(\underset{\sim}{r})$ such that any two of the small boxes are separated by a distance of at least R. Hence, by Lemma A2.2,

$$\begin{aligned} &V(\Lambda_{v'}(2j_1r_1,\ldots,j_\nu r_\nu),\ 2N(\underset{\sim}{r})N(\underset{\sim}{j}), f, \tfrac{1}{2}J_1 + \tfrac{1}{2}J_2) \\ &\ge [V(\Lambda_{v_1}(\underset{\sim}{r}), N(\underset{\sim}{r}), f, J_1)V(\Lambda_{v_2}(\underset{\sim}{r}), N(\underset{\sim}{r}), f, J_2)]^{N(\underset{\sim}{j})} \end{aligned}$$

for all sufficiently large $\underset{\sim}{j}$. Taking the logarithm, dividing by $2N(\underset{\sim}{r})N(\underset{\sim}{j})$, then passing to the limit $\underset{\sim}{j} \to \infty$, gives

$$s(v', f, \tfrac{1}{2}J_1 + \tfrac{1}{2}J_2) \ge \frac{\frac{1}{2}\log V_{v_1}(\underset{\sim}{r}, J_1)}{N(\underset{\sim}{r})} + \frac{\frac{1}{2}\log V_{v_2}(\underset{\sim}{r}, J_2)}{N(\underset{\sim}{r})}$$

If we now let $\underset{\sim}{r} \to \infty$ we get

$$s(v', f, \tfrac{1}{2}J_1 + J_2) \ge \tfrac{1}{2}s(v_1, f, J_1) + \tfrac{1}{2}s(v_2, f, J_2)$$

Letting J_1 and J_2 shrink to x_1 and x_2 respectively proves the lemma.

For each f,x, we define

$$v_{min} = v_{min}(f,x) = \inf\{v: s(v,f,x) > -\infty\} \quad .$$

$\frac{1}{v_{min}}$ is the largest density compatible with $\frac{f}{N} \approx x$, i.e., it is a sort of close-packing density. It may be zero or $+\infty$. Since s is increasing in v, $s(v,f,x) > -\infty$ if $v > v_{min}$.

Lemma A2.13

For fixed f,x, s(v,f,x) *is a continuous concave function of* v *on* (v_{min},∞).

Proof. We will prove that

$$s(v+,f,x) = \lim_{v'\downarrow v} s(v',f,x)$$

is concave (and hence continuous) in v. Since s is increasing in v, this implies s(v+,f,x) = s(v,f,x), so s(v,f,x) itself is concave and continuous. To prove s(v+,f,x) is concave, we choose v_1, v_2, and α between 0 and 1. We also let $v' > \alpha v_1 + (1-\alpha)v_2$, and we choose $v_1' > v_1$ and $v_2' > v_2$ such that $\alpha v_1' + (1-\alpha)v_2' < v'$. For any α' which is a dyadic rational sufficiently near to α, we have

$$\alpha' v_1' + (1-\alpha')v_2' < v',$$

and Lemma A2.12, together with a simple induction argument, gives

$$s(v',f,x) \geq \alpha' s(v_1',f,x) + (1-\alpha')s(v_2',f,x) \quad ;$$

letting $\alpha' \to \alpha$ gives

$$s(v',f,x) \geq \alpha s(v_1',f,x) + (1-\alpha)s(v_2',f,x) \quad ;$$

letting $v_1' \downarrow v_1'$ and $v_2' \downarrow v_2$ gives

$$s(v',f,x) \geq \alpha s(v+_1,f,x) + (1-\alpha)s(v+_2,f,x) \quad ;$$

finally, letting $v' \downarrow \alpha v_1 + (1-\alpha)v_2$ gives

$$s((\alpha v_1 + (1-\alpha)v_2)^+,f,x) \geq \alpha s(v+_1,f,x) + (1-\alpha)s(v+_2,f,x) \quad .$$

Thus, s(v+,f,x) is concave in v, so it is continuous in v on (v_{min},∞), so s(v,f,x) is concave and continuous in v on (v_{min},∞).

We still do not know very much about what happens to s at v_{min}. Since s is increasing in v, we know at least that

$$s(v_{min},f,s) \leq s(v_{min}^+,f,x) \quad ,$$

so s(v,f,x) is actually concave in v everywhere. Let

$$\Gamma(f) = \{(v,x)\colon s(v,f,x) > -\infty\} \quad .$$

We would like to be able to show that $\Gamma(f)$ is convex. Unfortunately, this seems to be hard to prove and may even be false, but we have something almost as good.

Proposition A2.14.

The interior of $\Gamma(f)$ *is non-empty, convex, and dense in* $\Gamma(f)$ *(i.e.,* $\Gamma(f)$ *is obtained by starting from an open convex set and adding part of its boundary).* $s(v,f,x)$ *is a continuous concave function of* (v,x) *on the interior of* $\Gamma(f)$.

Proof. By Lemma A2.11, if $(v,x) \in \Gamma(f)$ and $v' > v$, then $(v',x) \in \Gamma(f)$. Also, for any fixed v, $\{x: (v,x) \in \Gamma(f)\} = \{x: s(v,f,x) > -\infty\}$ has non-empty interior (by Proposition A2.10). Hence, $\Gamma(f)$ has non-empty interior. If (v_1,x_1) and (v_2,x_2) are in $\Gamma(f)$ and if $v' > \frac{1}{2} v_1 + \frac{1}{2} v_2$, then $(v', \frac{1}{2} x_1 + \frac{1}{2} x_2)$ is in $\Gamma(f)$ by Lemma A2.12. In particular, if (v_1,x_1) is in the interior of $\Gamma(f)$, so we can find $v_1' < v_1$ such that $(v_1',x_1) \in \Gamma(f)$, then

$$\frac{1}{2} v_1 + \frac{1}{2} v_2 > \frac{1}{2} v_1' + \frac{1}{2} v_2, \text{ so } (\frac{1}{2} v_1 + \frac{1}{2} v_2, \frac{1}{2} x_1 + \frac{1}{2} x_2) \text{ is in } \Gamma(f) .$$

In fact, since we may displace (v_1,x_1) a little without going outside $\Gamma(f)$, $(\frac{1}{2} v_1 + \frac{1}{2} v_2, \frac{1}{2} x_1 + \frac{1}{2} x_2)$ is actually in the interior of $\Gamma(f)$. By induction, if α is a dyadic rational, $0 < \alpha \le 1$, if (v_1,x_1) is in the interior of $\Gamma(f)$ and if (v_2,x_2) is in $\Gamma(f)$,

$$(\alpha v_1 + (1 - \alpha)v_2, \alpha x_1 + (1 - \alpha)x_2)$$

is in the interior of $\Gamma(f)$. Letting $\alpha \downarrow 0$ shows that (v_2,x_2) is in the closure of the interior of $\Gamma(f)$, i.e., the interior of $\Gamma(f)$ is dense in $\Gamma(f)$. Now if α is any real number between 0 and 1, we want to show that $(\alpha v_1 + (1 - \alpha)v_2, \alpha x_1 + (1 - \alpha)x_2)$ is in the interior of $\Gamma(f)$, provided that (v_1,x_1) is in the interior of $\Gamma(f)$ and (v_2,x_2) is in $\Gamma(f)$. We do this by remarking that we can write

$$(\alpha v_1 + (1 - \alpha)v_2, \alpha x_1 + (1 - \alpha)x_2) = (\alpha' v_1' + (1 - \alpha)v_2, \alpha' x_1' + (1 - \alpha')x_2) .$$

where α' is a dyadic rational ("near α") and (v_1',x_1') may be taken to be arbitrarily near to (v_1,x_1). Hence, if (v_1,x_1) is in the interior of $\Gamma(f)$ we may take (v_1',x_1') as to be in the interior of $\Gamma(f)$, so $(\alpha v_1 + (1 - \alpha)v_2, \alpha x_1 + (1 - \alpha)x_2)$ is in the interior of $\Gamma(f)$. In particular, the interior of $\Gamma(f)$ is convex.

If $(\frac{1}{2} v_1 + \frac{1}{2} v_2, \frac{1}{2} x_1 + \frac{1}{2} x_2)$ is in the interior of $\Gamma(f)$, then, by Lemma A2.13, $s(v,f, \frac{1}{2} x_1 + \frac{1}{2} x_2)$ is continuous in v at $\frac{1}{2} v_1 + \frac{1}{2} v_2$, so Lemma A2.12 implies

$$s(\frac{1}{2} v_1 + \frac{1}{2} v_2, f, \frac{1}{2} x_1 + \frac{1}{2} x_2) \ge \frac{1}{2} s(v_1,f,x_1) + \frac{1}{2} s(v_2,f,x_2) .$$

Thus, $s(v,f,x)$ is semi-concave* on the interior of $\Gamma(f)$.

---

* We say that a function h on a convex set is *semi-concave* if

$$h(\frac{1}{2} \xi_1 + \frac{1}{2} \xi_2) \ge \frac{1}{2} h(\xi_1) + \frac{1}{2} h(\xi_2)$$

for all ξ_1,ξ_2.

Now the argument given in the second half of the proof of Lemma A1.4 can easily be adapted to show that a locally bounded semi-concave function on an open convex set is continuous and hence also concave. (Just note that it suffices to make the argument with the integer n taken to be a power of 2.) Thus, to complete the proof, we have only to show that $s(v,f,x)$ is locally bounded in (v,x) on the interior of $\Gamma(f)$. We already have the upper bound.

$$s(v,f,x) \le 1 + \log(v) \quad .$$

To get a local lower bound, we let (v_0,x_0) be in the interior of $\Gamma(f)$, and we choose $v' < v_0$ such that (v',x_0) is still in the interior of $\Gamma(f)$. Then x_0 is in the interior of $\{x: s(v',f,x) > -\infty\}$. By Corollary A2.8, $s(v',f,x)$ is continuous in x' at x_0, so there is a constant C and a neighborhood N_{x_0} of x_0 such that

$$s(v',f,x\ ) > -C \quad \text{for} \quad x \in N_{x_0}$$

Since s is increasing in v,

$$s(v,f,x) > -C \quad \text{for all} \quad v \ge v';\ x \in N_{x_0}\ , \quad \text{so} \quad \{(v,x):\ v > v',\ x \in N_{x_0}\}$$

is a neighborhood of (v_0,x_0) on which $s(v,f,x)$ is bounded below, so $s(v,f,x)$ is locally bounded below at (v_0,x_0).

The difficulty is showing that $\Gamma(f)$ is actually convex has to do with some problems in taking thermodynamic limits with specific volume precisely equal to v_{min}. These problems will show up in a more serious way shortly. Nevertheless it seems to be fairly simple exercise in the techniques we have been using to prove that

$$\{(v,x):\ s(v+,f,x) > -\infty\}$$

is, in fact, convex (and differs from $\Gamma(f)$ only by adding some more boundary points).

So far, we have obtained a result on the concavity of s in $v = \frac{1}{\rho}$. There is a corresponding concavity in ρ; indeed $\rho s(\frac{1}{\rho},\ f,\ \frac{x}{\rho})$ is concave in (ρ,x), in the interior of the set on which it is greater than $-\infty$. We will give the argument for the concavity in ρ at fixed x; the changes necessary to prove joint concavity are trivial.

Lemma A2.15

Let $h(v)$ *be concave in* v *on some interval* $(v_1,v_2), v_1 \ge 0$. *Then* $h(\frac{1}{\rho})$ *is concave in* ρ *on* $(\frac{1}{v_2},\ \frac{1}{v_1})$.

Proof. Let ρ_1,ρ_2, and α, $0 \le \alpha \le 1$ be given; let $\rho = \alpha\rho_1 + (1-\alpha)\rho_2$. Consider $\alpha\rho_1 h(\frac{1}{\rho_1}) + (1-\alpha)\rho_2 h(\frac{1}{\rho_2}) = \rho[\beta h(\frac{1}{\rho_1}) + (1-\beta)h(\frac{1}{\rho_2})]$

$$\beta = \frac{\alpha\rho_1}{\rho};\quad 1-\beta = \frac{(1-\alpha)\rho_2}{\rho} \quad .$$

By the concavity of h, (and the positivity of ρ), this is not greater than

$$\rho h(\beta \frac{1}{\rho_1} + (1-\beta)\frac{1}{\rho_2}) \quad .$$

But

$$\frac{\beta}{\rho_1} + \frac{(1-\beta)}{\rho_2} = \frac{\alpha\rho_1}{\rho \cdot \rho_1} + \frac{(1-\alpha)\rho_2}{\rho \cdot \rho_2} = \frac{\alpha}{\rho} + \frac{(1-\alpha)}{\rho} = \frac{1}{\rho} \quad ,$$

so

$$\alpha\rho_1 h(\frac{1}{\rho_1}) + (1-\alpha)\rho_2 h(\frac{1}{\rho_2}) \leq \rho h(\frac{1}{\rho}) \quad .$$

A3. THERMODYNAMIC LIMITS ALONG MORE GENERAL SEQUENCES OF REGIONS

To investigate the limit of $\frac{\log V(\Lambda,N,f,J)}{N}$ as Λ becomes infinitely large, we need to specify in precisely what sense the region Λ becomes large. A first attempt, which will not quite suffice for our purposes, is the following: A sequence (Λ_n) of bounded open sets in $\mathbb{R}^\nu$ is said to *become infinitely large in the sense of van Hove* if, for all $r > 0$, the volume (Lebesgue measure) of the set of points within a distance r of the boundary of Λ_n, (both inside and outside Λ_n), divided by the volume of Λ_n itself, goes to zero as $n \to \infty$.* If (Λ_n) becomes infinitely large in the sense of van Hove, then $V(\Lambda_n)$ goes to infinity. A sequence of rectangular parallelepipeds becomes infinitely large in the sense of van Hove if and only if the lengths of all edges go to infinity with n. If Λ is a non-empty bounded open set whose boundary has Lebesgue measure zero, and if (λ_n) is a sequence of positive numbers going to infinity, then the sequence of dilates $(\lambda_n\Lambda)$ becomes infinitely large in the sense of van Hove.

Our first result on limits along general sequences of regions requires only convergence to infinity in the sense of van Hove:

Proposition A3.1.

Let Λ_n *be a sequence of bounded open sets becoming infinitely large in the sense of van Hove, and let*

$$\lim_{n\to\infty} \frac{V(\Lambda_n)}{N_n} = v > 0 \quad .$$

Then

$$\liminf_{n\to\infty} \frac{1}{N_n} \log V(\Lambda_n, N_n, f, J) \geq s(v,f,J) \tag{1}$$

---

* This definition of van Hove convergence is superficially different from that given in Ruelle [1], Definition 2.11. It is not hard to see that the two forms of the definition are in fact equivalent.

Proof. The idea is simple. We almost fill Λ_n, for large n, with translates of a given $\Lambda_v(\underset{\sim}{r})$, and distribute the N_n particles in these boxes. Then, using Lemma A2.2, we obtain a lower bound for $V(\Lambda_n, N_n, f, J)$ which, after passage to the limit $n \to \infty$ and then the limit $\underset{\sim}{r} \to \infty$, gives the inequality (1). To make the argument work, we must be able to fit approximately $N_n/N(\underset{\sim}{r})$ copies of $\Lambda_v(\underset{\sim}{r})$ into Λ_n; this requires a volume which is precisely of the order of that of Λ_n and makes the packing problem delicate. These difficulties disappear if we replace v by a slightly smaller specific volume, so we will prove an intermediate result:

Lemma A3.2

If $v' < v$, *and if* $\hat{J}$ *is a bounded convex open set with closure contained in* J, *then*

$$\liminf_n \frac{1}{N_n} \log V(\Lambda_n, N_n, f, J) \geq s(v', f, \hat{J})$$

We then recover our original assertion (1) by proving

Lemma A3.3

$$s(v,f,J) = \sup\{s(v',f,\hat{J}) : v' < v;\ \hat{J}\ \textit{bounded, convex open, with closure contained in}\ J\} .$$

These two lemmas immediately imply the proposition. We prove them in reverse order.

Proof of Lemma A3.3. We have:

$$s(v,f,J) = \sup_{x \in J}\{s(v,f,x)\} ,$$

so the lemma would be immediate if we knew that $s(v,f,x)$ was continuous in (v,x). We do know that it is continuous on the interior of $\Gamma(f)$, so we have only to prove

$$s(v,f,J) = \sup\{s(v,f,x) : x \in J : (v,x) \in \text{int}\,\Gamma(f)\} \tag{2}$$

To prove this, we first note that, since there exists $(v_1,x_1) \in \Gamma(f)$ with $v_1 < v$, $(v,x) \in \text{int}\,\Gamma(f)$ if $x \in \text{int}\{y : s(v,f,y) > -\infty\}$.

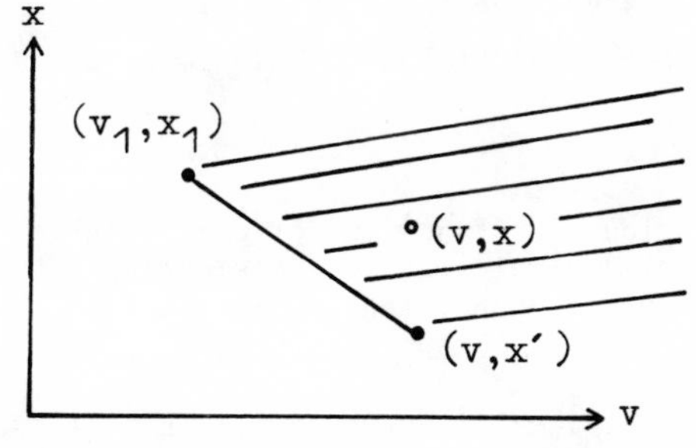

(If x' is chosen so that $s(v,f,x') > -\infty$, then everything to the right of the line joining (v_1,x_1) to (v,x') is in $\Gamma(f)$, so (v,x) is in the interior of $\Gamma(f)$.)

Thus, if I denotes the interior of $\{y: s(v,f,y) > -\infty\}$, we want to show that

$$s(v,f,J) = \sup\{s(v,f,x): x \in I \cap J\} \quad .$$

Now $I \cap J$ is dense in $\{y \in J: s(v,f,Y) > -\infty\}$, and the supremum of a concave function on a convex open set on which the function is finite-valued is the same as the supremum on the closure of that set,* so the lemma is proved.

Note that we use in the proof of the lemma the assumption that the observables do not take on the value $+\infty$. We can remove this assumption if we assume instead that v is strictly larger than the reciprocal of the close-packing density for the observable under consideration. If, on the other hand, v is strictly smaller than the reciprocal of the close-packing density, there is no problem with either the lemma or the proposition. For v exactly equal to the reciprocal of the close-packing density, our argument fails and the proposition could conceivably be false.

<u>Proof of Lemma A3.2</u>. It will suffice to prove:

$$\liminf_n \frac{1}{N_n} \log V(\Lambda_n,N_n,f,J) \geq \frac{\log V_{v'}(\underset{\sim}{r},\hat{J})}{N(\underset{\sim}{r})} \qquad (\dagger)$$

for all $\underset{\sim}{r}$. We will prove this by putting $[N_n/N(\underset{\sim}{r})]$** translates of $\Lambda_{v'}(\underset{\sim}{r})$ in Λ_n with each pair of translates separated by a distance of at least R, and putting $N(\underset{\sim}{r})$ particles in each translate. Since N_n need not be divisible by $N(\underset{\sim}{r})$, there may be some particles left over; these we deal with by arranging them in the unused volume of Λ_n in such a way that they do not "interact" with each other or with the particles in the translates of $\Lambda_{v'}(\underset{\sim}{r})$. Adding these left-over particles changes N without changing f; hence, changes f/N by a small fraction of its original value. In particular, if n is large enough and the particles in the translates of $\Lambda_{v'}(\underset{\sim}{r})$ have f/N in $\hat{J}$, then the whole configuration has f/N in J, so we get an approximate lower bound:

$$V(\Lambda_n,N_n,f,J) \gtrsim [V(\Lambda_{v'}(\underset{\sim}{r}),N(\underset{\sim}{r}),f,\hat{J})]^{[N_n/N(\underset{\sim}{r})]}$$

(The bound is only approximate since there should also be on the right-hand side a contribution from the left-over particles). This lower bound yields readily the as-

---

* This fact is nearly obvious for functions of one variable; we may reduce the many-variable case to the one-variable case by considering the behavior of the function along straight line segments connecting the interior of the set with points on the boundary.

** If x is a real number, $[x]$ will denote the largest integer not greater than x.

sertion in Eq. (†) and hence the proposition.

Turning to the technical details, we prove first a lemma about sequences of regions becoming infinitely large in the sense of van Hove.

Lemma A3.4

Let Λ_n *be a sequence of regions becoming infinitely large in the sense of van Hove, and let* Λ' *be a fixed rectangular parallelepiped. Let* k_n *denote the largest number of non-overlapping translates of* Λ' *which can be fitted into* Λ_n. *Then*

$$\lim_{n\to\infty} \frac{k_n V(\Lambda')}{V(\Lambda_n)} = 1$$

Proof. Let d denote the diameter of Λ', and let $\tilde{\Lambda}_n$ denote the set of points in Λ_n at a distance greater than d from the boundary of Λ_n. We cover $\tilde{\Lambda}_n$, up to sets of measure zero, by non-overlapping translates of Λ' (and we include in the covering only translates which actually intersect $\tilde{\Lambda}_n$.) Since each of the translates is entirely contained in Λ_n, the number of translates is not greater than k_n. Thus, the total volume of the family of translates is on the one hand not greater than $k_n V(\Lambda')$ and on the other hand not less than $V(\tilde{\Lambda}_n)$. Thus

$$V(\Lambda_n) \geq k_n V(\Lambda') \geq V(\tilde{\Lambda}_n) \quad .$$

But by the definition of becoming infinitely large in the sense of van Hove,

$$\lim_{n\to\infty} \frac{V(\tilde{\Lambda}_n)}{V(\Lambda_n)} = 1 \quad ,$$

so the lemma is proved.

We apply the lemma, keeping the same notation, with Λ' the rectangular parallelepiped $\{(q_1, \ldots, q_\nu): 0 < q_i < r_i \ell_\nu, 1 \leq i \leq \nu\}$. By centering a translate of $\Lambda_{v'}(\underset{\sim}{r})$ in each translate of Λ', we can put k_n translates of $\Lambda_{v'}(\underset{\sim}{r})$ in Λ_n in such a way that the distance between any two of these translates is at least R. Since

$$V(\Lambda') = v' N(\underset{\sim}{r}) \quad \text{and} \quad \lim_{n\to\infty} \frac{N_n}{V(\Lambda_n)} = v \quad ,$$

we have

$$\lim_{n\to\infty} \frac{k_n N(\underset{\sim}{r})}{N_n} = \frac{v}{v'} > 1$$

Writing

$$N_n = p_n N(\underset{\sim}{r}) + q_n \qquad 0 \leq q_n < N(\underset{\sim}{r}) \quad ,$$

we see that, for sufficiently large n,

$$k_n \geq p_n + q_n \quad .$$

For n large enough so this is true, we choose $p_n + q_n$ translates of $\Lambda_v(\underset{\sim}{r})$ in Λ_n any pair of which are separated by at least R. In each of p_n of these translates we put $N(\underset{\sim}{r})$ particles; in each of the remaining q_n we put a single particle. By lemma A2.2, we have:

$$V(\Lambda_n, N_n, f, \frac{p_n N(\underset{\sim}{r})}{N_n} \hat{J} + \frac{q_n}{N_n} J') \geq [V_{v'}(\underset{\sim}{r}, \hat{J})]^{p_n} [V(\Lambda_{v'}(\underset{\sim}{r}), 1, f, J')]^{q_n}$$

for any open convex J'. Since

$$f(q_1) \equiv 0$$

by the definition of observable,

$$V(\Lambda_{v'}(\underset{\sim}{r}), 1, f, J') = V(\Lambda_{v'}(\underset{\sim}{r}))$$

if J' contains 0. If we take n large enough and J' small enough (but containing 0), we have

$$\frac{p_n N(\underset{\sim}{r}) J}{N_n} \hat{J} + \frac{q_n}{N_n} J' \subset J$$

so, for large n,

$$V(\Lambda_n, N_n, f, J) \geq [V_{v'}(\underset{\sim}{r}, \hat{J})]^{p_n} [V(\Lambda_{v'}(\underset{\sim}{r}))]^{q_n} \quad ,$$

i.e.

$$\frac{1}{N_n} \log V(\Lambda_n, N_n, f, J) \geq \frac{p_n N(\underset{\sim}{r})}{N_n} \frac{1}{N(\underset{\sim}{r})} \log[V_{v'}(\underset{\sim}{r}, \hat{J})]$$
$$+ \frac{q_n}{N_n} \log[V(\Lambda_{v'}(\underset{\sim}{r}))] \quad .$$

As $n \to \infty$, $\frac{p_n N(\underset{\sim}{r})}{N_n} \to 1$ and $\frac{q_n}{N_n} \to 0$, so

$$\liminf_{n \to \infty} \frac{1}{N_n} \log V(\Lambda_n, N_n, f, J) \geq \frac{1}{N(\underset{\sim}{r})} \log[V_{v'}(\underset{\sim}{r}, \hat{J})]$$

and the proposition is proved.

We have thus obtained half of the proof of the existence of the limit of $\frac{1}{N} \log V$ for a general sequence of regions. For the second half, we need to impose another restriction on the sequence of regions. We say that a sequence (Λ_n) of regions becoming infinitely large in the sense of van Hove is *approximable by rectangles* if there is a sequence $\hat{\Lambda}_n$ of rectangular parallelepipeds with edges parallel to the coordinate axes such that

1. $\Lambda_n \subset \hat{\Lambda}_n$ for each n.
2. $\liminf_n \frac{V(\Lambda_n)}{V(\hat{\Lambda}_n)} \neq 0$.

(The sequence of spherical shells

$$\{q: n > |q| > n - \sqrt{n}\ \}$$

is an example of a sequence of regions becoming infinitely large in the sense of van Hove but not approximable by rectangles.)

Theorem A3.5

Let Λ_n *be a sequence of bounded open sets becoming infinitely large in the sense of van Hove and approximable by rectangles, and let* N_n *be a sequence of integers such that*

$$\lim_{n\to\infty} \frac{V(\Lambda_n)}{N_n} = v$$

Then:

(a) *If* $s(v,f,J) > -\infty$,

$$\lim_{n\to\infty} \frac{1}{N_n} \log \mathcal{V}(\Lambda_n, N_n, f, J) = s(v,f,J)$$

(b) *If* $s(v',f,J) = -\infty$ *for some* $v' > v$,

$$\mathcal{V}(\Lambda_n, N_n, f, J) = 0 \text{ for sufficiently large } n \ .$$

(c) $\limsup\limits_{n\to\infty} \frac{1}{N_n} \log \mathcal{V}(\Lambda_n, N_n, f, J) \le \lim\limits_{v'\downarrow v} s(v',f,J)$.

Proof. We consider first case (a). By Proposition A3.1 we know that

$$\liminf_{n} \frac{1}{N_n} \log \mathcal{V}(\Lambda_n, N_n, f, J) \ge s(v,f,J) \quad ,$$

so we have only to prove

$$\limsup_{n} \frac{1}{N_n} \log \mathcal{V}(\Lambda_n, N_n, f, J) \le s(v,f,J) \quad .$$

By translating the Λ_n's appropriately, we can assume that the $\tilde{\Lambda}_n$'s in the definition of approximability by rectangles have the form $\Lambda_V(\underset{\sim}{r}_n)$, and, enlarging $\underset{\sim}{r}_n$ if necessary, we can also assume

$$\frac{V(\Lambda_n)}{V(\tilde{\Lambda}_n)} \le \frac{1}{2} \ .$$

Again passing to a subsequence, we can assume that

$$\alpha = \lim_{n} \frac{V(\Lambda_n)}{V(\tilde{\Lambda}_n)} \quad \text{exists; we have}$$

$$0 < \alpha \le \frac{1}{2} \ .$$

Let Λ'_n be the set of points in $\tilde{\Lambda}_n$ at a distance greater than R from Λ_n, (see illustration)

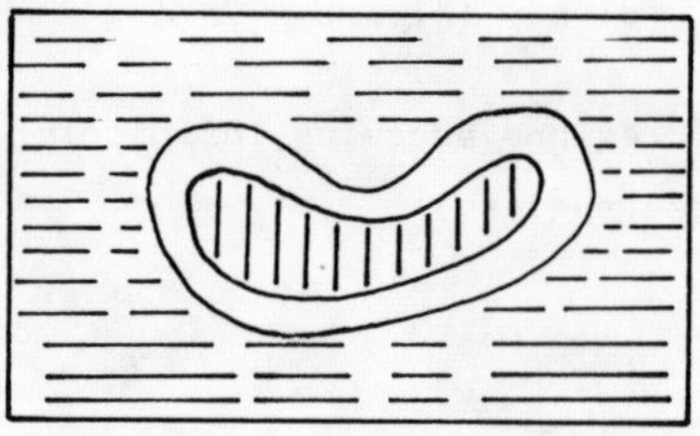

$\tilde{\Lambda}_n$ is the large rectangle; Λ_n the vertically shaded region, and Λ'_n the horizontally shaded region.

It follows readily from the definition that (Λ'_n) becomes infinitely large in the sense of van Hove. We define

$$N'_n = N(\underset{\sim}{r}_n) - N_n \quad ;$$

then

$$\lim_{n\to\infty} \frac{N'_n}{N(\underset{\sim}{r}_n)} = 1 - \lim_{n\to\infty} \frac{N_n}{N(\underset{\sim}{r}_n)} = 1 - \alpha, \quad \text{and}$$

$$\lim_{n\to\infty} \frac{V(\Lambda'_n)}{N'_n} = \lim_{n\to\infty} \frac{V(\Lambda'_n)}{V(\tilde{\Lambda}_n)} \cdot \frac{V(\tilde{\Lambda}_n)}{N(\underset{\sim}{r}_n)} \cdot \frac{N(\underset{\sim}{r}_n)}{N'_n}$$

$$= \frac{1}{1-\alpha} \cdot v \cdot (1-\alpha) = v \quad .$$

Now we apply Lemma A2.2:

$$V(\Lambda_v(\underset{\sim}{r}_n), N(\underset{\sim}{r}_n), f, J) \geq V(\Lambda_n, N_n, f, J) \cdot V(\Lambda'_n, N'_n, f, J) \quad .$$

Taking the logarithm, dividing by $N(\underset{\sim}{r}_n)$, and using

$$\liminf_{n\to\infty} \frac{1}{N'_n} \log V(\Lambda'_n, N'_n, f, J) \geq s(v,f,J), \quad \text{we get}$$

$$s(v,f,J) = \lim_{n\to\infty} \frac{1}{N(\underset{\sim}{r}_n)} \log V(\Lambda_v(\underset{\sim}{r}_n), N(\underset{\sim}{r}_n), f, J)$$

$$\geq \alpha[\limsup_{n\to\infty} \frac{1}{N_n} \log V(\Lambda_n, N_n, f, J)]$$

$$+ (1-\alpha) s(v,f,J) \quad .$$

Since $\alpha > 0$,

$$\limsup_{n\to\infty} \frac{1}{N_n} \log V(\Lambda_n, N_n, f, J) \leq s(v,f,J) \quad ,$$

so the assertion is proved.

We will not give the detailed proofs of (b) and (c); they are similar to the argument just given except that one assigns to $\tilde{\Lambda}_n$ a specific volume $\tilde{v}$ slightly larger than v.

This theorem says that

$$\lim_{n\to\infty} \frac{1}{N_n} \log V(\Lambda_n, N_n, f, J) = s(v,f,J)$$

whenever either

(a) $\{v\} \times J \cap \Gamma(f) \neq \emptyset$ (so $s(v,f,J) > -\infty$)

or

(b) the distance from $\{v\} \times J$ to $\Gamma(f)$ is non-zero. If, however, the distance from J to $\{x: s(v,f,x) > -\infty\}$ is zero, so J "touches" $\{x: s(v,f,x) > -\infty\}$, but does not actually intersect it, we may have

$$\lim_{v'\downarrow v} s(v',f,J) > s(v,f,J) \quad ,$$

and in this case we are unable to deduce the existence of

$$\lim_{n\to\infty} \frac{1}{N_n} \log V(\Lambda_n, N_n, f, J) \quad .$$

This small defect in our results does not seriously affect the physical conclusions which can be drawn; the reader should convince himself that the interpretation given in Section A1 for the behavior of

$$s(v,(u,f_2),(\varepsilon,x_2)) \quad ,$$

as a function of x_2, remains unchanged.

A4. A DIGRESSION: SUMS OF INDEPENDENT RANDOM VARIABLES

The ideas we have been developing can be modified easily to give results about sums of independent random variables, or, equivalently, about the statistical mechanics of collections of large numbers of identical non-interacting systems. These results are interesting both in their own right and for the insight they give into the general theory; they also supply a number of amusing exercises.

We start with a space Ω equipped with a measure μ. For applications to statistical mechanics, Ω may be, for example, a finite set representing some set of states of an atom or molecule, and the μ-measure of a subset of Ω is just the number of states in that set. Alternatively, Ω may be the configuration space for a continuous system, in which case μ is Lebesgue measure. These two examples show that there is no physical reason to assume μ is normalized. For reasons of technical convenience, we will want to assume that μ is finite (although a good deal of the theory can be carried through without this assumption), so we can always normalize it. The normalization constant plays no important role in anything we do, and it clutters up the formulas, so we put it equal to one, i.e., we assume μ is normalized. We let μ^N be the product of N copies of μ on Ω^N.

Let f be a measurable function on Ω; to allow us to consider several numerical functions simultaneously, we let f take values in $\mathbb{R}^t$. We will study the asymptotic properties of the distribution of

$$\frac{1}{N}\,[f(\omega_1) + \ldots + f(\omega_N)] \quad .$$

with respect to μ^N, for large N. In the language of probability theory, we are investigating the behavior of a sum of N independent random variables, all with the same distribution as f. From most points of view, this behavior is extremely well understood: If f is integrable, the weak law of large numbers says that, $\dfrac{f(\omega_1) + \ldots + f(\omega_N)}{N}$ converges in probability to $\int f d\mu$, the expected value of f, i.e., for any $\varepsilon > 0$,

$$\lim_{N\to\infty} \mu^N\{(\omega_1,\ldots,\omega_N): \left|\frac{f(\omega_1) + \ldots + f(\omega_N)}{N} - \int f d\mu\right| > \varepsilon\} = 0 \quad .$$

Moreover, if f is square-integrable, the central limit theorem gives very precise information about the probability of small fluctuations (of the order of $\frac{1}{\sqrt{N}}$) about the limiting value. We want to ask instead questions about large fluctuations, i.e., fluctuations of order 1. Specifically, we consider

$$\mu^N\{(\omega_1,\ldots,\omega_N): \frac{f(q_1) + \ldots + f(q_N)}{N} \in J\} \quad ,$$

where J is an open convex set in $\mathbb{R}^t$. This quantity is an analogue of the quantities $V(\Lambda,N,f,J)$ Considered earlier, and it can be analyzed in essentially the same way. Thus we have:

<u>Lemma A4.1</u>

$$\mu^{N+M}\{(\omega_1,\ldots,\omega_{N+M}): \frac{f(\omega_1) + \ldots + f(\omega_{N+M})}{N + M} \in J\}$$

$$\geq \mu^N\{(\omega_1,\ldots,\omega_N): \frac{f(\omega_1) + \ldots + f(\omega_N)}{N} \in J\}$$

$$\times\ \mu^M\{(\omega_1,\ldots,\omega_M): \frac{f(\omega_1) + \ldots + f(\omega_M)}{M} \in J\} \quad ,$$

i.e.

$$\log[\mu^N\{(\omega_1,\ldots,\omega_N): \frac{f(\omega_1) + \ldots + f(\omega_N)}{N} \in J\}]$$

is a superadditive function of N.

By Lemma A2.4, we thus get:

Proposition A4.2

$$\lim_{N\to\infty} \frac{1}{N} \log[\mu^N\{(\omega_1,\ldots,\omega_N): \frac{f(\omega_1) + \ldots + f(\omega_N)}{N} \in J\}]$$

exists and is equal to

$$\sup_N \frac{1}{N} \log[\mu^N\{(\omega_1,\ldots,\omega_N): \frac{f(\omega_1) + \ldots + f(\omega_N)}{N} \in J\}] \quad .$$

To be entirely precise, we must worry about the fact that

$$\mu^N\{(\omega_1,\ldots,\omega_N): \frac{f(\omega_1) + \ldots + f(\omega_N)}{N} \in J\}$$

may be zero for some N. An elementary argument (which we will omit) shows that if it is non-zero for one N, it is non-zero for all sufficiently large N and that the limit does in fact exist.

We define

$$s(f,J) = \lim_{N\to\infty} \frac{1}{N} \log[\mu^N\{(\omega_1,\ldots,\omega_N): \frac{f(\omega_1) + \ldots + f(\omega_N)}{N} \in J\})$$

and, for all $x \in \mathbb{R}^t$, we define

$$s(f,x) = \inf\{s(f,J): x \in J\} \quad .$$

In the same way as we proved Proposition A2.7, we prove:

Proposition A4.3

$s(f,x)$ *is upper semi-continuous in* x, *and*

$$s(f,J) = \sup\{s(f,x): x \in J\} \quad .$$

Theorem A4.4

$s(f,x)$ *is a concave function of* x.

Proof.

$$\begin{aligned}\mu^{2N}\{(\omega_1,\ldots,\omega_{2N}): \frac{f(\omega_1) + \ldots + f(\omega_{2N})}{2N} \in \frac{1}{2} J_1 + \frac{1}{2} J_2\} \\ \geq \mu^N\{(\omega_1,\ldots,\omega_N): \frac{f(\omega_1) + \ldots + f(\omega_N)}{N} \in J_1\} \\ \times \mu^N\{(\omega_1,\ldots,\omega_N): \frac{f(\omega_1) + \ldots + f(\omega_N)}{N} \in J_2\} \quad .\end{aligned}$$

Hence,

$$s(f, \frac{1}{2} J_1 + \frac{1}{2} J_2) \geq \frac{1}{2} s(f,J_1) + \frac{1}{2} s(f,J_2) \quad ;$$

hence

$$s(f, \frac{1}{2} x_1 + \frac{1}{2} x_2) \geq \frac{1}{2} s(f,x_1) + \frac{1}{2} s(f,x_2) \quad .$$

Proposition A4.5

The closure of $\{x: s(f,x) > -\infty\}$ *is the closed convex hull of the essential range of* f. *If the components of* f *are linearly independent modulo the constant functions, the interior of* $\{x: s(f,x) > -\infty\}$ *is non-empty.*

From now on, we will assume that the components of f are linearly independent modulo the constants.

In the present simple situation, unlike the more complicated one considered earlier, we can actually compute $s(f,x)$ in many cases.

Example: The coin-tossing problem. f takes on two values, say 0 and 1, and $\mu\{\omega: f(\omega) = 0\} = \frac{1}{2} = \mu\{\omega: f(\omega) = 1\}$.

Then

$$\mu^N\{(\omega_1,\dots,\omega_N): \alpha < \frac{f(\omega_1) + \dots + f(\omega_N)}{N} < \beta\}$$
$$= \frac{1}{2^N} \sum_{\alpha N<j<\beta N} \binom{N}{j} \quad .$$

If we temporarily denote

$$Q_N = \sup_{\alpha N<j<\beta N} \binom{N}{j} \quad ,$$

we have

$$\frac{Q_N}{2^N} \leq \frac{1}{2^N} \sum_{\alpha N<j<\beta N} \binom{N}{j} \leq \frac{(\beta - \alpha) \cdot N \cdot Q_N}{2^N}$$

Taking logarithms, dividing by N, and letting N go to infinity, and using the fact that

$$\frac{1}{N} \log(\beta - \alpha) \cdot N \cdot Q_N = \frac{1}{N} \log Q_N + \frac{1}{N} \log[(\beta - \alpha) \cdot N]$$

approaches $\frac{1}{N} \log Q_N$ as N goes to infinity, we see that

$$s(f,(\alpha,\beta)) = \lim_{N\to\infty} \frac{1}{N} \log Q_N - \log 2 \quad .$$

This is a simple example of a phenomenon which occurs again and again: only the largest contribution counts when computing the entropy.

We now use Stirling's formula: For $0 < \alpha < \beta < 1$, and $\alpha N < j < \beta N$, we have:

$$\frac{1}{N} \log\binom{N}{j} = \frac{1}{N} \log[\frac{N^N}{j^j (N - j)^{N-j}}] + o(1)$$

$$= -\log[(\tfrac{j}{N})^{j/N}(1 - j/N)^{(1-j/N)}] + o(1) \quad .$$

Hence:

$$\lim_{N\to\infty} \frac{1}{N} \log Q_N = \sup\{-x \log x - (1 - x)\log(1 - x) : \alpha < x < \beta\}$$

so

$$s(f,(\alpha,\beta)) = \sup\{-x \log x - (1 - x)\log(1 - x) - \log 2 :$$
$$\alpha < x < \beta\} \quad ,$$

so finally

$$s(f,x) = \begin{cases} -x \log x - (1 - x)\log(1 - x) - \log 2 & 0 \le x \le 1 \\ -\infty & \text{otherwise.} \end{cases}$$

This function has the expected qualitative properties; it is zero at $\frac{1}{2}$ (the mean value of f), and it is concave in x. Note that it approaches a finite value $(-\log 2)$ as x approaches zero or one; then jumps discontinuously to $-\infty$. This behavior is not so surprising since, for example,

$$\mu^N\{(\omega_1,\ldots,\omega_N) : f(\omega_1) + \ldots + f(\omega_N) = 0\} = \frac{1}{2^N}$$

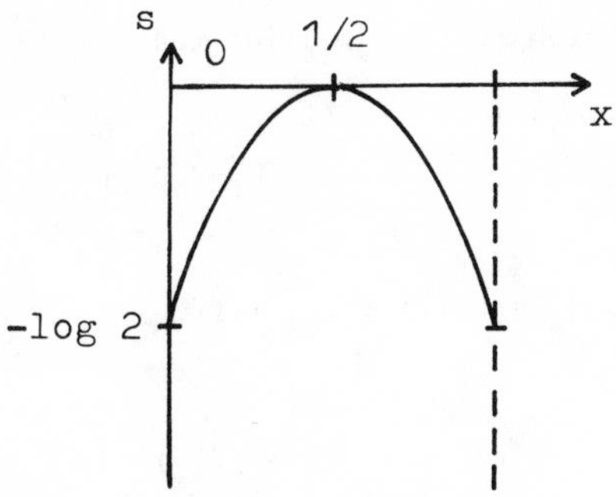

There are various other examples for which the entropy can be computed in this elementary way. We mention only one other: If f has distribution $\frac{1}{\pi} \frac{1}{1+x^2} dx$, it is easily verified that $\frac{f(\omega_1) + \ldots + f(\omega_N)}{N}$ has the same distribution, and hence that $s(f,x) \equiv 0$. (Such a function would seem to violate the weak law of large numbers; it doesn't because it is non-integrable.)

We will next give a general technique for computing $s(f,x)$. To avoid technical difficulties, we will assume for the moment that f is bounded and real-valued rather than vector-valued. For each real β, we define

$$Z(\beta) = \int e^{-\beta f(\omega)} \mu(d\omega) \quad ;$$

$$\mu_\beta(d\omega) = \frac{e^{-\beta f(\omega)}}{Z(\beta)} \mu(d\omega) \quad .$$

Note that

$$\frac{d}{d\beta}\log(Z(\beta)) = \frac{-\int f(\omega)e^{-\beta f(\omega)}\mu(d\omega)}{\int e^{-\beta f(\omega)}\mu(d\omega)} = -\int f(\omega)\mu_\beta(d\omega); = -E_\beta(f)$$

where $E_\beta(f)$ denotes $\int f(\omega)\mu_\beta(d\omega)$. Differentiating again we get

$$\frac{d^2}{d\beta^2}\log(Z(\beta)) = +\int[f(\omega) - E_\beta(f)]^2\mu_\beta(d\omega) \geq 0 \quad .$$

Thus, $\log Z(\beta)$ is a strictly convex function of β, and $E_\beta(f)$ is a strictly decreasing function of β. As β approaches ∞, $E_\beta(f)$ approaches the essential infimum of f, and, as β approaches $-\infty$, $E_\beta(f)$ approaches the essential supremum of f. Thus, if ess. $\inf\{f\} < x <$ ess. $\sup\{f\}$, there is a uniquely determined $\beta = \beta(x)$ such that

$$E_\beta(f) = x \quad .$$

Since $E_\beta(f)$ is strictly decreasing and real-analytic in β, the inverse function theorem implies that $\beta(x)$ is a real analytic function of x.

Now let μ_β^N denote the product of N copies of μ_β as a measure on Ω^N. If $\beta = \beta(x)$, then

$$\int f(\omega)\mu_\beta(d\omega) = x \quad ,$$

so, if $a < x < b$, the law of large numbers implies that

$$\lim_{N\to\infty} \mu_\beta^N \left\{\frac{f(\omega_1) + \ldots + f(\omega_N)}{N} \in (a,b)\right\} = 1 \quad . \tag{1}$$

But

$$\mu_\beta^N(d\omega_1 \ldots d\omega_N) \cdot Z(\beta)^N e^{\beta\cdot(f(\omega_1) + \ldots + f(\omega_N))} = \mu^N(d\omega_1 \ldots d\omega_N) \quad ,$$

so

$$\mu^N\{(\omega_1,\ldots,\omega_N): \frac{f(\omega_1) + \ldots + f(\omega_N)}{N} \in (a,b)\}$$

lies between

$$Z(\beta)^N e^{N\beta a}\, \mu_\beta^N\{(\omega_1,\ldots,\omega_N): \frac{f(\omega_1) + \ldots + f(\omega_N)}{N} \in (a,b)\}$$

and

$$Z(\beta)^N e^{N\beta b}\, \mu_\beta^N\{(\omega_1,\ldots,\omega_N): \frac{f(\omega_1) + \ldots + f(\omega_N)}{N} \in (a,b)\}$$

(We have simply replaced $e^{\beta(f(\omega_1) + \ldots + f(\omega_N))}$ by the upper and lower bounds $e^{N\beta a}$ and $e^{N\beta b}$ on the set in question.) Taking logarithms, dividing by N, and using (1), we get that $s(f,(a,b))$ is between $\beta a + \log Z(\beta)$ and $\beta b + \log Z(\beta)$ if $a < x < b$. Letting $a \uparrow x$, $b \downarrow x$, we find

$$s(f,x) = \beta x + \log Z(\beta) \quad , \tag{2}$$

where, we repeat, β is the unique solution of the equation

$$x = \int f(\omega)\mu_\beta(d\omega)$$

The formula may be rewritten in a more suggestive form: If we define

$$\rho_\beta(\omega) = \frac{e^{-\beta f(\omega)}}{Z(\beta)} ,$$

so

$$\mu_\beta(d\omega) = \rho_\beta(\omega)\mu(d\omega) ,$$

then

$$s(f,x) = -\int \rho_\beta(\omega)\log \rho_\beta(\omega)\mu(d\omega) ,$$

i.e., if we interpret f as the energy, then $s(f,x)$ is equal to the entropy computed in the canonical ensemble with the temperature chosen so that the mean energy is x. From the formula (2) and the fact that β is a real-analytic function of x, it follows that $s(f,x)$ is a real-analytic function of x. We have, moreover,

$$\frac{ds(f,x)}{dx} = \beta + x\frac{d\beta}{dx} + \frac{d}{d\beta}(\log Z(\beta))\frac{d\beta}{dx}$$

But $\frac{d}{d\beta}\log Z(\beta) = -E_\beta(f) = -x$, so

$$\frac{ds(f,x)}{dx} = \beta \tag{3}$$

Differentiating again we get

$$\frac{d^2s(f,x)}{dx^2} = \frac{d\beta}{dx} = \frac{1}{\frac{d}{d\beta}E_\beta(f)} < 0 ,$$

so $s(f,x)$ is a strictly concave function of x.

We note two other relations which follow from the above. For any x', we consider $s(f,x') - \beta x'$. Graphically, this is obtained by finding the y-intercept of the line through the point $(x',s(f,x'))$ with slope β.

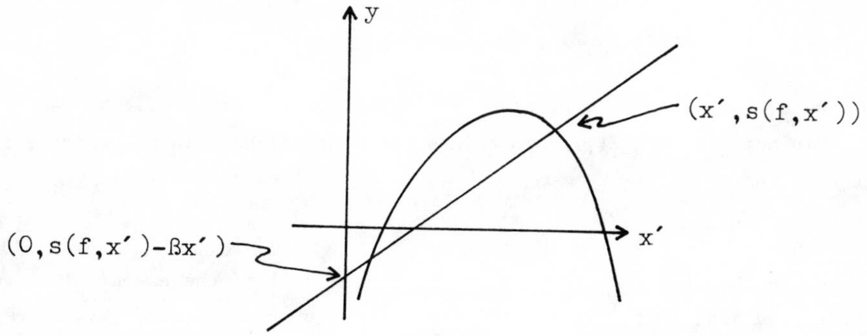

It is geometrically obvious, and easy to prove analytically, that this quantity, as a function of x', takes on its maximum when the line is tangent to the graph of $s(f,x)$, i.e., when

$$\beta = \frac{ds(f,x')}{dx'} \quad ,$$

i.e., $\beta = \beta(x)$. But, for this choice of x,

$$s(f,x) - \beta x = \log Z(\beta) \quad ,$$

so we have

$$\log Z(\beta) = \sup_{x'}\{s(f,x\ ) - \beta x'\} \tag{4}$$

In a similar way, one shows:

$$s(f,x) = \inf_{\beta'}\{\log Z(\beta') + \beta' x\} \tag{5}$$

We next turn to the consideration of a vector-valued function f. We continue to assume that f is (essentially) bounded. If we let $\beta = (\beta_1, \ldots, \beta_t) \in \mathbb{R}^t$,

$$\beta \cdot f(\omega) = \sum_{i=1}^{t} \beta_i f_i(\omega) \quad ,$$

then all the preceding analysis goes through, with only minor changes of notation, except for the monotonicity argument showing that any interesting value of x can be obtained as $\int f(\omega)\mu_\beta(d\omega)$ for appropriately chosen β. This monotonicity argument is replaced by the following:

<u>Lemma A4.6</u>

If x *is in the interior of the convex hull of the essential range of* f *then there is a unique* $\beta = \beta(x) \in \mathbb{R}^t$ *such that*

$$x = \int f(\omega)\mu_\beta(d\omega) \quad .$$

<u>Proof</u>. Subtracting a constant from f does not change $\mu_\beta(d\omega)$, so we may subtract x from f and therefore assume $x = 0$. Since

$$\int f(\omega)\mu_\beta(d\omega) = -\mathrm{grad}_\beta(\log Z(\beta))$$

we want to find a place where the gradient of $\log Z(\beta)$ is zero, i.e., we want to find a local minimum for $\log Z(\beta)$. To be assured that a local minimum exists, it suffices to show that

$$\lim_{|\beta|\to\infty} Z(\beta) = \infty \quad .$$

We now have to use the fact that zero is in the convex hull of the essential range of f. It is nearly evident that this implies that, for any $\beta \neq 0$ there exists $\varepsilon > 0$ such that

$$\mu\{\omega: \beta \cdot f(\omega) < -|\beta|\varepsilon\} > \varepsilon \quad .$$

A simple argument shows that ε can be chosen to be independent of the direction of β, i.e., there exists $\varepsilon > 0$ such that

$$\mu\{\omega: \beta \cdot f(\omega) < -|\beta|\varepsilon\} > \varepsilon$$

for all $\beta \neq 0$. Then

$$Z(\beta) = \int e^{-\beta \cdot f(\omega)} \mu(d\omega) \geq \int_{\{\omega:\beta\cdot f(\omega)<\varepsilon\beta\}} e^{-\beta f(\omega)} \mu(d\omega) \geq \varepsilon \cdot e^{\varepsilon|\beta|} ,$$

which goes to infinity as $|\beta| \to \infty$. This, then, proves the existence of the desired β. To show that there can be only one, we remark that $\log Z(\beta)$ is a strictly convex function of β and hence can have zero gradient at no more than one point.

Using this lemma, and the arguments developed above for real-valued f, it is easy to prove the following:

Theorem A4.7

Let f *be a bounded measurable function on* Ω *with values in* $\mathbb{R}^t$*; assume that the components of* f *are linearly independent modulo the constants. For* x *in the interior of the convex hull of the range of* f, *let* $\beta(x)$ *be the unique solution of*

$$x = \int f(\omega)\mu_\beta(d\omega)$$

Then

$$s(f,x) = \log Z(\beta(x)) + x \cdot \beta(x)$$

$s(f,x)$ *is a real-analytic and strictly concave function of* x,

$$\beta(x) = \operatorname{grad}_x s(f,x) ,$$

$$s(f,x) = \inf_{\beta'}\{\beta' x + \log Z(\beta')\} ,$$

$$\log Z(\beta) = \sup_{x'}\{s(f,x') - \beta \cdot x'\} .$$

The strict concavity of $s(f,x)$ in x implies in particular that, if we choose a particular bounded measurable function U on Ω, and if we consider

$$s((U,f), (\varepsilon,x)) ,$$

then for each ε this takes on its maximum as a function of x at a single point, which we denote by $\bar{f}(\varepsilon)$. The interpretation of $\bar{f}(\varepsilon)$ should by now be clear; if N is large, then $\frac{f(\omega_1) + \ldots + f(\omega_N)}{N}$ is very near $\bar{f}(\omega)$ for most $(\omega_1,\ldots,\omega_N)$ for which $\frac{U(\omega_1) + \ldots + U(\omega_N)}{N}$ is near ε. In particular, if we define a probability measure ${}^{N}\mu^{\text{m.c.}}_{(\varepsilon-\delta,\varepsilon)}$ on Ω^N by restricting μ^N to

$$\{(\omega_1,\ldots,\omega_N): \frac{U(\omega_1) + \ldots + U(\omega_N)}{N} \in (\varepsilon - \delta,\varepsilon)\}$$

and normalizing, then it is easy to show that

$$\bar{f}(\varepsilon) = \lim_{\delta\downarrow 0}\lim_{N\to\infty} \int {}^{N}\mu^{\text{m.c.}}_{(\varepsilon-\delta,\varepsilon)}(d\omega_1\ldots d\omega_N)\left|\frac{f(\omega_1) + \ldots + f(\omega_N)}{N}\right| \tag{6}$$

We now want to find a simpler way of evaluating $\bar{f}(\varepsilon)$. It will surely come as no surprise that we have:

Proposition A4.8

If β *is chosen so that*

$$\varepsilon = \frac{\int e^{-\beta U(\omega)} U(\omega)\,d\omega}{\int e^{-\beta U(\omega)}\,d\omega}$$

then

$$\bar{f}(\varepsilon) = \frac{\int e^{-\beta U(\omega)} f(\omega)\,d\omega}{\int e^{-\beta U(\omega)}\,d\omega} \tag{7}$$

Proof. We note first that, for all x,

$$s((U,f),\ (\varepsilon,x)) \le s(U,\varepsilon) \quad .$$

Since $\bar{f}(\varepsilon)$ was defined to be the unique value of x for which the left-hand side of this inequality takes on its supremum, it suffices to show that

$$s((U,f),\ (\varepsilon,\hat{f})) = s(U,\varepsilon) \quad ,$$

where $\hat{f}$ denotes, temporarily, the right-hand side of (7). Now, by Theorem A4.7 and the choice of β, $\hat{f}$, we have

$$s((U,f),\ (\varepsilon,\hat{f})) = \beta\cdot\varepsilon + 0\cdot\hat{f} + \log\left[\int e^{-\beta U(\omega)}\mu(d\omega)\right]$$

which, again by Theorem A4.7, is exactly $s(U,\varepsilon)$. Hence $\bar{f}(\varepsilon) = \hat{f}$, as desired.

Let us now put together in a more physical language what we have proved about the relation between the micro-canonical and canonical ensembles. Choosing an interaction U, we set up a relation between energy ε and inverse temperature β by

$$\varepsilon(\beta) = \frac{\int U(\omega)e^{-\beta U(\omega)}\mu(d\omega)}{\int e^{-\beta U(\omega)}\mu(d\omega)} \tag{8}$$

There is exactly one inverse temperature corresponding to each value of ε between ess. inf. (U) and ess. sup. (U). We may compute $s(\varepsilon)$ either by the microcanonical prescription

$$s(U,\varepsilon) = \lim_{\delta \downarrow 0} \lim_{N \to \infty} \frac{1}{N} \log \mu^N\{(\omega_1,\ldots,\omega_N) : \varepsilon - \delta < \frac{U(\omega_1) + \ldots + U(\omega_N)}{N} < \varepsilon \} \tag{9}$$

or by the canonical prescription:

$$\begin{aligned} s(\varepsilon) &= [\beta\varepsilon + \log Z(\beta)], \quad \beta = \beta(\varepsilon) \quad , \\ &= -\int \rho_\beta(\omega) \log(\rho_\beta(\omega)) \mu(d\omega) \quad , \end{aligned} \tag{10}$$

where $\rho_\beta(\omega) = \dfrac{e^{-\beta U(\omega)}}{Z(\beta)}$.

If s is a bounded measurable function on Ω, we may compute the expected value $\bar{f}$ of f either by the microcanonical prescription given in (6) or by the canonical prescription

$$\bar{f}(\varepsilon) = \int \mu_\beta(d\omega) f(\omega) \tag{11}$$

In physics, one is interested in positive values of β and hence in only part of the possible range of energy values - that part below the point where $s(U,\varepsilon)$ takes on its supremum:

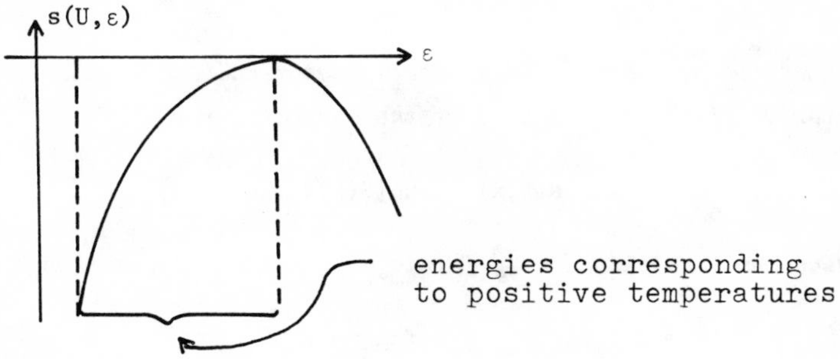

In this range, $s(U,\varepsilon)$ is strictly increasing in ε, so

$$\frac{\mu^N\{(\omega_1,\ldots,\omega_N) : \frac{U(\omega_1) + \ldots + U(\omega_N)}{N} < \varepsilon - \delta\}}{\mu^N\{(\omega_1,\ldots,\omega_N) : \frac{U(\omega_1) + \ldots + U(\omega_N)}{N} < \varepsilon\}}$$

goes to zero exponentially as N goes to infinity. Hence, if we consider microcanonical measures ${}^N\mu^{m.c.}_{(\varepsilon-\delta,\varepsilon)}$ and ${}^N\mu^{m.c.}_{(\varepsilon-\delta'\,\varepsilon)}$, we have

$$\lim_{N \to \infty} \left\| {}^N\mu^{m.c.}_{(\varepsilon-\delta,\varepsilon)} - {}^N\mu^{m.c.}_{(\varepsilon-\delta'\varepsilon)} \right\| = 0 \quad ,$$

so the choice of δ in the construction of the microcanonical measure makes very little difference for large N. In particular, the $\lim \delta \downarrow 0$ in the microcanonical prescription (6) for computing averages is unnecessary.

Note that, by symmetry

$$\int \frac{f(\omega_1) + \ldots + f(\omega_N)}{N} \, {}^N\mu^{m.c.}_{(\varepsilon-\delta,\varepsilon)}(d\omega_1 \ldots d\omega_N)$$

$$= \int f(\omega_1) \, {}^N\mu^{m.c.}_{(\varepsilon-\delta,\varepsilon)}(d\omega_1 \ldots d\omega_N)$$

Hence, if we project each ${}^N\mu^{m.c.}_{(\varepsilon-\delta,\varepsilon)}$ (which is a measure on $\underbrace{\Omega \times \ldots \times \Omega}_{N \text{ times}}$) onto the first factor, we get a sequence of measures on Ω which converges to μ_β in the sense that, if f is any bounded measurable function on Ω, the integral of f with respect to this sequence of measures converges to the integral of f with respect to μ_β.

We can give yet another expression for $s(f,x)$. For any probability measure σ on Ω, we define the *entropy of* σ *with respect to* μ, denoted by $S(\sigma)$, as

$$S(\sigma) \begin{cases} = -\int \left(\frac{d\sigma}{d\mu}\right) \log \left(\frac{d\sigma}{d\mu}\right) d\mu & \text{if } \sigma \text{ is absolutely continuous with respect to } \mu \\ = -\infty & \text{otherwise.} \end{cases}$$

Proposition A4.9

Let f *be a bounded measurable vector-valued function on* Ω*, and let* x *be in the interior of the convex hull of the essential range of* f*. Then*

$$s(f,x) = \sup\{S(\sigma) : \int f d\sigma = x\} \quad ,$$

and the supremum is attained only for $\sigma = \mu_\beta$ *where* β *is chosen so that*

$$\int f d\mu_\beta = x \quad .$$

In other words, $s(f,x)$ is the largest value of the entropy consistent with having the mean value of f equal to x.

Proof. We have only to show that, if $\int f d\sigma = x$, and if $\sigma \neq \mu_\beta$, then $S(\sigma) < s(f,x)$. We need only consider measures σ which are absolutely continuous with respect to μ. Then we can write:

$$S(\sigma) = -\int \frac{d\sigma}{d\mu} \log \frac{d\sigma}{d\mu} \, d\mu$$

$$= -\int \frac{d\sigma}{d\mu} \log \left(\frac{d\sigma}{d\mu} \cdot \frac{1}{\rho_\beta}\right) d\mu - \int \frac{d\sigma}{d\mu} \log \rho_\beta \, d\mu \quad ,$$

where, as before

$$\rho_\beta(\omega) = \frac{e^{-\beta f(\omega)}}{Z(\beta)}$$

Now

$$\int \frac{d\sigma}{d\mu} \log \rho_\beta d\mu = \int d\sigma[\beta \cdot f + \log Z(\beta)]$$
$$= \beta \cdot x + \log Z(\beta) = s(f,x) \quad ,$$

where the second equality follows from

$$\int f d\sigma = x \quad .$$

Thus

$$S(\sigma) = s(f,x) - \int (\frac{d\sigma}{d\mu} \cdot \frac{1}{\rho_\beta}) \log (\frac{d\sigma}{d\mu} \cdot \frac{1}{\rho_\beta}) \rho_\beta d\mu$$
$$= s(f,x) - \int d\mu_\beta \cdot (\frac{d\sigma}{d\mu} \cdot \frac{1}{\rho_\beta}) \log (\frac{d\sigma}{d\mu} \cdot \frac{1}{\rho_\beta}) \quad .$$

By the strict concavity of $-\lambda \log \lambda$,

$$\int d\mu_\beta [-(\frac{d\sigma}{d\mu} \cdot \frac{1}{\rho_\beta}) \log (\frac{d\sigma}{d\mu} \cdot \frac{1}{\rho_\beta})] < 0$$

unless $\frac{d\sigma}{d\mu} = \rho_\beta$, so the proposition is proved.

We next want to see what happens if we drop the assumption that f is bounded. For simplicity we restrict ourselves to a special case; we will consider a positive real-valued function f such that

$$\int e^{-\beta f} d\mu = \infty$$

for all $\beta < 0$. Since f is positive we have, however, that

$$\int e^{-\beta f} d\mu < \infty \quad \text{if} \quad \beta \geq 0 \quad .$$

If $x > 0$ is such that

$$x = E_\beta(f) = \frac{\int f(\omega) e^{-\beta f(\omega)} \mu(d\omega)}{\int e^{-\beta f(\omega)} \mu(d\omega)} \quad ,$$

with $\beta > 0$, then just as before we conclude that

$$s(f,x) = \beta x + \log[\int e^{-\beta f} d\mu] \quad .$$

By the monotone convergence theorem, and the fact that

$$E_\beta(f)$$

is a decreasing function of β, $E_\beta(f)$ increases to $\int f d\mu$ as β decreases to zero. Thus, on the interval

$$\text{ess inf } (f) < x < \int f d\mu \quad ,$$

$s(f,x)$ is, as before, a real-analytic and strictly concave function of x. If f is non-integrable, so $\int f d\mu = \infty$, there is nothing more to be said; the techniques we have been using give a formula for s valid for all interesting values of x.

We assume therefore that $\int f d\mu < \infty$. We are thus considering an intermediate situation: f increases rapidly enough so that $e^{-\beta f}$ is non-integrable for all negative β, but not fast enough so that f itself is non-integrable. In this case we claim that $s(f,x) = 0$ for all $x \geq \int f d\mu$. To see this, we note first that, as β decreases to zero, x increases to $\int f d\mu$ and $\int e^{-\beta f} d\mu$ decreases to $\int 1 d\mu = 1$. Hence,

$$s(f,x) = x\beta + \log[\int e^{-\beta f} d\mu]$$

approaches zero, so

$$s(f, \int f d\mu) = 0 \quad .$$

On the other hand, the condition that $e^{-\beta f}$ is non-integrable for all $\beta < 0$ implies that, if we define

$$F(\lambda) = \mu\{\omega: f(\omega) > \lambda\} \quad ,$$

then $F(\lambda)$ cannot go to zero exponentially as λ goes to infinity, and in particular for any $a > 0$, $b > 0$

$$\lim_n \sup e^{bn} F(an) > 0 \quad .$$

Thus, for any a,

$$\lim_n \sup \frac{1}{n} \log F(an) = 0 \quad .$$

Now

$$s(f,(a,\infty)) = \lim_{N\to\infty} \frac{1}{N} \log \mu^N\{f(\omega_1) + \ldots + f(\omega_N) > Na\} \quad .$$

But

$$\mu^N\{f(\omega_1) + \ldots + f(\omega_N) > Na\} \geq \mu^N\{f(\omega_1) > Na\} \quad (\text{since } f \geq 0)$$

$$= F(Na), \text{ so}$$

$$s(f,(a,\infty)) \geq \lim_N \sup \frac{1}{N} \log F(Na) = 0, \text{ so } s(f,(a,\infty)) = 0$$

for all a. Therefore

$$\sup_{x>a} s(f,x) = 0$$

for all a, i.e., $s(f,x)$ takes the value 0 for arbitrarily large x; since $s(f, \int f d\mu) = 0$, $s(f,x) \leq 0$ for all x, and $s(f,x)$ is concave in x, this implies

$$s(f,x) = 0$$

for all $x \geq \int f d\mu$, as asserted.

Note that, although the entropy remains identically zero for x larger than $\int fd_{\mu}$, we have

$$\lim_{N\to\infty} \mu^N\{(\omega_1,\ldots,\omega_N): \frac{f(\omega_1) + \ldots + f(\omega_N)}{N} > a\} = 0$$

for all $a > \int fd\mu$, by the law of large numbers. The above argument simply shows that, because of the possibility of large positive fluctuations in a single $f(\omega_i)$, this probability does not go to zero exponentially with N.

It may easily be seen that the relation:

$$s(f,x) = \sup\{S(\sigma): \int fd\sigma = x\}$$

remains true for unbounded positive f, but that the supremum is not attained for $x > \int fd\mu$.

We have worked out the above theory as an application of the techniques used in treating interacting systems. It is possible to get more detailed results (e.g., on fluctutions in the microcanonical ensemble) by noting that $Z(\beta)^N$, as a function of β, is the Laplace transform of the distribution of $f(\omega_1) + \ldots + f(\omega_N)$ with respect to μ^N, and then reading off the asymptotic properties of this distribution from the stationary phase approximation for the inverse Laplace transform.

A5. THE CANONICAL ENSEMBLE

The results of the preceding sections provide a conceptual framework for understanding the fact that macroscopic matter is describable by a small number of parameters. They are, however, almost entirely mechanical in character; i.e., are concerned with description in terms of such quantities as energy and density rather than in terms of more thermodynamic quantities such as temperature and entropy. We have therefore the problem of seeing how such quantities, and notably the temperature, can be built into the theory.

We will tackle the problem in an indirect way. We first introduce the canonical ensemble, containing a parameter to be identified eventually with the temperature, without attempting to justify it at the outset. We then work out the relation between the canonical ensemble and the microcanonical framework developed in the first section. After this has been done, we will try to make plausible the identification of the parameter appearing in the canonical framework with the temperature.

We start, then, by defining, for each bounded open Λ, each positive integer N, and each $\beta > 0$, a probability measure (the *canonical probability measure* or *canonical ensemble* on Λ^N (the configuration space for an assembly of N particles in Λ) by

$$\mu(\Lambda,N,\beta)(dq_1\ldots dq_N) = \frac{\exp\{-\beta U(q_1,\ldots,q_N)\} \dfrac{dq_1\ldots dq_N}{N!}}{Z(\Lambda,N,\beta)} \tag{1}$$

where the normalization factor $Z(\Lambda,N,\beta)$ is given by

$$Z(\Lambda,N,\beta) = \frac{1}{N!} \int dq_1, \ldots, dq_N \exp\{-\beta U(q_1,\ldots,q_N)\} \quad .$$

The parameter β is of course to be identified with the reciprocal of the temperature (in appropriate units); we will anticipate this identification by referring to β as the *inverse temperature*. The heuristic basis of the analysis we will give is the following: One wants to show that, except for exceptional values of β, the function $\frac{U(q_1,\ldots,q_N)}{N}$ on Λ^N is for large Λ,N, nearly constant with respect to the canonical probability measure μ_β. This means that, except for some set of configuration of small probability, all configurations have approximately the same energy. For these configurations, the weight factor $e^{-\beta U}$ is nearly constant, so one would except that the canonical probability measure is much the same as the one obtained by assigning equal weight to all configurations with approximately the right value for $\frac{U}{N}$ and weight zero to all other configuration, i.e., the canonical probability measure should be in some sense equivalent to the one obtained by normalizing the restriction of Lebesgue measure to be appropriately chosen "thickened energy surface." The main task of this section will be to see to what extent these heuristic ideas can be made precise.

In order to be able to use our previous results, we will continue to assume that the interaction is a finite-range observable as defined in Section A1. We will also need another assumption about the interaction, the stability condition, which was not needed in our discussion of the microcanonical ensemble.* We say that an observable U is *stable* if there exists a constant B such that

$$U(q_1,\ldots,q_N) \geq - NB \tag{2}$$

for all $N,q_1,\ldots,q_N$. We will use the term *interaction* to denote a stable observable. We will not stop to make a detailed investigation of what observables are stable, but we will make a few remarks; these remarks will apply to more general observables than those we have been considering:

(a) Any non-negative observable is stable.

(b) If a stable observable U is obtained from a two-body potential Φ (i.e.,

$$U(q_1,\ldots,q_N) = \frac{1}{2} \sum_{i\neq j} \Phi(q_i - q_j) \tag{3}$$

then Φ must be bounded below.

(c) A two-body potential which is continuous and of positive type defines a stable interaction:

---

* We would have needed the stability condition, or something like it, if we had defined our microcanonical ensemble by fixing the total energy rather than just the interaction energy.

$$U(q_1,\dots,q_N) = \frac{1}{2}\sum_{i\neq j}\Phi(q_i - q_j) = -\frac{1}{2}N\Phi(0) + \frac{1}{2}\sum_{i,j}\Phi(q_i - q_j) \quad ,$$

and $\sum_{i,j}\Phi(q_i - q_j) \geq 0$ since Φ is of positive type.

(d) The sum of two stable observables is stable; in particular, any two-body potential which is the sum of a continuous function of positive type and a non-negative function defines a stable observable.

With a little work, this remark implies the following stability criterion for two-body potentials:

(e) Let Φ be bounded below. Assume there exists a decreasing function φ on $(0,\infty)$, with values in $[0,\infty]$, such that $\int t^{\nu-1}\varphi(t)dt$ diverges at 0 and converges at $+\infty$, and such that

$$\Phi(q) \geq \varphi(|q|) \qquad \text{for small } q$$

$$\Phi(q) \geq -\varphi(|q|) \qquad \text{for large } q$$

Then Φ defines a stable interaction.

The proof of this criterion is obtained by constructing a function of positive type which is everywhere less than Φ. See Ruelle [9]. Proposition 3.2.8, p. 38 for the details.

There is a slightly stronger condition than stability which is often useful. We will say that an interaction U is *superstable* if, for every continuous Φ of compact support on $\mathbf{R}^\nu$,

$$U(q_1,\dots,q_N) + \lambda \cdot \frac{1}{2}\sum_{i\neq j}\Phi(q_i - q_j)$$

is stable for sufficiently small λ. In other words, an interaction is superstable if it is stable and if it remains stable when perturbed slightly by any continuous finite-range two-body observable. Although this condition looks rather restrictive, interactions which appear in practice are nearly always superstable. For example, it is easily shown that an interaction is superstable if it can be obtained by adding to a stable interaction another interaction defined by a continuous non-negative two-body potential strictly positive at the origin. (The converse of this assertion follows at once from the definition of superstability, so this is in fact a characterization of superstable interactions.) For a sample of the results obtainable through the use of superstability rather than ordinary stability, see Ruelle [11].

We will now fix a stable interaction U, and we will let

$$s(v,\varepsilon) = s(v,U,\varepsilon)$$

$$s(v,J) = \sup_{\varepsilon\in J} s(v,\varepsilon) \tag{4}$$

We will also make the following convention: Whenever we consider a vector-valued observable f, the first component of f is always assumed to be U. Also, when we consider $x \in \mathbb{R}^t$, the first component of x is denoted by ε.

We now define, for any observable f with values in $\mathbb{R}^t$, any bounded open Λ, any positive integer N, and any open convex set J in $\mathbb{R}^t$,

$$Z(\Lambda,N,\beta,f,J) = \int_\Delta \frac{dq_1\ldots dq_N}{N!} e^{-\beta U(q_1,\ldots,q_N)}, \tag{5}$$

where the region Δ of integration is

$$\Delta = \{(q_1,\ldots,q_N) \in \Lambda^N : \frac{f(q_1,\ldots,q_N)}{N} \in J\}$$

Thus, for example, the probability in the canonical ensemble of finding $\frac{f}{N}$ in J is just

$$\frac{Z(\Lambda,N,\beta,f,J)}{Z(\Lambda,N,\beta)},$$

so, by analyzing the asymptotic behavior of $Z(\Lambda,N,\beta,f,J)$ as Λ and N become large simultaneously, we can obtain information about the distribution of f with respect to the canonical probability measure, in the same limit. The main result in this direction is the following:

Theorem A5.1

Let Λ_n *be a sequence of regions becoming infinitely large in the sense of van Hove and approximable by rectangles, and let* N_n *be a sequence of integers such that*

$$\lim_{n\to\infty} \frac{V(\Lambda_n)}{N_n} = v .$$

Let f *be an* $\mathbb{R}^t$*-valued observable whose first component is a stable interaction* U, *and let* $\beta > 0$. *Let* J *be a convex open set in* $\mathbb{R}^t$ *such that either*

(a) $J \cap \{x: s(v,f,x) > -\infty\} \neq \emptyset$ *or*

(b) *The distance from* J *to* $\{x: s(v,f,x) > -\infty\}$ *is strictly positive.*

Then

$$\lim_{n\to\infty} \frac{1}{N_n} \log Z(\Lambda_n,N_n,\beta,f,J) = \sup_{x\in J} s(v,f,x) - \beta\varepsilon\} , \tag{6}$$

where ε *denotes the first component of* x.

From this theorem, we can immediately deduce some useful corollaries:

Corollary A5.2

If Λ_n, N_n *are as in the preceding theorem:*

$$\lim_{n\to\infty} \frac{1}{N_n} \log Z(\Lambda_n,N_n,\beta) = \sup_\varepsilon \{s(v,\varepsilon) - \beta\varepsilon\} \tag{7}$$

For reasons to be explained later, we will denote

$$\lim_{n\to\infty} \frac{1}{N_n} \log Z(\Lambda_n,N_n,\beta) = -\beta f(v,\beta) \qquad (8)$$

Corollary A5.3

Let Λ_n, N_n, f *be as in Theorem* A5.1, *and let* $W \subset \mathbb{R}^t$ *be an open set such that*

$$\sup\{s(v,f,x) - \beta\varepsilon \colon x \notin W\}$$
is strictly less than $-\beta f(v,\beta)$.

Then the probability with respect to the canonical ensemble of having $\frac{f}{N_n}$ *outside* W *goes to zero exponentially with the size of the system as* n *goes to infinity.*

Proof of Theorem A5.1. To simplify the notation, we will treat only the case of the real-valued observable U, instead of a more general vector-valued observable. The idea of the proof is as follows: If J is a very small interval centered at ε_0, then

$$Z(\Lambda_n,N_n,\beta,U,J) = \int_\Delta \frac{dq_1\ldots dq_{N_n}}{N_n!} \exp\{-\beta U(q_1,\ldots,q_{N_n})\} \quad ,$$

$$(\Delta = \{(q_1,\ldots,q_{N_n}) \in \Lambda_n^{N_n} \colon \frac{U(q_1,\ldots,q_N)}{N_n} \in J\})$$

$$\approx \exp\{-\beta N_n\varepsilon_0\} \cdot \int_\Delta \frac{dq_1\ldots dq_{N_n}}{N_n!} = \exp\{-\beta N_n\varepsilon\} \cdot V(\Lambda_n,N_n,U,J) \quad .$$

But for large n,

$$V(\Lambda_n,N_n,U,J) \approx \exp\{N_n s(v,\varepsilon_0)\}, \quad \text{so}$$

$$\frac{1}{N_n} \log Z(\Lambda_n,N_n,U,J) \approx s(v,\varepsilon_0) - \beta\varepsilon_0 \quad .$$

This simple argument shows that the theorem is at least approximately true for very small intervals. To deal with a general interval we break it up into small intervals, make the above argument for each of these, and then show that in the limit only the largest contribution matters. There is no need to consider values of $\varepsilon < -B$, since these are rules out by the stability condition. We do need, however, a separate argument to deal with large values of ε.

To simplify the notation, we write $Z_n(J)$ for $Z(\Lambda_n,N_n,U,J)$. Also, we will let

$$\varepsilon_{min} = \inf\{\varepsilon \colon s(v,\varepsilon) > -\infty\}$$

$$\varepsilon_{max} = \sup\{\varepsilon \colon s(v,\varepsilon) > -\infty\}$$

Whenever we consider an interval $J = (\varepsilon_1,\varepsilon_2)$, we will always assume $\varepsilon_1 \neq \varepsilon_{max}$;

$\varepsilon_2 \neq \varepsilon_{min}$, and we will not repeat this at each step. (The restriction is necessary to allow us to use Theorem A3.5 to conclude

$$\lim_{n\to\infty} \frac{1}{N_n} \log V(\Lambda_n, N_n, U, J) = s(v,J) \quad .)$$

We will also write, provisionally,

$$g(J) = \sup_{\varepsilon\in J} [s(v,\varepsilon) - \beta\varepsilon] \quad .$$

<u>Lemma A5.4</u>

Let $J = (\varepsilon_1, \varepsilon_2)$. *Then*

$$\begin{aligned} s(v,J) - \beta\varepsilon_2 &\le \liminf_n \frac{1}{N_n} \log Z_n(J) \\ &\le \limsup_n \frac{1}{N_n} \log Z_n(J) \le s(v,J) - \beta\varepsilon_1 \quad . \end{aligned} \tag{9}$$

and

$$\begin{aligned} g(J) - \beta(\varepsilon_2 - \varepsilon_1) &\le \liminf_n \frac{1}{N_n} \log Z_n(J) \\ &\le \limsup_n \frac{1}{N_n} \log Z_n(J) \le g(J) + \beta(\varepsilon_2 - \varepsilon_1) \quad . \end{aligned} \tag{10}$$

<u>Proof</u>. To obtain (9), we note that, in the definition (5) of $Z_n(J)$, the integrand $\exp\{-\beta U\}$ lies between $\exp\{-\beta N_n \varepsilon_1\}$ and $\exp\{-\beta N_n \varepsilon_2\}$ in the region of configuration space over which we integrate, and the volume of this region, including the factor $1/N_n!$, is just what we have called $V(\Lambda_n, N_n, U, J)$. Hence we have

$$\exp\{-\beta N_n \varepsilon_2\} V(\Lambda_n, N_n, U, J) \le Z_n(J) \le \exp\{-\beta N_n \varepsilon_1\} V(\Lambda_n, N_n, U, J) \quad .$$

Taking the logarithm of this inequality, dividing by N_n, and using

$$\lim_{n\to\infty} \frac{1}{N_n} \log V(\Lambda_n, N_n, U, J) = s(v,J) \quad ,$$

we get (9). To deduce (10) from (9), we note that

$$\begin{aligned} s(v,J) - \beta\varepsilon_2 &= \sup_{\varepsilon\in J} [s(v,\varepsilon) - \beta\varepsilon_2] \\ &= \sup_{\varepsilon\in J} [s(v,\varepsilon) - \beta\varepsilon - \beta(\varepsilon_2 - \varepsilon)] \\ &\ge \sup_{\varepsilon\in J} [s(v,\varepsilon) - \beta\varepsilon] - \beta(\varepsilon_2 - \varepsilon_1) = g(J) - \beta(\varepsilon_2 - \varepsilon_1); \end{aligned}$$

similarly,

$$s(v,J) - \beta\varepsilon_1 \le g(J) + \beta(\varepsilon_2 - \varepsilon_1) \quad .$$

We next prove Theorem A5.1 for bounded intervals J:

Lemma A5.5

Let $J = (\varepsilon_1, \varepsilon_2)$ *be bounded. Then*

$$\lim_{n\to\infty} \frac{1}{N_n} \log Z_n(J) = \sup_{\varepsilon \in J}\{s(v,\varepsilon) - \beta\varepsilon\}$$

Proof. We choose $\delta > 0$, and write

$$J = J_1 \cup \dots \cup J_k \quad .$$

where the J_i are open intervals each having length less than δ. We have

$$\sup_i\{Z_n(J_i)\} \le Z_n(J) \le k \cdot \sup_i\{Z_n(J_i)\}, \quad \text{so}$$

for large n,

$$\frac{1}{N_n} \log Z_n(J) \approx \sup_i \left\{\frac{1}{N_n} \log Z_n(J_i)\right\}$$

By the preceding lemma, $\lim_n \sup \frac{1}{N_n} \log Z_n(J_i)$ and $\lim_n \inf \frac{1}{N_n} \log Z_n(J_i)$ both differ from $g(J_i)$ by no more than $\beta \cdot \delta$ for each i. Since

$$\sup_i g(J_i) = \sup_i \sup_{\varepsilon \in J_i}\{s(v,\varepsilon) - \beta\varepsilon\} = \sup_{\varepsilon \in J}\{s(v,\varepsilon) - \beta\varepsilon\} = g(J),$$

we have

$$g(J) - \beta\delta \le \lim_n \inf \frac{1}{N_n} \log Z_n(J) \le \lim_n \sup \frac{1}{N_n} \log Z_n(J) \le g(J) + \beta\delta \ .$$

Letting δ decrease to zero gives the lemma.

The proof of Theorem A5.1 is now nearly immediate. We want to split the interval J into a bounded part to which we can apply Lemma A5.5 and an unbounded part whose contribution is negligible. Thus, we will write:

$$J = J_1 \cup J_2 \cup J_3 \quad ,$$

with

$$J_1 = J \cap (-B - 2, \varepsilon' + 1)$$

$$J_2 = J \cap (\varepsilon', \infty)$$

$$J_3 = J \cap (-\infty, -B - 1)$$

and where ε' will be chosen later. By the stability condition

$$\frac{U}{N} \ge -B \quad ,$$

$$Z_n(J) = Z_n(J_1 \cup J_2) \quad .$$

Note that, since

$$s(v,\varepsilon) - \beta\varepsilon \leq 1 + \log v - \beta\varepsilon$$

goes to $-\infty$ as ε goes to infinity, we may make

$$\sup_{\varepsilon \in J}[s(v,\varepsilon) - \beta\varepsilon] = \sup_{\varepsilon \in J_1}[s(v,\varepsilon) - \beta\varepsilon] \tag{11}$$

by making ε' large enough. Similarly, by Lemma A5.4,

$$\limsup_{n\to\infty} \frac{1}{N_n} \log Z_n(J_2) \leq s(v,(\varepsilon',\infty)) - \beta\varepsilon' \leq 1 + \log v - \beta\varepsilon'$$

Hence, if $\sup_{\varepsilon \in J}\{s(v,\varepsilon) - \beta\varepsilon\} > -\infty$,

$$\lim_{n\to\infty} \frac{1}{N_n} \log Z_n(J) = \lim_{n\to\infty} \frac{1}{N_n} \log Z_n(J_1) = \sup_{\varepsilon \in J_1}\{s(v,\varepsilon) - \beta\varepsilon\}$$
$$= \sup_{\varepsilon \in J}\{s(v,\varepsilon) - \beta\varepsilon\}$$

provided ε' is taken large enough so that (11) holds and so that

$$1 + \log v - \beta\varepsilon' < \sup_{\varepsilon \in J}\{s(v,\varepsilon) - \beta\varepsilon\} .$$

On the other hand, if

$$\sup_{\varepsilon \in J}\{s(v,\varepsilon) - \beta\varepsilon\} = -\infty \quad ,$$

then for any choice of ε'

$$\lim_{n\to\infty} \frac{1}{N_n} \log Z_n(J_1) = -\infty, \text{ so}$$

$$\limsup_n \frac{1}{N_n} \log Z_n(J) \leq \limsup_n \frac{1}{N_n} \log Z_n(J_2) \leq 1 + \log v - \beta\varepsilon' ;$$

since the right-hand side can be made as close as we like to $-\infty$, the left-hand side must equal $-\infty$, i.e.,

$$\lim_{n\to\infty} \frac{1}{N_n} \log Z_n(J) = -\infty = \sup_{\varepsilon \in J}\{s(v,\varepsilon) - \beta\varepsilon\},$$

so the proof of the theorem is complete.

Let us now see what we can say about the distribution of the energy with respect to the canonical probability measure by looking at the set of points where $s(v,\varepsilon) - \beta\varepsilon$ takes on its supremum. (Since $s(v,\varepsilon) - \beta\varepsilon$ is an upper semi-continuous function of ε, which is $-\infty$ for large negative ε and which goes to $-\infty$ as $\varepsilon \to \infty$, it necessarily takes on its supremum at least once.) Now $s(v,\varepsilon) - \beta\varepsilon$ is a concave function of ε; hence, takes on its supremum at a given point ε^* if and only if there is a horizontal line tangent to the graph of $s(v,\varepsilon) - \beta\varepsilon$ at ε^*, i.e., if and only if there is a line with slope β tangent to the graph of $s(v,\varepsilon)$ at ε^*. Let us examine first the case in which $s(v,\varepsilon) - \beta\varepsilon$ takes on its supremum exactly once:

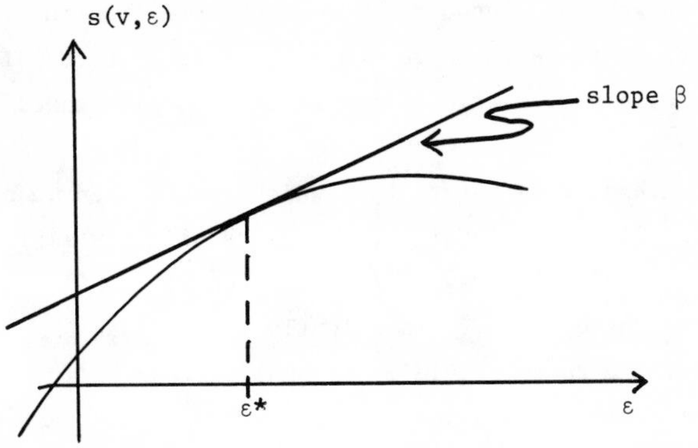

If $\delta > 0$, then Theorem A.5.1 implies that the canonical probability of finding $\frac{U(q_1,\ldots,q_N)}{N}$ less than $\varepsilon^* - \delta$:

$$\frac{Z(\Lambda,N,\beta,U,(-\infty,\varepsilon^* - \delta))}{Z(\Lambda,N,\beta,U,(-\infty,\infty))}$$

goes to zero exponentially as Λ,N go to infinity. Similarly, the canonical probability of finding $\frac{U(q_1,\ldots,q_N)}{N}$ larger than $\varepsilon^* + \delta$ goes to zero exponentially. Thus, for large Λ,N, the distribution of $\frac{U}{N}$ in the canonical ensemble becomes very sharply peaked about ε^*. This suggests that the mean value of $\frac{U}{N}$ also approaches ε^*, and a simple argument, which we will omit, shows that this is indeed the case, i.e., that very large values of $\frac{U}{N}$ do not contribute inordinately much to the mean value. Thus, we are well justified in speaking of $\varepsilon^* = \varepsilon^*(v,\beta)$ as canonical energy per particle corresponding to (v,β); it should be kept in mind that, for large systems, not only is the mean energy per particle approximately equal to ε^* but large fluctuations (i.e., fluctuations of order 1) about this value are extremely improbable.

Next suppose that $s(v,\varepsilon) - \beta\varepsilon$ takes on its supremum on an interval of non-zero length. This means that the graph of $s(v,\varepsilon)$ as a function of ε contains a linear portion with slope β:

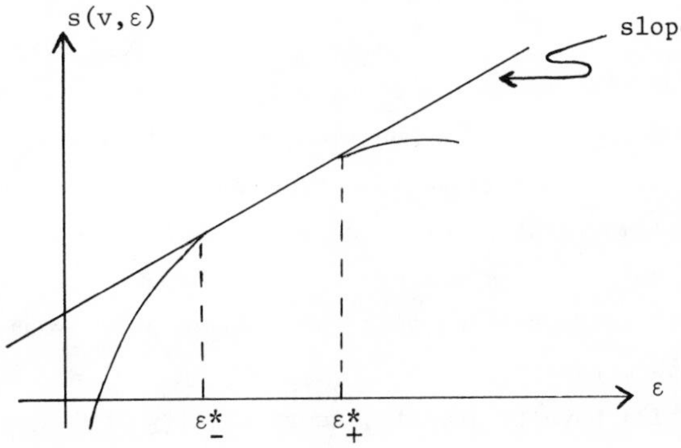

As before, we can argue that the canonical probability of finding $\frac{U}{N}$ larger than $\varepsilon_+^* + \delta$ or less than $\varepsilon_-^* - \delta$ goes to zero exponentially with the size of the system and that the mean energy per particle lies in or very near to the interval $[\varepsilon_-^*,\varepsilon_+^*]$

when the system is large. We cannot say very much more than this. On the basis of the above considerations there is no reason to suppose that the distribution of the energy per particle either does or does not become sharply peaked about some value in $[\varepsilon_-^*, \varepsilon_+^*]$.

Notice something very interesting which occurs when we are in the situation described above. If β' is an inverse temperature slightly smaller than β, and which corresponds to a unique energy $\varepsilon^*(v,\beta')$, then $\varepsilon^*(v,\beta')$ is larger than $\varepsilon_+^*(v,\beta)$. On the other hand, if β' is slightly larger than β and corresponds to a unique energy $\varepsilon^*(v,\beta')$, then $\varepsilon^*(v,\beta')$ is smaller than $\varepsilon_-^*(v,\beta)$. Hence, as β' decreases past β, $\varepsilon^*(v,\beta')$ has a jump discontinuity of at least $\varepsilon_+^* - \varepsilon_-^*$. In other words, the system we are investigating undergoes a phase transition at inverse temperature β and this phase transition has a latent heat per particle of at least $\varepsilon_+^* - \varepsilon_-^*$. Note that there can be at most countably many values of β at which such phase transitions occur (for each value of the density), but that nothing we have said prevents the set of inverse temperatures at which such phase transitions occur from being dense in the positive real axis.

It should also be added that phase transitions of the type considered above are not quite the ones which usually occur in nature. The point is that typical phase transitions (e.g., the freezing of water) occur at a well-defined temperature and have a non-zero latent heat when carried out at constant *pressure*. The situation we have been describing refers to systems where the volume, rather than the pressure, is held fixed. It seems that, if one tries to freeze water in a container of fixed size (i.e., at constant volume) one finds a range of temperatures over which both ice and water are present (in proportions depending on the temperature), and one finds that the internal energy of the system varies continuously with the temperature. In order to see how phase transitions with non-zero latent heats occur in realistic situations, we need a formalism which enables us to see how the energy depends on the temperature at constant pressure. The necessary machinery will appear when we consider the grand-canonical ensemble. (I am indebted to R. B. Griffiths for pointing the above facts out to me.)

In addition to having linear portions, it is possible at least in principle for the graph of $s(v,\varepsilon)$, as a function of ε, to have corners i.e., discontinuities in its slope. Such a corner would correspond to a situation where the energy is independent of the temperature over a range of temperatures, i.e., where the specific heat vanishes for a range of temperatures. Such phenomena do not seem to occur in nature and it is therefore reassuring that they can be proved not to occur in lattice systems (see Griffiths and Ruelle [9]).

We still have to justify, heuristically at least, the identification of β with the inverse temperature. We will do this by arguing that two systems correspond to the same value of β if and only if no heat flows from one to the other when they are placed in thermal contact. To make this idea precise, we consider two cubic boxes

Λ_1 and Λ_2 placed side by side. We place N_1 particles in Λ_1 and N_2 particles in Λ_2. We do not want to assume that the particles in Λ_1 are identical with those in Λ_2, so we do not require that the interaction among the particles in Λ_1 is the same as the interaction among the particles in Λ_2, and we assume that the interaction between particles in Λ_1 and those in Λ_2 is in some appropriate sense short-range, i.e., only depends on the particles near the boundary separating the two boxes. We fix the overall energy per particle in the two boxes to lie in a small interval about ε, and we ask how the energy is distributed between the two boxes with respect to normalized Lebesgue measure on the thickened energy surface in the configuration space $\Lambda_1^{N_1} \times \Lambda_2^{N_2}$. To try to answer this question we proceed as follows: If the energy per particle in Λ_1 is ε_1 and the energy per particle in Λ_2 is ε_2, then the overall energy per particle is approximately

$$\frac{N_1\varepsilon_1 + N_2\varepsilon_2}{N_1 + N_2} ;$$

the approximation involved here is that we drop the interaction between the two boxes. We make this approximation because we expect the interaction energy to be a surface term, of the order of $(N_1 + N_2)^{\frac{\nu-1}{\nu}}$, and hence to be negligible in comparison with the total energy which is of order $(N_1 + N_2)$. We thus want to consider values $(\varepsilon_1,\varepsilon_2)$ of the energy per particle in (Λ_1,Λ_2) such that

$$N_1\varepsilon_1 + N_2\varepsilon_2 \approx (N_1 + N_2)\varepsilon \tag{12}$$

Now the configurations with energy per particle in $\Lambda_1 \approx \varepsilon_1$ and energy per particle in $\Lambda_2 \approx \varepsilon_2$ gives a contribution to the configuration space volume in $\Lambda_1^{N_1} \times \Lambda_2^{N_2}$ of approximately

$$N_1!N_2! \exp\{N_1 s_1(v_1,\varepsilon_1) + N_2 s_2(v_2,\varepsilon_2)\}$$

and what we have to do is find the value(s) of $(\varepsilon_1,\varepsilon_2)$ which maximize the quantity subject to the constraint (12); all other values of $(\varepsilon_1,\varepsilon_2)$ have a probability which is exponentially small if N_1,N_2 are large. We therefore look for the values of $(\varepsilon_1,\varepsilon_2)$ which maximize

$$\frac{N_1}{N_1 + N_2} s_1(v_1,\varepsilon_1) + \frac{N_2}{N_1 + N_2} s_2(v_2,\varepsilon_2)$$

subject to the constraint

$$\frac{N_1}{N_1 + N_2}\varepsilon_1 + \frac{N_2}{N_1 + N_2}\varepsilon_2 = \varepsilon \quad .$$

This is easy to do; if we put $\alpha = \frac{N_1}{N_1 + N_2}$, the constraint gives

$$\alpha d\varepsilon_1 + (1 - \alpha)d\varepsilon_2 \quad ,$$

and the condition

$$0 = \frac{\partial}{\partial \varepsilon_1} [\alpha s_1(v_1,\varepsilon_1) + (1-\alpha)s_2(v_2,\varepsilon_2)]$$

becomes

$$\frac{\partial s_1(v_1,\varepsilon_1)}{\partial \varepsilon_1} = \frac{\partial s_2(v_2,\varepsilon_2)}{\partial \varepsilon_2} .$$

We now give a physical interpretation of the above result. We have considered two systems, in the boxes Λ_1 and Λ_2, and we have placed them in thermal contact, i.e., we have allowed them to interchange energy but not particles. We have found that, with probability very close to one, the energy will be distributed between the systems in such a way as to make

$$\frac{\partial s_1(v_1,\varepsilon_1)}{\partial \varepsilon_1} = \frac{\partial s_2(v_2,\varepsilon_2)}{\partial \varepsilon_2} \tag{13}$$

We interpret this as meaning that, when two systems are placed in thermal contact, there will be no net heat flow from one to the other if and only if (13) holds. In other words, two systems are at the same temperature if and only if they have the same value of $\frac{\partial s}{\partial \varepsilon}$. But we saw that, starting with the canonical ensemble, the energy per particle corresponding to β is found by solving the equation

$$\frac{\partial s(v,\varepsilon)}{\partial \varepsilon} = \beta .$$

Thus, two systems are at the same temperature if and only if they have the same value of β. This by itself does not show that β is the reciprocal of the temperature, but it shows at least that β is a function of the temperature alone and that conversely the temperature is uniquely determined by β. A more careful examination of the above argument shows that β is a decreasing function of the temperature; if

$$\frac{\partial s_1(v_1,\varepsilon_1)}{\partial \varepsilon_1} < \frac{\partial s_2(v_2,\varepsilon_2)}{\partial \varepsilon_2} ,$$

then ε_1 is larger than the equilibrium value, so heat will flow from system 1 to system 2 if they are placed in thermal contact.

This is about as far as we can get in identifying β with the reciprocal of the temperature by using just the above ideas. Indeed, one needs more information about the temperature than the simple fact that heat flows from bodies at high temperatures to bodies at low temperatures in order to establish a temperature scale. One method of completing the identification of β with the reciprocal of the temperature is to prove that, for any given system, the product $p \cdot v \cdot \beta$ (p the pressure, v the specific volume) approaches a constant as v goes to infinity (i.e., as the density goes to zero). This would show at the same time that the system behaves like an ideal gas at low densities and that β is a constant times the reciprocal of the temperature. There is no particular difficulty in carrying out this program, but it would involve a considerable amount of technical work; we will not do it.

The necessary results can be read out of Theorem 4.3.1 of Ruelle [9].

The above argument is evidently not completely satisfactory on two counts. In the first place, we did not really prove that, for a composite system of N_1 particles in Λ_1 and N_2 particles in Λ_2, if we consider Lebesgue measure on a thickened energy surface in the configuration space $\Lambda_1^{N_1} \times \Lambda_2^{N_2}$, and if the system is large, then with probability very near to one the energy will be distributed between the two systems in such a way as to make $N_1 s_1(v_1,\varepsilon_1) + N_2 s_2(v_2,\varepsilon_2)$ a maximum. This is a purely technical question, and it should be more or less straightforward to give a precise proof (although, to the best of my knowledge, this has never been done). The main difficulty would appear to lie in proving that the interaction energy between the two boxes is negligible in comparison with the total energy.

The second problem is a much more fundamental one: we have assumed that, if the systems in Λ_1 and Λ_2 are prepared separately with fixed energies, and if the systems are then allowed to interact, energy will be transferred from one to the other in such a way as to maximize the probability (with respect to Lebesgue measure) of the composite system. Proving that this is the case is part of the general problem, alluded to in the introduction, of understanding why events which are very improbable with respect to Lebesgue measure really do not occur. As with the general question, we will have nothing to say about this problem.

Once we have identified β with the reciprocal of the temperature, we can at least partially justify the identification of s with the entropy. We do this by remarking that

$$\frac{\partial s(v,\varepsilon)}{\partial \varepsilon} = \beta \quad ,$$

so, for an infinitesimal change in the state of the system at constant specific volume, we have

$$T ds = d\varepsilon$$

Since, by its very definition, the entropy per particle must satisfy the same relation, the difference between s and the entropy can at most depend on the specific volume. We must postpone the argument showing that this difference actually is a constant.

If we accept provisionally the identification of s with the entropy per particle, we can also interpret

$$\lim_{\substack{\Lambda\to\infty \\ N\to\infty \\ \frac{V(\Lambda)}{N}\to v}} \frac{1}{N} \log Z(\Lambda,N,\beta)$$

By Corollary A5.2, this limit is equal to

$$s(v,\varepsilon^*(v,\beta)) - \beta\varepsilon^*(v,\beta) = -\beta[\varepsilon^*(v,\beta) - \frac{1}{\beta} s(v,\varepsilon^*(v,\beta))]$$

Now $\varepsilon^* - \frac{1}{\beta} s$ is what is called the *Helmholtz free energy* (per particle); hence, the function $f(v,\beta)$ which we defined by

$$\lim \frac{1}{N} \log Z(\Lambda,N,\beta) = -\beta \mathfrak{f}(v,\beta)$$

is to be interpreted as the Helmholtz free energy. Since

$$\mathfrak{f}(v,\beta) = \inf_{\varepsilon}\{\varepsilon - \frac{1}{\beta} s(v,\varepsilon)\} \quad ,$$

$\mathfrak{f}(v,\beta)$ is a concave function of $1/\beta$. (It is the infimum of a family of affine function of $1/\beta$.) In its dependence on v, it is the infimum over ε of the function $\varepsilon - \frac{1}{\beta} s(v,\varepsilon)$ which is jointly convex in (v,ε). Although the infimum of a family of convex functions is not generally convex, the following proposition shows that $\mathfrak{f}(v,\beta)$ is convex in v:

Proposition A5.6

Let $h(\xi,\eta)$ *be a jointly convex function of* (ξ,η). *Then* $\inf_{\eta}\{h(\xi,\eta)\}$ *is convex in* ξ.

Proof. For any (ξ_1,η_1), (ξ_2,η_2), for any α between 0 and 1,

$$h(\alpha\xi_1 + (1-\alpha)\xi_2, \alpha\eta_1 + (1-\alpha)\eta_2) \le \alpha h(\xi_1,\eta_1) + (1-\alpha)h(\xi_2,\eta_2) \quad .$$

Hence

$$\inf_{\eta}\{h(\alpha\xi_1 + (1-\alpha)\xi_2,\eta)\} \le \alpha h(\xi_1,\eta_1) + (1-\alpha)h(\xi_2,\eta_2)$$

for all η_1,η_2. Taking the infimum of the right-hand side with respect to η_1,η_2 gives

$$\inf_{\eta}\{h(\alpha\xi_1 + (1-\alpha)\xi_2,\eta)\} \le \alpha \inf_{\eta}\{h(\xi_1,\eta)\} + (1-\alpha)\inf_{\eta}\{h(\xi_2,\eta)\} \quad ,$$

proving the proposition.

The formula

$$\lim_{\substack{\Lambda\to\infty \frac{V(\Lambda)}{N}\to v \\ N\to\infty}} \frac{1}{N} \log Z(\Lambda,N,\beta) = s(v,\varepsilon^*) - \beta\varepsilon^*$$

gives another way of calculating $s(v,\varepsilon^*)$ which relates s to the notion of entropy appearing in information theory. The canonical probability measure is defined by

$$\mu^c(\Lambda,N,\beta)(dq_1\ldots dq_N) = \frac{\exp[-\beta U(q_1,\ldots,q_N)]}{Z(\Lambda,N,\beta)} \frac{dq_1\ldots dq_N}{N!}$$

$$= \rho(q_1,\ldots,q_N) \frac{dq_1\ldots dq_N}{N!} \quad ,$$

where the second equality is the definition of ρ. Now consider

$$\frac{1}{N}\int \frac{dq_1\ldots dq_N}{N!}\,[-\rho(q_1,\ldots,q_N)\log \rho(q_1,\ldots,q_N)]$$

$$= \frac{1}{N}\int \frac{dq_1\ldots dq_N}{N!}\,\rho(q_1,\ldots,q_N)[\beta U(q_1,\ldots,q_N) + \log Z(\Lambda,N,\beta)]$$

$$= \int \mu^c(\Lambda,N,\beta)(dq_1\ldots dq_N)[\beta\,\frac{U(q_1,\ldots,q_N)}{N}] + \frac{1}{N}\log Z(\Lambda,N,\beta) \quad .$$

Now let $\Lambda, N \to \infty$ in such a way that $\frac{V(\Lambda)}{N} \to v$. We have already remarked that, if we are not at a phase-transition point,

$$\lim_{\substack{\Lambda\to\infty \\ N\to\infty}} \int \mu^c(\Lambda,N,\beta)(dq_1,\ldots,dq_N)\,\frac{U(q_1,\ldots,q_N)}{N} = \varepsilon^*(v,\beta) \quad ,$$

(with $\frac{V(\Lambda)}{N}$ under the limit), and we also have

$$\lim_{\substack{\Lambda\to\infty \\ N\to\infty \\ \frac{V(\Lambda)}{N}\to v}} \frac{1}{N}\log Z(\Lambda,N,\beta) = s(v,\varepsilon^*) - \beta\varepsilon^* \quad .$$

Hence

$$\lim_{\substack{\Lambda\to\infty \\ N\to\infty \\ \frac{V(\Lambda)}{N}\to v}} \frac{1}{N}\int \frac{dq_1\ldots dq_N}{N}\,[-\rho(q_1,\ldots,q_N)\log \rho(q_1,\ldots,q_N)] = s(v,\varepsilon^*) \quad , \qquad (14)$$

so the entropy per particle $s(v,\varepsilon^*(v,\beta))$ may be obtained as the infinite-volume limit of the information-theoretic entropy per particle of the canonical probability measure.

Many of the results we have obtained so far in this section come under the general heading of equivalence of ensembles for thermodynamic functions. That is, we have seen how to replace the independent variable ε by β; we have seen the relation between the free energy as a function of (v,β) and the entropy as a function of (v,ε), and we have seen that there are two distinct ways of computing the entropy, one giving it as a function of (v,β), the other giving it as a function of (v,ε). We turn next to the consideration of the equivalence of ensembles for observables, i.e., we want to see what can be said about the relation between the distribution of a given observable f/N with respect to the canonical probability measure for (Λ,N,β) and its distribution with respect to the micro-canonical probability measure for some very thin energy surface about ε, where ε and β are related as above. The machinery for analyzing the situation is provided by Theorem A5.1.

We choose, then, a scalar observable f which we may as well assume not to be a constant multiple of U, and we define:

$$\eta^c(f,x|v,\beta) = \sup_{\varepsilon}\{s(v,(U,f),(\varepsilon,x)) - \beta\varepsilon\} - [-\beta\mathfrak{f}(v,\beta)] \qquad (15)$$

We collect the properties of η^c in the following proposition:

Proposition A5.7

(a) *For fixed* (v,β), η^c *is a concave function of* x.

(b) *Let*

$$x_{min}(f,v,\beta) = \inf\{x: \eta^c(f,x|v,\beta) > -\infty\}$$

$$x_{max}(f,v,\beta) = \sup\{x: \eta^c(f,x|v,\beta) > -\infty\}$$

Then, if $x_1 \neq x_{max}$, $x_2 \neq x_{min}$, *we have*

$$\lim_{\substack{\Lambda\to\infty \\ N\to\infty \\ \frac{V(\Lambda)}{N}\to v}} \frac{1}{N} \log \mu^c(\Lambda,N,\beta)(\{\tfrac{f}{N} \in (x_1,x_2)\})$$

$$= \sup_{x_1<x<x_2} \{\eta^c(f,x|v,\beta)\}$$

Proof. (a) Since $s(v,(U,f),(\varepsilon,x)) - \beta\varepsilon$ is jointly concave in (ε,x), $\sup_\varepsilon s(v,(U,f),(\varepsilon,x)) - \beta\varepsilon\}$ is concave in x by proposition A5.6.

(b) This follows at once from Theorem A5.1. Note that (b) implies

$$\sup_x\{\eta^c(f,x|v,\beta\} = 0 \quad .$$

The interpretation of $\eta^c(f,x|v,\beta)$ should be fairly clear from the above. The essential fact is that, for large Λ,N with $\frac{V(\Lambda)}{N} \approx v$, all values x for f/N are extremely improbable (with respect to $\mu^c(\Lambda,N,\beta)$) except those where $\eta^c(f,x|v,\beta)$ is very near to zero.

In order to compare the canonical with the microcanonical ensemble, we will introduce as the microcanonical analogue of η^c the following:

$$\eta^{m.c.}(f,x|v,\varepsilon) = s(v,(U,f),(\varepsilon,x)) - s(v,\varepsilon) \tag{16}$$

$\eta^{m.c.}$ satisfies part (a) of Proposition A5.7 and a more or less obvious analogue of part (b), with $\mu^c(\Lambda,N,\beta)$ replaced by normalized Lebesgue measure on a very slightly thickened energy surface with energy per particle ε. Again, only values x of f/N such that $\eta^{m.c.}(f,x|v,\beta) \approx 0$ actually occur with any reasonable probability. Our main result on equivalence of ensembles for observables is the following:

Theorem A5.8

$\eta^c(f,x_0|v,\beta) = 0$ *if and only if there exists* ε^* *with*

$$-\beta\mathfrak{f}(v,\beta) = s(v,\varepsilon^*) - \beta\varepsilon^* \tag{17}$$

such that

$$\eta^{m.c.}(f,x_0|v,\varepsilon^*) = 0$$

In particular, if $\eta^c(f,x|v,\beta) = 0$ *for only one value* x_0 *of* x, *then for any* ε^* *satisfying* (17), x_0 *is the unique value of* x *at which* $\eta^{m.c.}(f,x|v,\varepsilon^*) = 0$. *If*

$$\eta^c(f,x|v,\beta) < 0$$

for all x, *but*

$$\lim_{x\to\infty} \eta^c(f,x|v,\beta) = 0$$

(so f/N *approaches* $+\infty$ *in probability with respect to the canonical probability measure), then for all* ε^* *satisfying* (17),

$$\eta^{m.c.}(f,x\ v,\varepsilon^*) < 0 \ \textit{for all}\ x,\ \lim_{x\to\infty} \eta^c(f,x\ v,\varepsilon^*) = 0 \quad .$$

To see in more concrete terms what this means, suppose we take values of (v,β) such that there is only one corresponding ε^* (i.e., only one ε^* satisfying (17)). Then, for any observable f, f/N converges in probability to a constant with respect to the canonical probability measure in such a way that the probability of finding a large fluctuation about the limiting value decreases exponentially with the size of the system if and only if it converges in the same way to the same constant with respect to the microcanonical ensemble with energy per particle ε^*.

Proof of Theorem A5.8. Assume first that

$$-\beta \mathfrak{f}(v,\beta) = s(v,\varepsilon^*) - \beta\varepsilon^* \tag{18}$$

and

$$\eta^{m.c.}(f,x_0|v,\varepsilon^*) = 0 \quad . \tag{19}$$

Then (19) implies

$$s(v,\varepsilon^*) = s(v,(U,f),(\varepsilon^*,x_0)), \quad \text{so}$$

$$-\beta \mathfrak{f}(v,\beta) = s(v,(U,f),(\varepsilon^*,x_0)) - \beta\varepsilon^*, \quad \text{so}$$

$$\sup_{\varepsilon}\{s(v,(U,f),(\varepsilon,x_0)) - \beta\varepsilon\} \geq -\beta \mathfrak{f}(v,\beta) \quad ,$$

so

$$\eta^c(f,x_0|v,\beta) \geq 0 \quad .$$

Since

$$\eta^c(f,x|v,\beta) \leq 0$$

for all x,

$$\eta^{c}(f,x_0|v,\beta) = 0 \quad .$$

Conversely, suppose

$$\eta^{c}(f,x_0|v,\beta) = 0 \quad .$$

Then

$$\sup_{\varepsilon}\{s(v,(U,f),(\varepsilon,x_0)) - \beta\varepsilon\} = -\beta\mathfrak{f}(v,\beta) \quad .$$

Since

$$\varepsilon \to s(v,(U,f),(\varepsilon,x_0)) - \beta\varepsilon$$

is upper semicontinuous, goes to $-\infty$ as $\varepsilon \to -\infty$ (by stability), and goes to $-\infty$ as $\varepsilon \to \infty$ (since it is not greater than $1 + \log(v) - \beta\varepsilon$), it must take on its supremum for some value ε^* of ε. Thus

$$s(v,(U,f),(\varepsilon^*,x_0)) - \beta\varepsilon^* = -\beta\mathfrak{f}(v,\beta) \quad .$$

Since

$$s(v,(U,f),(\varepsilon,x_0)) - \beta\varepsilon^* \leq s(v,\varepsilon^*) - \beta\varepsilon^* \leq -\beta\mathfrak{f}(v,\beta) \ ,$$

we must have

$$s(v,\varepsilon^*) - \beta\varepsilon^* = -\beta\mathfrak{f}(v,\beta) \quad \text{and}$$

$$s(v,(U,f),(\varepsilon^*,x_0)) = s(v,\varepsilon^*), \quad \text{i.e.,} \ \eta^{m.c.}(f,x_0|v,\varepsilon^*) = 0 \quad .$$

The second assertion of the theorem is an immediate consequence of the first. The proof of the final assertion is very similar to the proof of the first assertion; we omit the details.

Note that we have not shown that, even if $\eta^{m.c.}(f,x|v,\varepsilon)$ or $\eta^{c.}(f,x|v,\beta)$ takes on the value zero for only one value x_0 of x, the mean values of f/N in the corresponding microcanonical or canonical ensembles converge as the system becomes large to x_0. This is presumably true, but proving it would require estimates on the contribution to the mean coming from very large values of f/N; these estimates are not quite obvious.

We next want to outline a technique for proving that, in certain cases, $\eta^{m.c.}(f,x|v,\varepsilon)$ takes on the value zero exactly once. In order not to have to interrupt the argument, it is convenient to prove first a technical property of $s(v,f,x)$ which we will need later:

Proposition A5.9

(a) *Let* f *be an* $\mathbb{R}^t$*-valued observable, and let* A *be an invertible* $t \times t$ *matrix. Let* A *define a mapping of* $\mathbb{R}^t$ *into itself by* $(Ax)_i = \sum_j A_{ij}x_j$*. Then*

$$s(v,Af,Ax) = s(v,f,x) \quad \textit{for all} \quad x \in \mathbb{R}^t$$

(b) *Let* f_1, f_2 *be scalar-valued observables; let*

$$x_{1,\min} = \inf\{x_1 : s(v,f_1,x_1) > -\infty\}$$

$$x_{1,\max} = \sup\{x_1 : s(v,f_1,x_1) > -\infty\}$$

Then, if $x_1 \neq x_{1,\min}, x_{1,\max}$,

$$s(v,f_1,x_1) = \sup_{x_2}\{s|v,(f_1,f_2),(x_1,x_2)\} \quad .$$

Proof. Part (a) is immediate from the construction of $s(v,f,x)$. To prove (b), we note that if $J = (a,b)$, where $a \neq x_{1,\max}$, $b \neq x_{1,\min}$,

$$\lim_{\substack{\Lambda\to\infty \\ N\to\infty \\ \frac{N(\Lambda)}{N}\to v}} \frac{1}{N} \log V(\Lambda,N,(f_1,f_2),J\times\mathbb{R})$$

$$= \sup_{\substack{x_1\in J \\ x_2\in\mathbb{R}}} s(v,(f_1,f_2),(x_1,x_2)) = \sup_{x_1\in J} s(v,f_1,x_1))$$

Thus, if we let $\varphi(x_1)$ temporarily denote $\sup_{x_2} s(v,(f_1,f_2),(x_1,x_2))$, we have

$$\sup_{x_1\in J} s(v,f_1,x_1) = \sup_{x_1\in J} \varphi(x_1) \quad .$$

Since this is true for nearly every J,

$$\varphi(x_1) = s(v,f_1,x_1)$$

wherever both sides are continuous, and the interior of $\{x_1 : \varphi(x_1) > -\infty\}$ is the same as the interior of $\{x_1 : s(v,f_1,x_1) > -\infty\}$. But $\varphi(x_1)$ is concave; hence, continuous except possibly at $x_{1,\min}$, $x_{1,\max}$, so the proposition is proved.

We now turn to the problem of finding a condition which guarantees that $\eta^{m.c.}(f,x|v,\varepsilon)$ is zero for only one value of x, i.e., that $s(v,(U,f),(\varepsilon,x))$ takes on its maximum (as a function of x) only once. The general idea of the argument is to show that a horizontal segment in the graph of $s(v,(U,f),(\varepsilon,x))$ manifests itself in non-differentiability of thermodynamic functions in their dependence on the interaction. We will consider the free energy $\mathfrak{f}(v,\beta)$ as a function of the interaction; to indicate what interaction we are considering we will write $\mathfrak{f}_U(v,\beta)$. Suppose that, for given values of (v,β), ε^* is chosen so that

$$\varepsilon^* - \frac{1}{\beta} s(v,\varepsilon^*) = \mathfrak{f}_U(v,\beta) \quad ,$$

and assume that, for some observable f,

$$s(v,(U,f),(\varepsilon^*,x)) \quad ,$$

as a function of x, takes on its supremum $s(v,\varepsilon^*)$ on an interval $[x_1,x_2]$ with $x_1 < x_2$.

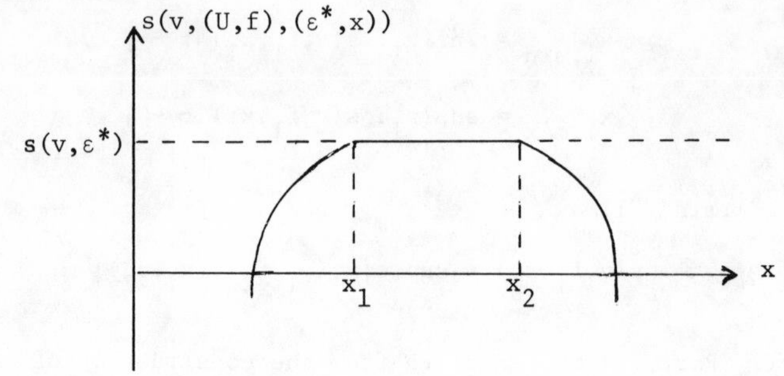

Assume also that the observable f is such that $U + \lambda f$ is stable for all sufficiently small λ, so $\mathscr{f}_{U+\lambda f}(v,\beta)$ makes sense for all λ in an interval about zero. (This will be the case, for example, if U is a superstable interaction and f is a continuous finite-range two-particle observable.)

Then, from what we have shown above,

$$\mathscr{f}_{U+\lambda f}(v,\beta) = \inf_{\varepsilon}\{\varepsilon - \frac{1}{\beta}\, s(v,U + \lambda f,\varepsilon)\} \quad .$$

On the other hand, by Proposition A5.9,

$$s(v,U + \lambda f,\varepsilon) = \sup\{s(v,(U,f),(\varepsilon',x)) \colon \varepsilon' + \lambda x = \varepsilon\}$$

for all but at most two values of ε, so

$$\mathscr{f}_{U+\lambda f}(v,\beta) = \inf_{\varepsilon',x}\{\varepsilon' + \lambda x - \frac{1}{\beta}\, s(v,(U,f),(\varepsilon',x))\} \quad .$$

We may therefore insert any values we like for (ε',x) in $\varepsilon' + \lambda x - \frac{1}{\beta}\, s(v,(U,f),(\varepsilon',x))$ and get an upper bound for $\mathscr{f}$. We insert $\varepsilon' = \varepsilon^*$ and $x = x_1$ or x_2. Using

$$s(v,(U,f),(\varepsilon^*,x_1)) = s(v,(U,f),(\varepsilon^*,x_2)) = s(v,\varepsilon^*)$$

and

$$\varepsilon^* - \frac{1}{\beta}\, s(v,\varepsilon^*) = \mathscr{f}_U(v,\beta), \quad \text{we get}$$

$$\mathscr{f}_{U+\lambda x}(v,\beta) \le \varepsilon^* + \lambda x_i - \frac{1}{\beta}\, s(v,(U,f),(\varepsilon^*,x_i))$$

$$= \mathscr{f}_U(v,\beta) + \lambda x_i, \quad i = 1,2.$$

Thus, for $\lambda > 0$.

$$\frac{\mathscr{f}_{U+\lambda f}(v,\beta) - \mathscr{f}_U(v,\beta)}{\lambda} \le x_1 \quad ,$$

and for $\lambda < 0$

$$\frac{\mathfrak{f}_{U+\lambda f}(v,\beta) - \mathfrak{f}_U(v,\beta)}{\lambda} \geq x_2 \quad .$$

Since $x_1 < x_2$, the right and left derivatives of $\mathfrak{f}_{U+\lambda f}(v,\beta)$ with respect to λ at $\lambda = 0$ must be different, so $\mathfrak{f}_{U+\lambda f}(v,\beta)$ cannot be differentiable with respect to λ at $\lambda = 0$. Equivalently: If $\mathfrak{f}_{U+\lambda f}(v,\beta)$ is differentiable in λ at $\lambda = 0$, then $\eta^{m.c.}(f,x|v,\varepsilon^*)$ is zero for only one value of x.

There is a general situation in which this criterion can be applied. If U is a superstable two-body interaction and f is a continuous finite-range two-particle observable, one shows by special techniques (the Kirkwood-Salsburg equations and expansions in the activity) that, for sufficiently large v (depending on β), $\mathfrak{f}_{U+\lambda f}(v,\beta)$ is actually analytic in λ for small λ. This implies, then, that every two-particle f converges in probability to a constant in the microcanonical ensemble with specific volume v and the energy per particle corresponding to (v,β).

We will also sketch roughly a second application of the criterion. Note that for finite Λ,N,

$$\log Z(\Lambda,N,\beta) = \log \left\{\frac{1}{N!} \int_\Lambda \cdots \int_\Lambda dq_1 \ldots dq_N \exp[-\beta U(q_1,\ldots,q_N)]\right\}$$

is convex in U. The same is therefore true of

$$-\beta\mathfrak{f}_U(v,\beta) = \lim_{\substack{\Lambda\to\infty \\ N\to\infty \\ \frac{V(\Lambda)}{N}\to\infty}} \frac{1}{N} \log Z(\Lambda,N,\beta) \quad ,$$

so $\mathfrak{f}_U(v,\beta)$ is concave in U. Now concave functions tend to be continuous, and continuous concave functions are differentiable "almost everywhere" (e.g., on the complement of a set of first category for a continuous concave function on an open subset of a separable Banach space.) Hence, for any given (v,β) and "almost all" superstable interactions U, $\mathfrak{f}_{U+\lambda f}(v,\beta)$ is differentiable with respect to λ at $\lambda = 0$ for every continuous two-particle observable f, so for such interactions we have convergence in probability to constants for all continuous two-particle observables as the system becomes large. The above somewhat vague argument, and in particular the meaning of "almost all" superstable interactions, can be made precise, although not in a very elegant way. We will see later a version of this argument for lattice systems.

A6. THE GRAND CANONICAL ENSEMBLE

In passing from the microcanonical point of view from which we started to the canonical point of view developed in the preceding section, we allowed the energy, originally fixed to lie in some small interval, to vary, but controlled its variation through the introduction of the temperature. We now want to do much the same thing with the particle number. We introduce a new parameter, z, called the *activity* or *fugacity*, whose logarithm plays the same role with respect to the particle number as $-\beta$ plays with respect to the energy. For each bounded open Λ, and each positive β, z, we define a probability measure $\mu^{g.c.}(\Lambda, z,\beta)$ on $\bigcup_{N=0}^{\infty} \Lambda^N$, called the *grand canonical probability measure*, by

$$\mu^{g.c.}(\Lambda, z,\beta)(N;dq_1\ldots dq_N) = \frac{1}{\Xi(\Lambda,z,\beta)} z^N \exp[-\beta U(q_1,\ldots,q_N)] \frac{dq_1\ldots dq_N}{N!} , \quad (1)$$

where

$$\begin{aligned}\Xi(\Lambda,z,\beta) &= \sum_{N=0}^{\infty} z^N \int_\Lambda \ldots \int_\Lambda \frac{dq_1\ldots dq_N}{N!} \exp[-\beta U(q_1,\ldots,q_N)] \\ &= \sum_{N=0}^{\infty} z^N Z(\Lambda,N,\beta)\end{aligned} \quad (2)$$

In other words, $\mu^{g.c.}(\Lambda, z,\beta)$ is a weighted average over N of $\mu^{c.}(\Lambda,N,\beta)$, with the weight assigned to N proportional to $z^N Z(\Lambda,N,\beta)$.

We want to investigate, as before, the behavior of the distribution of observables, and also of the particle number, in the grand canonical probability measure for large Λ. Because particle number is now variable, it is convenient to normalize observables and thermodynamic functions by dividing by the volume rather than the particle number as we have done up to now. We therefore transform our notation as follows: We let:

$$\hat{s}(\rho,f,x) = \rho s(\tfrac{1}{\rho},f,\tfrac{x}{\rho}) \quad (3)$$

$$\left(= \lim_{\substack{\Lambda\to\infty \\ N\to\infty \\ \frac{N}{V(\Lambda)}\to\rho}} \frac{1}{V(\Lambda)} \log\left[\frac{1}{N!} \cdot \text{configuration space volume with } \frac{f}{V(\Lambda)} \approx x\right]\right)$$

$$\hat{s}(\rho,\varepsilon) = \rho s(\tfrac{1}{\rho}, \tfrac{\varepsilon}{\rho}) \quad (4)$$

$$\hat{f}(\rho,\beta) = \rho f(\tfrac{1}{\rho}, \beta) \quad . \quad (5)$$

We have already remarked (at the end of Section A2) that $\hat{s}(\rho,f,x)$ is jointly concave in (ρ,x); similarly, $\hat{f}(\rho,\beta)$ is a convex function of ρ. We let

$$\hat{\Gamma}(f) \subset \mathbb{R}^{t+1} = \{(\rho,x): \hat{s}(\rho,f,x) > -\infty\} \quad . \quad (6)$$

We continue to use the convention, introduced in the preceding section, that the first component of an $\mathbb{R}^t$-valued observable f is the interaction U and that the first component of a point x of $\mathbb{R}^t$ is called ε.

We now define, for any convex open $J \subset \mathbb{R}^{t+1}$,

$$\Xi(\Lambda, z,\beta,f,J) = \sum_{N=0}^{\infty} \int_{\Delta_N} \ldots \int \frac{dq_1\ldots dq_N}{N!} z^N \exp[-\beta U(q_1,\ldots,q_N)]$$

where the region of integration Δ_N is given by:

$$\Delta_N = \{(q_1,\dots,q_N) \in \Lambda^N : (\frac{N}{V(\Lambda)}, \frac{f(q_1,\dots,q_N)}{V(\Lambda)}) \in J\} \tag{8}$$

Then the grand-canonical probability of finding $(\frac{N}{V(\Lambda)}, \frac{f}{V(\Lambda)})$ in J is just

$$\frac{\Xi(\Lambda,z,\beta,f,J)}{\Xi(\Lambda,z,\beta)}$$

The following theorem is proved using techniques which are essentially the same as those used in the proof of Theorem A5.1; we will omit the proof:

Theorem A6.1

Let Λ_n *be a sequence of bounded open sets becoming infinitely large in the sense of van Hove and approximable by rectangles. Let* J *be a convex open set in* $\mathbf{R}^{t+1}$*; assume that either:*

(a) J *contains an interior point of* $\hat{\Gamma}(f)$ *or*

(b) *The distance from* J *to* $\hat{\Gamma}(f)$ *is strictly positive. Then*

$$\lim_{n\to\infty} \frac{1}{V(\Lambda_n)} \log \Xi(\Lambda,z,\beta,f,J) = \sup\{\hat{s}(\rho,f,x) - \beta\varepsilon + \rho \log z : (\rho,x) \in J\} .$$

Corollary A6.2

If Λ_n *is as in the preceding theorem,*

$$\lim_{n\to\infty} \frac{1}{V(\Lambda_n)} \log \Xi(\Lambda_n,z,\beta) = \sup_{\rho,\varepsilon}\{\hat{s}(\beta,\varepsilon) - \beta\varepsilon + \rho \log z\}$$

$$= \sup_{\rho}\{-\beta\hat{\mathfrak{f}}(\rho,\beta) + \rho \log z\} .$$

The implications of Theorem A6.1 for convergence in probability of N/V and f/V in the grand canonical probability measure should by now be clear: in the limit of large systems, only values of (N/V, f/V) near to those values of (ρ,x) where $\hat{s}(\rho,f,x) - \beta\varepsilon + \rho \log z$ takes on its supremum occur. In particular, if $-\beta\hat{\mathfrak{f}}(\rho,\beta) + \beta \log z$ takes on its supremum as a function of ρ at a single point, there is a well-defined density corresponding to the given values of (z,β). Otherwise, $\hat{\mathfrak{f}}(\rho,\beta)$ has a linear portion with slope $\frac{\log z}{\beta}$ in its graph as a function of ρ, and ρ changes discontinuously as $\log z$ passes through the value in question with β held fixed. It may be shown that, for fairly general interactions, the graph of $\hat{\mathfrak{f}}(\rho,\beta)$ as a function of ρ does not have any corners (i.e., discontinuities in its derivative). This result is known, for complicated reasons, as "continuity of the pressure"; for proofs, see Ruelle [9], pp. 58-60 or Ruelle [11], section 4. For lattice systems, there is a more general result, due to Griffiths and Ruelle [5], which implies that the lattice analogue of

$$\lim_{\Lambda\to\infty} \frac{1}{V(\Lambda)} \log \Xi(\Lambda,z,\beta)$$

is a strictly convex function of β, $\log z$.

We still have to interpret

$$\lim_{\Lambda\to\infty} \frac{1}{V(\Lambda)} \log \Xi(\Lambda,z,\beta) \quad .$$

It is well known that this limit is to be identified with β times the pressure, but the justification for this identification is sometimes a bit mysterious. There are in fact two justifications, one based on the thermodynamic relations satisfied by the limit and the other on a purely mechanical definition of the pressure. We will give a heuristic version of the second justification; it has the advantage of being very down-to-earth and of not requiring any prior identification of β with the inverse temperature and s with the entropy.

We start by considering a finite system of particles in a region Λ with a smooth boundary which we take to be perfectly reflecting, and we set out to calculate the force exerted by the particles bouncing off the walls in a time interval of length T. We want to consider only very small values of T, so we have only to consider the effect of particles initially very near the wall. From the point of view of each such particle, the wall appears to be flat. Consider, then, a single particle, at a distance x from the wall. On the time-scale we are considering, we may ignore the force exerted on our particle by the other particles in the system, i.e., the particle moves as if it were free except when it collides with the wall. Thus, the particle will collide with the wall if and only if the component of its velocity in the direction of the wall, which we will denote by v_x, is larger than x/T. If it does collide with the wall, the momentum it imparts to the wall is $2v_x$. (We are putting the mass of the particles equal to one.) Now, instead of considering a single configuration of particles, we are going to average over a probability distribution on the set of all configurations of particles in Λ. We will assume that this probability distribution is Maxwellian in the velocities, i.e., has the form

$$\rho(N,dq_1,\ldots,dq_N) \cdot \exp[\tfrac{-\beta}{2}(v_1^2 + \ldots + v_N^2)](\tfrac{\beta}{2\pi})^{N\nu/2}$$

where $\rho(N; dq_1\ldots dq_N)$ is a probability measure on $\bigcup_{N=0}^{\infty} \Lambda^N$. Returning to the problem of calculating the momentum imparted to the wall by a single particle during a time interval of length T, we find, by averaging over that particle's velocity distribution, the average value

$$\sqrt{\frac{\beta}{2\pi}} \int_{x/T}^{\infty} 2v_x \exp[\tfrac{-\beta}{2} v_x^2]dv_x$$

$$= \frac{2}{\sqrt{2\pi\beta}} \exp[-\beta \frac{x^2}{T^2}] \quad ,$$

where x is the distance from the initial position of the particle to the wall.

Dividing by T gives the average force per unit time on the walls of the container due to the particular particle we are considering. We now sum up the contributions from all the particles and average over the configuration probability distribution; this gives the following expression for the average total force per unit time on the boundary of the container:

$$F = \int_{\Lambda} dq\ \rho(q) \frac{2}{\sqrt{2\pi\beta}} \cdot \frac{1}{T} e^{-\frac{\beta}{2} \cdot \frac{x(q)^2}{T^2}} , \tag{9}$$

where $\rho(q)$ is the average density at the point q and $x(q)$ is the distance from q to the boundary of Λ. Now

$$\int_0^{\infty} dx \frac{2}{\sqrt{2\pi\beta}} \cdot \frac{1}{T} e^{-\frac{\beta}{2} \cdot \frac{x^2}{T^2}} = \frac{1}{\beta} ,$$

so, if we write locally, near the boundary of Λ,

$$dq = dx \cdot d\sigma;\ d\sigma \text{ surface area on the boundary of } \Lambda,$$

then

$$F = \int d\sigma \cdot \frac{1}{\beta} \cdot [\text{an average over } x \text{ of } \rho(q)] ,$$

and the average over x is sharply peaked at $x = 0$ for small T. Hence, if we assume that the density $\rho(q)$ approaches a boundary value continuously as q approaches the boundary of Λ, we get, for the mean instantaneous force on the walls of the box:

$$\lim_{T \to 0} \int_{\Lambda} dq\ \rho(q) \frac{2}{\sqrt{2\pi\beta}} \cdot \frac{1}{T} e^{-\frac{\beta}{2} \cdot \frac{x(q)^2}{T^2}} = \frac{1}{\beta} \cdot \int_{\partial\Lambda} d\sigma(q) \cdot \rho(q) ,$$

and thus the instantaneous pressure is given by

$$p = \frac{1}{\beta} \frac{\int_{\partial\Lambda} \rho(q) d\sigma(q)}{\int_{\partial\Lambda} d\sigma(q)} \tag{10}$$

i.e., by the total force divided by the surface area.

We next rework the above expression for the pressure in the special case in which the probability distribution of particles in Λ is given by $\mu^{g.c.}(\Lambda,z,\beta)$. Let Λ' denote the region obtained by deleting from Λ a thin shell of uniform thickness around the surface. We may write :

$$\Xi(\Lambda,z,\beta) = \sum_{n,m=0} z^n z^m \int_{\Lambda'} \cdots \int \frac{dq_1 \ldots dq_n}{n!} \int_{\Lambda \setminus \Lambda'} \cdots \int \frac{dq_1 \ldots dq_m}{m!}$$

$$\exp[-\beta U(q_1',\ldots,q_n',q_1,\ldots,q_m)] ,$$

and $\Xi(\Lambda',z,\beta)$ is exactly the sum of all terms with $m = 0$. Thus

$$\frac{\Xi(\Lambda) - \Xi(\Lambda')}{\Xi(\Lambda)}$$

is the probability, with respect to $\mu^{g.c.}(\Lambda,z,\beta)$, of finding at least one particle in $\Lambda\backslash\Lambda'$. If the thickness of the shell $\Lambda\backslash\Lambda'$ is very small, the probability of finding more than one particle in $\Lambda\backslash\Lambda'$ is negligible, so

$$\frac{1}{V(\Lambda\backslash\Lambda')}\,\frac{\Xi(\Lambda)-\Xi(\Lambda')}{\Xi(\Lambda)} \approx \text{average density in } \Lambda\backslash\Lambda'$$

and, as the thickness of the shell goes to zero,

$$\frac{1}{V(\Lambda\backslash\Lambda')}\,\frac{\Xi(\Lambda)-\Xi(\Lambda')}{\Xi\Lambda} \to \frac{\int_{\partial\Lambda} d\sigma(q)\cdot\rho(q)}{\int_{\partial\Lambda} d\sigma(q)} = \beta\cdot p \quad .$$

Thus we have

$$\frac{d}{d(V(\Lambda))}\log\Xi(\Lambda,z,\beta) = \beta\cdot p \quad . \tag{11}$$

Now what happens as Λ becomes large? We have not shown that

$$\lim_{\Lambda\to\infty}\frac{d}{d(V(\Lambda))}\log\Xi(\Lambda,z,\beta)$$

exists. We do know, however, that

$$\lim_{\Lambda\to\infty}\frac{1}{V(\Lambda)}\log\Xi(\Lambda,z,\beta)$$

exists. In general, if we have a function $\psi(V)$ such that

$$\lim_{V\to\infty}\frac{d\psi(V)}{dV} = \alpha \quad ,$$

then

$$\lim_{V\to\infty}\frac{\psi(V)}{V} = \alpha$$

also, by l'Hôpital's rule (or by a simple geometric argument). Unfortunately, the argument does not work in the other direction, so we are left in the following slightly unfortunate situation: We don't know that

$$\lim_{\Lambda\to\infty} p(\Lambda,z,\beta)$$

exists, but we do know that, if the limit exists, it is equal to

$$\frac{1}{\beta}\cdot\lim_{\Lambda\to\infty}\frac{1}{V(\Lambda)}\log\Xi(\Lambda,z,\beta)$$

and this latter limit does exist. Despite the gap in the argument, we will identify

$$\frac{1}{\beta}\lim_{\Lambda\to\infty}\frac{1}{V(\Lambda)}\log\Xi(\Lambda,z,\beta)$$

with the pressure of the infinite system. Thus we have

$$p(z,\beta) = \sup_{\varepsilon,\rho}\left\{\frac{1}{\beta}\,\hat{s}(\rho,\varepsilon) - \varepsilon + \frac{\rho}{\beta}\log z\right\} \quad . \tag{12}$$

We can use this expression to identify z. In doing this, it will be convenient to introduce the kinetic energy and the corresponding entropy into our formulas. Recall that ε is the potential energy per unit volume and $\hat{s}$ the configurational entropy per unit volume. As a function of β, the mean kinetic energy per particle is $\frac{\nu}{2} \cdot \frac{1}{\beta}$ and the kinetic entropy per particle is $\frac{\nu}{2}\,[1 + \log(\frac{2\pi}{\beta})]$. Thus we define $s_t = \hat{s} + \rho \cdot \frac{\nu}{2}\,[1 + \log(\frac{2\pi}{\beta})]$ = total entropy per unit volume

$$\varepsilon_t = \varepsilon + \rho \cdot \frac{\nu}{2} \cdot \frac{1}{\beta} = \text{total energy per unit volume.}$$

We also introduce a new parameter μ, called the *chemical potential*, defined by

$$z = e^{\mu\beta} \cdot \left(\frac{2\pi}{\beta}\right)^{\nu/2} \tag{13}$$

so, for example,

$$\Xi(\Lambda,z,\beta) = \sum_{N=0}^{\infty} e^{\beta\cdot\mu\cdot N} \int_{\Lambda} \cdots \int \frac{dq_1 \ldots dq_N}{N!} \int_{\mathbb{R}^\nu} \cdots \int dv_1 \ldots dv_N$$
$$\times \exp\left[-\beta\left(\sum_{j=1}^{N} \frac{v_j^2}{2} + U(q_1,\ldots,q_N)\right)\right] .$$

In terms of these new variables, we have:

$$p = \sup_{\varepsilon_t,\rho}\left\{\frac{1}{\beta} s_t - \varepsilon_t + \rho\cdot\mu\right\}, \quad \text{i.e.}$$

$$p(\mu,\beta) = \frac{1}{\beta} s_t^* - \varepsilon_t^* + \rho^*\mu, \quad \text{or}$$

$$\rho^*\mu = \varepsilon_t^* - \frac{1}{\beta} s_t^* + p \quad .$$

Thus, $\rho^*\mu$ is what is called in thermodynamics the Gibbs free energy per unit volume, so the chemical potential μ is the Gibbs free energy per particle.

We are now able to do a better job of justifying the identification of s_t with the thermodynamic entropy. (Recall that all we have shown so far is that it differes from the thermodynamic entropy at most by a function of the density.) The argument we will give is semi-formal, in that we will argue in terms of differentials and "infinitesimal changes" in the parameters of the system. The task of making the argument precise (notably, taking into account the occurrence of phase transitions) will be left as an exercise.

We start from the thermodynamic relation

$$dE = TdS - pdV \tag{14}$$

which is frequently taken as the definition of the entropy. Here, E is the energy, T the temperature, S the entropy, p the pressure, and V the volume, and the relation (14) holds for an infinitesimal change made in a fixed quantity of matter. Taking the unit of matter to be a single particle, we see that what we want to prove is

$$d(\varepsilon_t/\rho) = \frac{1}{\beta}\, d(s_t/\rho) + p\, \frac{d\rho}{\rho^2} \tag{15}$$

If we prove the relation (15), then the identification of s_t as the entropy per unit volume (up to an undetermined additive constant) follows from our previous identifications of $\varepsilon_t, \rho, \beta$, and p.

From the formula

$$p = \sup_{\varepsilon_t, \rho} \left\{ \frac{1}{\beta}\, s_t(\rho, \varepsilon_t) - \varepsilon_t + \rho\mu \right\}$$

and the fact that the values of (ρ, ε_t) corresponding to the given values of (μ, β) are exactly those for which the supremum is attained, we see that we have:

$$0 = \frac{\partial}{\partial \varepsilon_t} \left\{ \frac{1}{\beta}\, s_t(\rho, \varepsilon_t) - \varepsilon_t + \rho\mu_t \right\} = \frac{\partial}{\partial \rho} \left\{ \frac{1}{\beta}\, s_t(\rho, \varepsilon_t) - \varepsilon_t + \rho\mu_t \right\} .$$

(In computing the partial derivatives, the independent variables are ρ, ε_t; the dependent variable is s_t, and β, μ are treated as constants.) Simplifying the expressions, we get

$$\frac{\partial}{\partial \varepsilon_t}\, s_t(\rho, \varepsilon_t) = \beta; \quad \frac{\partial}{\partial \rho}\, s_t(\rho, \varepsilon_t) = -\mu\beta \tag{16}$$

Thus, we have,

$$\frac{1}{\beta}\, ds_t = d\varepsilon_t - \mu d\rho \tag{17}$$

Dividing by ρ and using the formulas:

$$\frac{ds_t}{\rho} = d\left(\frac{s_t}{\rho}\right) + s_t\, \frac{d\rho}{\rho^2}$$

$$\frac{d\varepsilon_t}{\rho} = d\left(\frac{\varepsilon_t}{\rho}\right) + \varepsilon_t\, \frac{d\rho}{\rho^2} ,$$

we get

$$\frac{1}{\beta}\, d(s_t/\rho) = d(\varepsilon_t/\rho) + \frac{d\rho}{\rho^2} \left(\varepsilon_t - \frac{s_t}{\beta} - \mu\rho\right)$$

Finally, inserting

$$p = \frac{1}{\beta}\, s_t(\rho, \varepsilon_t) - \varepsilon_t + \rho\mu ,$$

we obtain

$$\frac{1}{\beta}\, d(s_t/\rho) = d(\varepsilon_t/\rho) - p d\rho/\rho^2$$

which is exactly the formula (15) we wanted to prove.

For the sake of completeness, we will state formally some results expressing the equivalence of the grand canonical and microcanonical ensembles for observables. We will omit the proofs and interpretations of these results; they are essentially the same as those for the results on the equivalence of the canonical and microcanonical ensembles given in the preceding section (see Theorem A5.8.) Let f denote a scalar-valued observable which is not a constant multiple of U, and define

$$\eta^{g.c.}(f,x|z,\beta) = \sup_{\rho,\varepsilon}\{\hat{s}(\rho,(U,f),(\varepsilon,x)) - \beta\varepsilon + \rho \log z\} - \beta p(z,\beta) \tag{18}$$

Then we have:

Proposition A6.3

(a) *For all* x, $\eta^{g.c.}(f,x|z,\beta) \leq 0$, *and*

$$\sup_{x} \eta^{g.c.}(f,x|z,\beta) = 0 \quad .$$

(b) $\eta^{g.c.}(f,x|z,\beta)$ *is a concave function of* x .

(c) *Let*

$$x_{min}(f,z,\beta) = \inf\{x: \eta^{g.c.}(f,x|z,\beta) > -\infty\}$$

$$x_{max}(f,z,\beta) = \sup\{x: \eta^{g.c.}(f,x|z,\beta) > -\infty\} \quad .$$

Then, if $x_1 \neq x_{max}$, $x_2 \neq x_{min}$

$$\lim_{\Lambda\to\infty} \frac{1}{V(\Lambda)} \log \mu^{g.c.}(\Lambda,z,\beta)(\{\frac{f}{V(\Lambda)} \in (x_1,x_2)\})$$

$$= \sup_{x_1<x<x_2} \{\eta^{g.c.}(f,x|z,\beta)\} \quad .$$

Theorem A6.4

$\eta^{g.c.}(f,x_0|z,\beta) = 0$ *if and only if there exist* ρ^*, ε^* *such that:*

(a) $\beta p(z,\beta) = \hat{s}(\rho^*,\varepsilon^*) - \beta\varepsilon^* + \rho^* \log z$

(b) $\eta^{m.c.}(f,x_0/\rho^*|1/\rho^*,\varepsilon^*/\rho^*) = 0.$

(The factors of $1/\rho^*$ appear because, for example, ε^* is to be interpreted as the energy per unit volume and $\eta^{m.c.}$ is defined as a function of the energy per particle.)

B. INVARIANT EQUILIBRIUM STATES

B1. PRELIMINARIES

Our next step is to investigate the relation between the analysis we have given of infinite volume limits of ensembles and the theory of states of actually infinite systems. We have been working up to now with continuous systems of particles.

For the ideas we now want to develop, the technicalities for continuous systems become very difficult and have not been worked out in detail. For this reason, we will now investigate instead lattice systems. Such systems can be interpreted either as discrete approximations to continuous system (lattice gases) or as models for magnetic materials (spin systems). To maintain contact with our earlier results on continuous systems, we will discuss lattice systems in lattice-gas language rather than spin-system language.

A lattice gas is an approximation of a continuous system in which the particles are constrained to lie on the points of a lattice, which in appropriate units we can take to be the lattice of integral points $\mathbb{Z}^\nu$ in $\mathbb{R}^\nu$. We also require that there be no more than one particle at each lattice site. Thus, if Λ is a finite subset of $\mathbb{Z}^\nu$ (the analogue of a bounded open set in $\mathbb{R}^\nu$ in the previous chapter), we can define a configuration of particles in Λ completely by giving the subset of Λ of lattice points at which there are particles. Thus, the analogue of configuration space for systems of an indefinite number of particles in Λ is just the set of all subsets of Λ, and the analogue of the set of all configurations of finitely many particles is just the set of all finite subsets of $\mathbb{Z}^\nu$. (From this remark, it is a short step to the remark that any subset of $\mathbb{Z}^\nu$ may be interpreted as a configuration of possibly infinitely many particles. Thus, the set of subsets of $\mathbb{Z}^\nu$ is the configuration space for an infinite lattice gas. The continuous analogue of this configuration space is a much more complicated object.)

The analogue of what we called in the preceding chapter a finite-range observable is a mapping f from finite subsets of $\mathbb{Z}^\nu$ to real numbers such that

(a) $f(X + a) = f(X)$ for all $a \in \mathbb{Z}^\nu$; X finite $\subset \mathbb{Z}^\nu$.

(b) $f(X \cup Y) = f(X) + f(Y)$ if the distance from X to Y is greater than R.

(c) $f(\phi) = 0 = f(\{x\})$ for all $x \in \mathbb{Z}^\nu$.

Given such an f, we define recursively another function Φ, again sending finite subsets to numbers, by

$$f(X) = \sum_{Y \subset X} \Phi(Y). \tag{1}$$

Φ again satisfies (a) and (c), but (b) is replaced by

$$(b') \ \Phi(X) = 0 \quad \text{if} \quad \operatorname{diam}(X) > R \quad .$$

Conversely, given Φ satisfying (a), (b′), and (c), we can reconstruct f by (1). Thus, we have a one-to-one correspondence between observables f and potentials Φ; we will from now on describe observables by giving the corresponding Φ's. The observable f associated with Φ will be denoted by $\Sigma\Phi$.

We make two technical changes in our definition of observable. In the last chapter, we gave ourselves considerable extra trouble by not treating the particle number as an observable. This had the advantage of allowing us to prove results about microcanonical and canonical ensembles with precisely defined numbers of particles.

We could do the same thing here, but the improvement in results does not seem worth the trouble, so we incorporate the particle number in the observable. In terms of potentials, this means we drop the requirement $\Phi(\{x\}) = 0$.

On the other hand, it will be important to be able to deal with observables which are not of strictly finite range. Note that the finite range condition (b′) implies that there are only finitely many X's containing 0 with $\Phi(X) \neq 0$. We therefore have, trivially,

$$\|\Phi\| = \sum_{0 \in X} \frac{|\Phi(X)|}{N(X)} < \infty \quad . \tag{2}$$

We will take this condition to replace the condition of strictly finite range. (The significance of the $\frac{1}{N(X)}$ will become clear later.) Thus, we define a *potential* to be a mapping from finite subsets of $\mathbb{Z}^\nu$ to $\mathbb{R}$ (or $\mathbb{R}^t$ to allow us to consider several potentials simultaneously) such that

(a) $\Phi(X) = \Phi(X + a)$ for $X \subset \mathbb{Z}^\nu$, $a \in \mathbb{Z}^\nu$

(b) $\Phi(\emptyset) = 0$

(c) $\|\Phi\| = \sum_{0 \in X} \frac{|\Phi(X)|}{N(X)} < \infty$.

We will denote the observable associated with such a potential by $\Sigma\Phi$. With the norm defined above, the set of real-valued potentials becomes a separable Banach space, $\mathcal{B}$, in which the set $\mathcal{B}_0$ of finite-range potentials is dense.

Examples: (a) $\Phi\{X\} = 1$ if X contains only one point, $= 0$ otherwise. Then $(\Sigma\Phi)(X) = N(X)$.

(b) $\Phi(\{x,y\}) = \varphi(x - y)$ if $x \neq y$, $\Phi(X) = 0$ if $N(X) \neq 2$. The $(\Sigma\Phi)(X)$ is just the potential energy of the configuration of particles X interacting by the two-body potential φ. The condition $\|\Phi\| < \infty$ becomes $\sum_{x \in \mathbb{Z}^\nu} |\varphi(x)| < \infty$.

We next want to give analogues of the results of Chapter A for the microcanonical entropy. We start by defining

$V(\Lambda,\Phi,J)$, where $\Lambda \subset \mathbb{Z}^\nu$ is finite

Φ is an $\mathbb{R}^t$-valued potential

J is a convex open set in $\mathbb{R}^t$,

to be the number of subsets X of Λ such that

$$\frac{\Sigma\Phi(X)}{N(\Lambda)} \in J \quad ;$$

$$V(\Lambda,\Phi,J) = N\{X \subset \Lambda : \frac{\Sigma\Phi(X)}{N(\Lambda)} \in J\} \quad .$$

This V is not quite an exact analogue of the V of the preceding chapter, since it is defined by specifying values of $\Sigma\Phi$ per lattice site, rather than per particle.

Using methods essentially the same as those used in Chapter A, I believe that it is possible to prove the following

Theorem B1.1

Let Φ *be an* $\mathbb{R}^t$*-valued potential whose components are linearly independent (in* $\mathcal{B}$*). Then there exists a concave upper semi-continuous function* $s(\Phi,x)$ *defined on* $\mathbb{R}^t$ *such that*

(a) $s(\Phi,x) \leq \log(2)$ *for all* x.

(b) $\Gamma(\Phi) = \{x\colon s(\Phi,x) > -\infty\}$ *has non-empty interior;* $s(\Phi,x)$ *is continuous on the interior of* $\Gamma(\Phi)$.

(c) *Let* J *be a convex open set in* $\mathbb{R}^t$ *which either contains an interior point of* $\Gamma(\Phi)$ *or is at a non-zero distance from* $\Gamma(\Phi)$*, and let* Λ_n *be a sequence of finite subsets of* $\mathbb{Z}^\nu$ *becoming infinitely large in the sense of van Hove. Then*

$$\lim_{n\to\infty} \frac{1}{N(\Lambda_n)} \log V(\Lambda_n,\Phi,J) = \sup_{x\in J} s(\Phi,x) \quad .$$

This theorem should be taken with a small grain of salt; I have never written out all the details of its proof. It is surely true if the potential is assumed to be of finite range and the condition of approximability by rectangles is added to the requirement that Λ_n becomes infinitely large in the sense of van Hove, and I believe it is possible to remove these restrictions.

The interpretation of $s(\Phi,x)$ is essentially the same as that of the function $s(v,f,x)$ of the preceding chapter. In particular, if Φ_1 is the potential such that $\Sigma\Phi_1 = N$, and if Φ_2 is the potential defining the interaction energy, then

$$s((\Phi_1,\Phi_2),\ (\rho,\varepsilon))$$

is the microcanonical entropy per lattice site as a function of the density ρ and the energy per lattice site ε.

For any scalar-valued potential Φ, and any finite $\Lambda \subset \mathbb{Z}^\nu$, we define the grand canonical partition function

$$\Xi(\Lambda,\Phi) = \sum_{X\subset\Lambda} e^{-\Sigma\Phi(X)} \tag{3}$$

(For notational convenience, we have absorbed the inverse temperature β and the activity into the interaction. Thus, if the interaction we really want to consider is given by a potential $\hat{\Phi}$, and if Φ_1 denotes the potential corresponding to the number of particles, we should put the potential Φ in (3) equal to $\beta\hat{\Phi} - \log z\Phi_1$). We now have:

Theorem B1.2

Let $\Phi \in \mathcal{B}$, *and let* Λ_n *be a sequence of finite subsets of* $\mathbb{Z}^\nu$ *becoming infinitely large in the sense of van Hove. Then*

$$\lim_{n\to\infty} \frac{1}{N(\Lambda_n)} \log \Xi(\Lambda_n,\Phi) = \sup_x [s(\Phi,x) - x] \quad .$$

We will denote the limit by $p(\Phi)$, and refer to it as the pressure. This theorem, except for the relation between $p(\Phi)$ and $s(\Phi,x)$ is contained in Theorem 2.3.3 of Ruelle [9].

We will now collect a few results which will be useful later.

Proposition B1.3

(a) *For* $\Phi \in \mathcal{B}$, X *a finite subset of* $\mathbb{Z}^\nu$, *we have* $|\Sigma\Phi(X)| \le \|\Phi\| \cdot N(X)$.

(b) $|p(\Phi_1) - p(\Phi_2)| \le \|\Phi_1 - \Phi_2\|$ *for all* $\Phi_1,\Phi_2 \in \mathcal{B}$.

(c) $p(\Phi)$ *is a convex function of* Φ *on* $\mathcal{B}$.

Proof. (a) $\Sigma\Phi(X) = \sum_{Y\subset X} \Phi(Y)$

$$= \sum_{x\in X} \sum_{x\in Y\subset X} \frac{\Phi(Y)}{N(Y)}$$

(The factor $\frac{1}{N(Y)}$ arises because each $Y \subset X$ contributes to the sum for each of $N(Y)$ different x's). Thus,

$$|\Sigma\Phi(X)| \le \sum_{x\in X} \sum_{x\in Y} \frac{|\Phi(Y)|}{N(Y)} = \sum_{x\in X} \|\Phi\| = N(X) \cdot \|\Phi\| \quad .$$

(b) We consider

$$\frac{d}{d\alpha} \log \Xi(\Lambda,\Phi_1 + \alpha(\Phi_2 - \Phi_1))$$

$$= \frac{\sum_{X\subset\Lambda} \Sigma(\Phi_2 - \Phi_1)(X) e^{-\Sigma(\Phi_1 + \alpha(\Phi_2 - \Phi_1))(X)}}{\Xi(\Lambda,\Phi_1 + \alpha(\Phi_2 - \Phi_1))} ,$$

which is just an average over $X \subset \Lambda$ of $\Sigma(\Phi_2 - \Phi_1)(X)$. By (a)

$$|\Sigma(\Phi_1 - \Phi_2)| \le N(X) \cdot \|\Phi_1 - \Phi_2\| \le N(\Lambda)\|\Phi_1 - \Phi_2\| \quad .$$

Thus,

$$\left|\frac{d}{d\alpha} \log \Xi(\Lambda,\Phi_1 + \alpha(\Phi_2 - \Phi_1))\right| \le N(\Lambda) \cdot \|\Phi_1 - \Phi_2\| \quad .$$

Integrating from $\alpha = 0$ to $\alpha = 1$ and dividing for $N(\Lambda)$ gives

$$\left|\frac{1}{N(\Lambda)} \log \Xi(\Lambda,\Phi_1) - \frac{1}{N(\Lambda)} \log \Xi(\Lambda,\Phi_2)\right| \le \|\Phi_1 - \Phi_2\| \quad .$$

Passing to the limit $\Lambda \to \infty$ proves (b).

(c) follows because $\frac{1}{V(\Lambda)} \log \Xi(\Lambda,\Phi)$ is a convex function of Φ (as may be proved by computing the second derivative of $\log \Xi(\Lambda,\Phi + \alpha\Phi')$ with respect to α) and a point-wise limit of a sequence of convex functions is convex.

B2. STATES OF INFINITE SYSTEMS

Everything we have done so far has been concerned with the properties of finite systems and especially with their asymptotic behavior as the system becomes large. We now want to introduce a formalism for investigating the statistical mechanics of actually infinite systems. The connection between this formalism and the limiting properties of finite systems will become apparent as we go along.

The statistical mechanics of finite systems is concerned with probability measures on the configuration space (or phase space) for finite numbers of particles. For infinite systems, therefore, we want to study probability measures on the configuration space of infinitely many particles. We have already remarked that, for the lattice gases we are considering, the configuration space of the infinite system is just the set of all subsets of $\mathbb{Z}^\nu$. We will denote this set by $\mathcal{P}(\mathbb{Z}^\nu)$ (and, in general, if X is a set, we will denote the set of subsets of X by $\mathcal{P}(X)$).

For finite $\Lambda \subset \mathbb{Z}^\nu$, $\mathcal{P}(\Lambda)$ is, of course, finite, and the theory of probability distributions on $\mathcal{P}(\Lambda)$ is completely free of measure-theoretic technicalities. This is not true for $\mathcal{P}(\mathbb{Z}^\nu)$, but $\mathcal{P}(\mathbb{Z}^\nu)$ may in a natural way be made into a compact topological space so we can use the theory of Borel measures on such spaces. The topology is defined by identifying $\mathcal{P}(\mathbb{Z}^\nu)$ with $\{0,1\}^{\mathbb{Z}^\nu}$, by identifying each subset of $\mathbb{Z}^\nu$ with its characteristic function, giving $\{0,1\}$ the discrete topology, and giving $\mathcal{P}(\mathbb{Z}^\nu)$ the product topology. Equivalently, we can define the topology by saying that a sequence X_n in $\mathcal{P}(\mathbb{Z}^\nu)$ converges to X if and only if, for each finite $\Lambda \subset \mathbb{Z}^\nu$, $X_n \cap \Lambda = X \cap \Lambda$ for all sufficiently large n.

A function f on $\mathbb{Z}^\nu$ will be said to be *measurable in* Λ ($\Lambda \subset \mathbb{Z}^\nu$) if $f(X) = f(X \cap \Lambda)$ for all $X \subset \mathbb{Z}^\nu$. A function measurable in some finite Λ is called a *cylinder function*. All cylinder functions are continuous, and any continuous function on $\mathcal{P}(\mathbb{Z}^\nu)$ is a uniform limit of cylinder functions.

When we speak of a measure on $\mathcal{P}(\mathbb{Z}^\nu)$, we will always mean a Borel measure (or the completion of such a measure). By the Riesz representation theorem, Borel measures are in one-one correspondence with positive linear functionals on the space $C(\mathcal{P}(\mathbb{Z}^\nu))$ of all continuous functions on $\mathcal{P}(\mathbb{Z}^\nu)$, the correspondence being defined by associating with the measure μ the functional $\int f d\mu$. If μ is a measure on $\mathcal{P}(\mathbb{Z}^\nu)$, if Λ is a finite subset of $\mathbb{Z}^\nu$, and if $X \subset \Lambda$, we define a measure μ_Λ on $\mathcal{P}(\Lambda)$ by

$$\mu_\Lambda(\{X\}) = \mu\{Y \in \mathbb{Z}^\nu : Y \cap \Lambda = X\} \quad \text{for all } X \subset \Lambda \ . \qquad (1)$$

These measures on different $P(\Lambda)$'s are related by a consistency condition: If $\Lambda \subset M$, then

$$\mu_\Lambda(\{X\}) = \Sigma\ \{\mu_M(\{Y\}):\ Y \subset M;\ Y \cap \Lambda = X\}\ . \tag{2}$$

Thus, a measure on $P(\mathbb{Z}^\nu)$ defines a consistent family of measures $\{\mu_\Lambda\}$. Conversely, it is easily verified that such a consistent family of measures defines a positive linear functional on the set of all cylinder functions on $P(\mathbb{Z}^\nu)$ which extends by continuity to a positive linear functional on $C(P(\mathbb{Z}^\nu))$ and hence uniquely defines a measure on $P(\mathbb{Z}^\nu)$. Thus, instead of mesures on $P(\mathbb{Z}^\nu)$, we may investigate consistent families $\{\mu_\Lambda\}$ of measures on the $P(\Lambda)$.

The additive group of $\mathbb{Z}^\nu$ (the *translation group*) acts on $P(\mathbb{Z}^\nu)$ by

$$\tau_a X = X - a$$

for $a \in \mathbb{Z}^\nu$; $X \in P(\mathbb{Z}^\nu)$. A measure on $P(\mathbb{Z}^\nu)$ is *translation-invariant* if it is invariant under this group of mappings. In this chapter, we will be concerned almost entirely with translation invariant probability measures on $P(\mathbb{Z}^\nu)$.

We will frequently have occasion to speak of measures on $P(\mathbb{Z}^\nu)$ obtained as limits of measures on $P(\Lambda_n)$'s, where Λ_n is a sequence of finite subsets of $\mathbb{Z}^\nu$ becoming large in one sense or another. By this we will mean the following: Let Λ_n be an increasing sequence of finite subsets of $\mathbb{Z}^\nu$ such that every finite $M \subset \mathbb{Z}^\nu$ is contained in Λ_n for sufficiently large n. Let $\mu^{(\Lambda_n)}$ be a measure on $P(\Lambda_n)$ for each n. We will say that the sequence $\mu^{(\Lambda_n)}$ converges to the measure μ on $P(\mathbb{Z}^\nu)$ if, for every cylinder function f on $P(\mathbb{Z}^\nu)$ (or, equivalently, for every continuous f)

$$\lim_{n\to\infty} \sum_{X\subset\Lambda_n} \mu^{(\Lambda_n)}(\{X\})f(X) = \int f d\mu\ . \tag{3}$$

An elementary compactness argument shows that, given any sequence Λ_n as above and any sequence $\mu^{(\Lambda_n)}$ of probability measures on $P(\Lambda_n)$, there is a subsequence n_j such that $\mu^{(\Lambda_{n_j})}$ converges to a probability measure μ. It is to be emphasized that we do *not* assume that the measures $\mu^{(\Lambda_n)}$ satisfy the consistency conditions of the family (μ_Λ) of measures on $P(\Lambda)$ corresponding to a given μ on $P(\mathbb{Z}^\nu)$, so the process we are describing is not just a question of the reconstruction of a measure on $P(\mathbb{Z}^\nu)$ from its projections on the $P(\Lambda)$.

We will also sometimes want to average the $\mu^{(\Lambda_n)}$ over translations before passing to the limit. To do this in a satisfactory way, we need to assume that the Λ_n become infinitely large in the sense of van Hove. Thus, if Λ_n is an increasing sequence of finite sets in $\mathbb{Z}^\nu$ such that each finite set is eventually contained in Λ_n, if (Λ_n) becomes infinitely large in the sense of van Hove, and if for each n $\mu^{(\Lambda_n)}$ is a probability measure on $P(\Lambda_n)$, we will say that $\mu^{(\Lambda_n)}$, averaged over translations, converges to a measure μ on $P(\mathbb{Z}^\nu)$ if, for every continuous function f on $P(\mathbb{Z}^\nu)$,

$$\lim_{n\to\infty} \frac{1}{N(\Lambda_n)} \sum_{a\in\Lambda_n} \sum_{X\subset\Lambda_n} \mu^{(\Lambda_n)}(\{X\}) f(X-a) = \int f d\mu \quad . \tag{4}$$

In this case, we will write $\overline{\lim_{n\to\infty} \mu^{(\Lambda_n)}} = \mu$. A measure μ obtained in this way is translation invariant. It suffices to assume (4) for cylinder functions f; if f is a cylinder function measurable in Λ, the left-hand side of (4) may be replaced by

$$\lim_{n\to\infty} \frac{1}{N(\Lambda'_n)} \sum_{a\in\Lambda'_n} \int d\mu^{(\Lambda_n)}(\tau_a f)$$

where $\Lambda'_n = \{a: a + \Lambda \subset \Lambda_n\}$; $(\tau_a f)(X) = f(X-a)$, and we are regarding $\tau_a f$ as defined on $P(\Lambda_n) \subset P(\mathbb{Z}^\nu)$. As an example of the use of this construction, we will want to use measures μ obtained as limits of translation averages of

$$\mu^{(\Lambda_n)}(\{X\}) = \frac{1}{\Xi_{\Lambda_n}} e^{-\beta(\Sigma\Phi)(X) + \log z\cdot N(X)}$$

to describe infinite volume equilibrium states corresponding to the interaction potential Φ, inverse temperature β, and activity z.

If Φ is a potential, and if μ is a translation invariant probability measure on $P(\mathbb{Z}^\nu)$, we would like to be able to speak of

$$\lim_{\Lambda\to\infty} \int d\mu \frac{(\Sigma\Phi)_\Lambda}{N(\Lambda)}, \text{ where } (\Sigma\Phi)_\Lambda(X) = (\Sigma\Phi)(X \cap \Lambda) \quad ,$$

as the mean value of $\Sigma\Phi$ per lattice site. We must, therefore, investigate this limit. To do this, it is convenient to map the space $\mathcal{B}$ of potentials into $C(P(\mathbb{Z}^\nu))$ by defining

$$A_\Phi(X) = \sum_{0\in Y\subset X} \frac{\Phi(Y)}{N(Y)} \quad . \tag{5}$$

Then

$$\Sigma\Phi_\Lambda(X) = \sum_{Y\subset X\cap\Lambda} \Phi(Y) = \sum_{a\in\Lambda} \sum_{a\in Y\subset X\cap\Lambda} \frac{\Phi(Y)}{N(Y)}$$

$$= \sum_{a\in\Lambda} (\tau_a A_\Phi)(X) - \sum_{a\in\Lambda} \sum_{\substack{a\in Y\subset X \\ Y\cap X\setminus\Lambda\neq\phi}} \frac{\Phi(Y)}{N(Y)} \quad .$$

Thus, we have:

$$\frac{(\Sigma\Phi)_\Lambda(X)}{N(\Lambda)} - \frac{1}{N(\Lambda)} \sum_{a\in\Lambda} \tau_a A_\Phi(X) = \frac{\text{boundary terms}}{N(\Lambda)} \quad ,$$

and it is easy to show that, if Λ_n is a sequence of finite subsets of $\mathbb{Z}^\nu$ becoming infinitely large in the sense of van Hove,

$$\lim_{n\to\infty} \sup_X \left| \frac{\Sigma\Phi_{\Lambda n}(X)}{N(\Lambda_n)} - \frac{1}{N(\Lambda_n)} \sum_{a\in\Lambda_n} \tau_a A_\Phi(X) \right| = 0 \quad . \tag{6}$$

Hence,

$$\lim_{n\to\infty} \int d\mu \frac{\Sigma\Phi_{\Lambda_n}}{N(\Lambda_n)} = \int d\mu A_\Phi(X) \tag{7}$$

so A_Φ is to be interpreted as the contribution of the origin to $\Sigma\Phi_\Lambda$ for large Λ. Note also that, by (6) and the pointwise ergodic theorem, if Λ_n is a sequence of rectangles with shortest side going to infinity

$$\lim_{n\to\infty} \frac{1}{N(\Lambda_n)} (\Sigma\Phi)\Lambda_n(X)$$

exists almost everywhere with respect to any invariant μ.

We note for future use two facts about A_Φ. First, it follows trivially from the definition that

$$\|A_\Phi\| \leq \|\Phi\|$$

(where the norm on the left is the supremum norm on $C(P(\mathbb{Z}^\nu))$. Second, if μ_1 and μ_2 are invariant probability measures on $P(\mathbb{Z}^\nu)$ such that

$$\int A_\Phi d\mu_1 = \int A_\Phi d\mu_2$$

for all $\Phi \in \mathcal{B}$, then $\mu_1 = \mu_2$. We leave the verification of this fact as an exercise.

We still need one more piece of machinery before we can seriously begin investigating equilibrium states. This is the definition of the entropy per lattice site of a general translation-invariant probability measure μ on $P(\mathbb{Z}^\nu)$. This definition goes as follows: For any finite subset Λ of $\mathbb{Z}^\nu$, and any probability measure μ on $P(\mathbb{Z}^\nu)$, we define

$$S(\mu,\Lambda) = \sum_{X\subset\Lambda} - \mu_\Lambda(\{X\}) \log \mu_\Lambda(\{X\}) \quad .$$

Then the mean entropy of μ per lattice site should be

$$\lim_{\Lambda\to\infty} \frac{1}{N(\Lambda)} S(\mu,\Lambda) \quad .$$

The existence of the limit requires proving. In order to do this, it is convenient to introduce the conditional entropy

$$S(\mu,\Lambda|M) = S(\mu|\Lambda \cup M) - S(\mu,M) \quad .$$

We now have the following.

<u>Lemma B2.1</u>

$S(\mu,\Lambda|M)$ *is decreasing in* M. (This result is generally expressed by saying that $\Lambda \to S(\mu,\Lambda)$ is *strongly subadditive.)*

The proof is by now fairly well-known, so we will only state the relevant result; see Robinson and Ruelle [8] for the details.

Theorem B2.2

Let μ *be invariant under translations by a subgroup of* $\mathbb{Z}^\nu$ *of finite index (i.e.,* μ *is periodic), and let* Λ_n *be a sequence of finite subsets of* $\mathbb{Z}^\nu$ *becoming infinitely large in the sense of van Hove. Then*

$$\lim_{n\to\infty} \frac{1}{N(\Lambda_n)} S(\mu,\Lambda_n)$$

exists. If μ *is invariant under the group of all translations and if* Λ_n *is a sequence of rectangles, then*

$$\lim_{n\to\infty} \frac{1}{N(\Lambda_n)} S(\mu,\Lambda_n) = \inf \frac{1}{N(\Lambda_n)} S(\mu,\Lambda_n) \quad .$$

We call the limit the *mean entropy of* μ *per lattice site,* and denote it by $s(\mu)$.

Corollary B2.3

(a) $s(\mu)$ *is an affine function of* μ, i.e., *if* μ_1 *and* μ_2 *are periodic with the same period, then*

$$s(\alpha\mu_1 + (1-\alpha)\mu_2) = \alpha s(\mu_1) + (1-\alpha)s(\mu_2) \quad .$$

(b) *The mapping* $\mu \to s(\mu)$ *is upper semicontinuous on the set of all translation-invariant* μ *(in the weak-* topology for measures.)*

Proof. (a) by the concavity of $-x \log x$,

$$\begin{aligned} s(\alpha\mu_1 + (1-\alpha)\mu_2,\Lambda) &= \sum_{X\subset\Lambda} -[\alpha\mu_{1,\Lambda}(\{X\}) \\ &\quad + (1-\alpha)\mu_{2,\Lambda}(\{X\})]\log[\alpha\mu_{1,\Lambda}(\{X\}) \\ &\quad + (1-\alpha)\mu_{2,\Lambda}(\{X\})] \\ &\geq \alpha S(\mu_1,\Lambda) + (1-\alpha)S(\mu_2,\Lambda) \quad . \end{aligned}$$

On the other hand,

$$\begin{aligned} &\sum_{X\subset\Lambda} -\alpha\mu_{1,\Lambda}(\{X\})\log[\alpha\mu_{1,\Lambda}(\{X\}) + (1-\alpha)\mu_{2,\Lambda}(\{X\})] \\ &\quad \leq \sum_{X\subset\Lambda} -\alpha\mu_{1,\Lambda}(\{X\})\log[\alpha\mu_{1,\Lambda}(\{X\})] \quad . \\ &\quad = \alpha[S(\mu_1,\Lambda) - \log\alpha] \quad . \end{aligned}$$

By making a similar argument we get:

$$\begin{aligned} &\sum_{X\subset\Lambda} -(1-\alpha)\mu_{2,\Lambda}(\{X\})\log[\alpha\mu_{1,\Lambda}(\{X\}) + (1-\alpha)\mu_{2,\Lambda}(\{X\})] \\ &\quad \leq (1-\alpha)S(\mu_2,\Lambda) - (1-\alpha)\log(1-\alpha) \quad . \end{aligned}$$

Combining these two equalities with the simple concavity inequality we get

$$\alpha S(\mu_1,\Lambda) + (1-\alpha)S(\mu_2,\Lambda) \le$$
$$S(\alpha\mu_1 + (1-\alpha)\mu_2,\Lambda) \le \alpha S(\mu_1,\Lambda) + (1-\alpha)S(\mu_2,\Lambda)$$
$$- [\alpha \log \alpha + (1-\alpha)\log(1-\alpha)] \quad .$$

Dividing by $N(\Lambda)$ and letting Λ become large, we get

$$s(\alpha\mu_1 + (1-\alpha)\mu_2) = \alpha s(\mu_1) + (1-\alpha)s(\mu_2) \quad .$$

(b) For any fixed Λ, and any $X \subset \Lambda$, $\mu_\Lambda(\{X\})$ is continuous in μ. Hence, $S(\mu,\Lambda)$ is continuous in μ. But, if μ is invariant, $s(\mu)$ is the infimum over all rectangles Λ of $\frac{S(\mu,\Lambda)}{N(\Lambda)}$; hence, is upper semi-continuous.

B 3. MAXIMIZATION OF THE ENTROPY

We have made two distinct uses of the term "entropy"; one as $\lim \frac{1}{N} \log V$, the other as a parameter associated with an invariant probability measure on $P(\mathbb{Z}^\nu)$. These two notions are related in a reasonably simple way; we have:

Theorem B3.1

Let Φ *be a vector-valued potential whose components are linearly independent in* $\mathcal{B}$, *and let* $x \in \mathbb{R}^t$. *Then* $s(\Phi,x) = \sup\{s(\mu): \mu$ *an invariant probability measure on* $P(\mathbb{Z}^\nu)$ *such that* $\int A_\Phi d\mu = x\}$. *Moreover, if* $s(\Phi,x) > -\infty$, *there is at least one probability measure* μ *such that* $\int A_\Phi d\mu = x$ *with* $s(\mu) = s(\Phi,x)$. *Conversely, for any invariant probability measure* μ *on* $P(\mathbb{Z}^\nu)$,

$$s(\mu) = \inf\{s(\Phi, \textstyle\int A_\Phi d\mu): \Phi \text{ a vector-valued potential}\} \quad .$$

Proof. To prove the first assertion, we need two distinct arguments: we first show that, for any probability measure μ, $s(\mu) \le s(\Phi, \int A_\Phi d\mu)$. We then have to give a procedure for constructing a measure μ such that $\int A_\Phi d\mu = x$ and such that $s(\mu) = s(\Phi,x)$.

To prove the first assertion, we consider first *ergodic* measures μ, i.e., those such that any invariant function is constant almost everywhere. For such a measure μ, and any sequence Λ_n of rectangles with shortest side going to infinity,

$$\frac{\Sigma\Phi_{\Lambda_n}}{N(\Lambda_n)}$$

converges almost everywhere with respect to μ to $\int A_\Phi d\mu$, by the pointwise ergodic theorem and the remark already made that

$$\left\| \frac{\Sigma\Phi_{\Lambda_n}}{N(\Lambda_n)} - \frac{1}{N(\Lambda_n)} \sum_{a\in\Lambda} T_a \cdot A_\Phi \right\|_\infty \to 0 \ .$$

Choose an open convex J containing $\int A_\Phi d\mu$, and let

$$A_n = \{X \subset \Lambda_n : \frac{\Sigma\Phi(X)}{N(\Lambda_n)} \in J\} \ .$$

By what we have already said,

$$\mu_{\Lambda_n}(A_n) \to 1 \text{ and } n \to \infty; \text{ on the other hand,}$$

$$\lim_{n\to\infty} \frac{1}{N(\Lambda_n)} \log N(A_n) = \sup_{y\in J} s(\Phi,y) \ .$$

We can write

$$S(\mu,\Lambda_n) = -\sum_{X\in\Lambda_n} \mu_{\Lambda_n}(\{X\}) \log \mu_{\Lambda_n}(\{X\})$$

$$= \sum_{X\in A_n} \mu_{\Lambda_n}(\{X\}) \log\left(\frac{1}{\mu_{\Lambda_n}(\{X\})}\right) + \sum_{X\notin A_n} \mu_{\Lambda_n}(\{X\}) \log\left(\frac{1}{\mu_{\Lambda_n}(\{X\})}\right) \ .$$

We want to estimate each of these sums separately, using the concavity of the logarithm. If, for example, we put

$$\alpha(X) = \frac{\mu_{\Lambda_n}(\{X\})}{\mu_{\Lambda_n}(A_n)} \ ,$$

we have

$$\sum_{X\in A_n} \mu_{\Lambda_n}(\{X\}) \log \frac{1}{\mu_{\Lambda_n}(\{X\})} = \mu_{\Lambda_n}(A_n) \cdot \sum_{X\in A_n} \alpha(X) \log \frac{1}{\mu_{\Lambda_n}(\{X\})}$$

$$\leq \mu_{\Lambda_n}(A_n) \log \sum_{X\in A_n} \frac{\alpha(X)}{\mu_{\Lambda_n}(\{X\})} = \mu_{\Lambda_n}(A_n) \log\left[\frac{N(A_n)}{\mu_{\Lambda_n}(A_n)}\right] \ .$$

Similarly,

$$\sum_{X\notin A_n} \mu_{\Lambda_n}(\{X\}) \log \mu_{\Lambda_n}(\{X\}) \leq [1 - \mu_{\Lambda_n}(A_n)] \log\left[\frac{2^{N(\Lambda_n)} - N(A_n)}{1 - \mu_{\Lambda_n}(A_n)}\right]$$

so

$$S(\mu,\Lambda_n) \leq -\mu_{\Lambda_n}(A_n) \log \mu_{\Lambda_n}(A_n) - (1 - \mu_{\Lambda_n}(A_n) \log[1 - \mu_{\Lambda_n}(A_n)]$$

$$+ \mu_{\Lambda_n}(A_n) \log N(A_n) + [1 - \mu_{\Lambda_n}(A_n)] \log[2^{N(\Lambda_n)}] \ .$$

Now divide by $N(\Lambda_n)$ and let $n \to \infty$. Using $\mu_{\Lambda_n}(A_n) \to 1$; $\frac{1}{N(\Lambda_n)} \log N(A_n) \to \sup_{y\in J} s(\Phi,y)$, we get

$$s(\mu) = \lim_{n\to\infty} \frac{1}{N(\Lambda_n)} S(\mu,\Lambda_n) \leq \sup_{y\in J} s(\Phi,y) \ .$$

This is true for all convex open J containing $\int A_\Phi d\mu$; hence, by the upper semi-continuity of $s(\Phi,x)$ in x,

$$s(\mu) \le s(\Phi, \int A_\Phi d\mu) \quad .$$

To extend this inequality from ergodic measures μ to general invariant probability measures, we use the theory of extremal points of compact convex sets. The ergodic measures are exactly the extremal points of the set of all invariant probability measures on $\mathcal{P}(\mathbb{Z}^\nu)$. We have two functions of μ,

$$\mu \to s(\mu) \quad \text{and} \quad \mu \to s(\Phi, \int A_\Phi d\mu)$$

the first affine and the second concave, such that

$$s(\Phi, \int A_\Phi d\mu) \ge s(\mu)$$

for every extremal μ. If the two functions were continuous, the Krein-Milman Theorem would imply immediately that the inequality holds for all invariant μ. Unfortunately, the functions are only upper semi-continuous, so we need a slightly more refined argument. The key remark is that, given an invariant probability measure μ_0 and $\varepsilon > 0$, there exists a continuous affine function $h(\mu)$ such that

$$h(\mu) \ge s(\Phi, \int A_\Phi d\mu) \quad \text{for all} \quad \mu$$

$$h(\mu_0) \le s(\Phi, \int A_\Phi d\mu_0) + \varepsilon$$

In other words, a concave upper semi-continuous function can be approximated from above by continuous affine functions. This fact is proved by using the Hahn-Banach Theorem to construct a hyperplane separating the point

$$(\mu_0, s(\Phi, \int A_\Phi d\mu_0) + \varepsilon)$$

from the region below the graph of $\mu \to s(\Phi, \int A_\Phi d\mu)$; h is then the function with this hyperplane as its graph. See, for example, Lemma II.C.2 of Lanford [7].

Now consider $s(\mu) - h(\mu)$; this is affine and upper semi-continuous and therefore must take on its supremum at some extremal point of the set of invariant probability measures, i.e. at some ergodic μ. But

$$h(\mu) \ge s(\Phi, \int A_\Phi d\mu) \ge s(\mu)$$

for all ergodic μ, so the supremum is negative, i.e.

$$s(\mu) \le h(\mu)$$

for all μ. Hence

$$s(\mu_0) \le h(\mu_0) \le s(\Phi, \int A_\Phi d\mu_0) + \varepsilon \quad .$$

This is true for all μ_0 and all $\varepsilon > 0$, so

$$s(\mu) \le s(\Phi, \int A_\Phi d\mu)$$

for all μ, as desired. (I am indebted to A. Connes for suggesting the above argument to me.)

The next step in the proof is the construction of a measure μ with $\int A_\Phi d\mu = x$ and $s(\mu) = s(\Phi,x)$, assuming $s(\Phi,x) > -\infty$. We take a sequence (Λ_n) of cubes becoming infinitely large as $n \to \infty$ and we choose the nth cube large enough so that

$$\frac{1}{N(\Lambda_n)} \log[N\{X \subset \Lambda_n : |\frac{\Sigma\Phi(X)}{N(\Lambda_n)} - x| < \frac{1}{n}\}] \ge s(\Phi,x) - \frac{1}{n} .$$

We construct a probability measure μ_{n,Λ_n} on $P(\Lambda_n)$ by assigning probability zero to any X with $|\frac{\Sigma\Phi(X)}{N(\Lambda_n)} - x| \ge \frac{1}{n}$ and equal probability to each of the other X's. Then

$$-\sum_{X\subset\Lambda} \mu_{n,\Lambda_n}(X)\log \mu_{n,\Lambda_n}(X) = \log[N\{X \subset \Lambda_n : |\frac{\Sigma\Phi(x)}{N(\Lambda_n)} - x| < \frac{1}{n}]$$

$$\ge N(\Lambda_n)[s(\Phi,x) - \frac{1}{n}]; \text{ also, clearly,}$$

$$|\int \mu_{n,\Lambda_n}(dX) \frac{\Sigma\Phi(X)}{N(\Lambda_n)} - x| \le \frac{1}{n} .$$

Now, to construct a measure on $P(\mathbb{Z}^\nu)$, we choose a subgroup $G^{(n)}$ of the additive group $\mathbb{Z}^\nu$ such that

$$\{\Lambda_n + a : a \in G^{(n)}\}$$

is a decomposition of $\mathbb{Z}^\nu$ into disjoint cubes. We may identify $P(\mathbb{Z}^\nu)$ with $\prod_{a\in G^{(n)}} P(\Lambda_n + a)$. Transporting μ_{n,Λ_n} to each $P(\Lambda_n + a)$ by translation, then taking the product measure, gives a probability measure on $P(\mathbb{Z}^\nu)$ which we denote by μ_n. μ_n is not necessarily invariant under translations, but it is certainly invariant under $G^{(n)}$, i.e., is periodic. Its mean entropy therefore exists and is easily calculated to be equal to

$$-\frac{1}{N(\Lambda_n)} \sum_{X\subset\Lambda_n} \mu_{n,\Lambda_n}(\{X\})\log \mu_{n,\Lambda_n}(\{X\}) \ge s(\Phi,x) - \frac{1}{n} .$$

We may now average μ_n over the finite group $\mathbb{Z}^\nu/G^{(n)}$ to obtain an invariant measure $\bar{\mu}_n$; since the entropy is an affine function of μ, we have

$$s(\bar{\mu}_n) = s(\mu_n) \ge s(\Phi,x) - \frac{1}{n} .$$

It is easy to see, but tedious to prove in detail, that

$$|\int A_\Phi d\bar{\mu}_n - \int \mu_{n,\Lambda_n}(dX) \frac{\Sigma\Phi(X)}{N(\Lambda_n)}| \to 0 \quad \text{as} \quad n \to \infty .$$

Thus, we have

(a) $s(\bar{\mu}_n) \geq s(\Phi,x) - \frac{1}{n}$ for all n.

(b) $\lim_{n\to\infty} \int A_\Phi d\bar{\mu}_n = x$.

Now let μ be the limit of a weak-* convergent subsequence of $(\bar{\mu}_n)$. Then $\int A_\Phi d\mu = \lim_{n\to\infty} \int A_\Phi d\bar{\mu}_n = x$, and, by the upper semi-continuity of the entropy

$$s(\mu) \geq \limsup_n s(\bar{\mu}_n) \geq s(\Phi,x) \quad .$$

We still have to prove the converse part of Theorem B3.1., i.e., we have to show that, for any invariant probability measure μ on $P(\mathbb{Z}^\nu)$,

$$s(\mu) = \inf\{s(\Phi, \int A_\Phi d\mu\}$$

where the infimum is taken over all vector-valued potentials. From the first part of the theorem, we have

$$s(\mu) \leq s(\Phi, \int A_\Phi d\mu)$$

for any vector-valued Φ. Hence, we assume

$$s(\mu) + \varepsilon < s(\Phi, \int A_\Phi d\mu)$$

for some $\varepsilon > 0$, and all vector-valued potentials Φ, and we attempt to derive a contradiction. Let $(\Phi^{(1)},\Phi^{(2)},\ldots)$ be a linearly independent sequence of elements of $\mathcal{B}$ whose linear span is dense in $\mathcal{B}$, and let Φ_n be the n-component vector valued potential $(\Phi^{(1)},\ldots,\Phi^{(n)})$. For each n, choose an invariant probability measure μ_n such that

$$s(\mu_n) > s(\mu) + \varepsilon \quad \text{and} \quad \int A_{\Phi_n} d\mu_n = \int A_\Phi d\mu \quad .$$

(This is possible since we know we can take $s(\mu_n)$ as large as $s(\Phi, \int A_{\Phi_n} d\mu)$, which by assumption is larger than $s(\mu) + \varepsilon$). Now

$$\lim_{n\to\infty} \int A_\Phi d\mu_n = \int A_\Phi d\mu$$

for all Φ in the finite linear span of $\{\Phi^{(1)},\Phi^{(2)},\ldots\}$, and hence for all Φ in $\mathcal{B}$. This implies that μ_n converges to μ in the weak-* topology for measures on $P(\mathbb{Z}^\nu)$. The upper semi-continuity of s implies $s(\mu) \geq \limsup_n s(\mu_n)$, but the μ_n's were all chosen so that $s(\mu_n) > s(\mu) + \varepsilon$, so we have the desired contradiction, and the proof of the theorem is complete.

B4. INVARIANT EQUILIBRIUM STATES

Suppose now we want to consider a system interacting with a potential $\hat{\Phi}$, and we let Φ_N denote the potential such that $\Sigma\Phi_N(X) = N(X)$. We will say that an invariant probability measure μ on $P(\mathbb{Z}^\nu)$ is an *invariant equilibrium state* with mean density ρ and mean energy per lattice site ε if $s(\mu) > -\infty$, $\int A_{\Phi_N} d\mu = \rho$; $\int A_{\hat{\Phi}} d\mu = \varepsilon$, and

$$s(\mu) = s((\Phi_N, \hat{\Phi}), (\rho, \varepsilon)) \quad .$$

In other words, an invariant equilibrium state is one which has the largest possible entropy consistent with its mean density and mean energy. We collect some facts about equilibrium states in the following theorem.

Theorem B4.1

For a given interaction $\hat{\Phi}$, *there exists an invariant equilibrium state with mean density* ρ *and mean energy per lattice site* ε *if and only if* $s((\Phi_N, \hat{\Phi}), (\rho, \varepsilon)) > -\infty$. *For fixed* (ρ, ε), *the set of invariant equilibrium states is a weak-* closed convex set. If* Φ' *is a potential which is not linearly dependent on* $\Phi_N, \hat{\Phi}$, *then there is an invariant equilibrium state* μ *with mean density* ρ, *and mean energy per lattice site* ε, *and with*

$$\int A_{\Phi}' d\mu = x$$

if and only if

$$s((\Phi_N, \hat{\Phi}, \Phi'), (\rho, \varepsilon, x)) = \sup_y \{ s((\Phi_N, \hat{\Phi}, \Phi'), (\rho, \varepsilon, y)) \quad .$$

There is only one invariant equilibrium state corresponding to (ρ, ε) *if and only if, for every* Φ', $s((\Phi_N, \hat{\Phi}, \Phi'), (\rho, \varepsilon, y))$ *takes on its supremum as a function of* y *at a single point.*

The proof of the theorem is a straightforward matter of applying Theorem B3.1 and the definition of equilibrium state. We omit the details.

We now digress briefly to mention another, more axiomatic, way of introducing the notion of "state" or "equilibrium state". Instead of talking about invariant probability measures on $P(\mathbb{Z}^\nu)$ we could define a state to be a mapping $\Phi \to \mu(\Phi)$ from real-valued potentials to real numbers, with the interpretation that $\mu(\Phi)$ is the mean value of $\frac{\Sigma\Phi}{N}$. This mapping should be subjected to some restrictions, e.g., $\mu(\Phi)$ should be linear in Φ and, if $\Sigma\Phi \geq 0$, $\mu(\Phi)$ should be non-negative. It is at this point probably not hard to impose a few more technical restrictions which guarantee that μ is obtained as above from a translation-invariant probability measure on $P(\mathbb{Z}^\nu)$. Suppose we refrain from doing so, but we still want to be able to speak of the entropy of μ. This we can do by introducing $s(\Phi, x)$ for vector-valued potentials as in Theorem B1.1, i.e., by taking thermodynamic limits, and defining

$$s(\mu) = \inf\{s(\Phi, \mu(\Phi)) : \Phi \text{ vector valued potentials}\} \quad .$$

We can thus define an equilibrium state as one such that

$$s(\mu) = s((\Phi_N,\hat{\Phi}),(\mu(\Phi_N),\mu(\hat{\Phi}))) \quad .$$

There remains the problem of proving that $s(\mu)$ is an affine function of μ (corresponding to the fact that the thermodynamic entropy is an extensive quantity); this seems to be relatively difficult to do directly, i.e., without introducing the measure on $P(\mathbb{Z}^\nu)$ corresponding to μ. This point of view about equilibrium states was developed by Ruelle in [10], and it precedes the serious investigation of the notion of a state as a probability measure on the configuration space for the infinite system.

Returning to the investigation of equilibrium states, we want next to change from parametrization of equilibrium states by energy and density to parametrization by temperature and activity. Recall how the change of variables from (ρ,ε) to $(\log z,\beta)$ went: We showed that, if $p(z,\beta)$ denotes the pressure as a function of z and β, then

$$\beta p(z,\beta) = \sup_{\rho,\varepsilon}[s(\rho,\varepsilon) - \beta\varepsilon + \rho \log z] \quad .$$

The values of (ρ,ε) for which the supremum is taken on are the values corresponding to the parameters (z,β); in particular, if the supremum is taken on at a single point only, then the mean density and mean energy per lattice site in the grand canonical ensemble converge in probability to the corresponding constants as the system becomes large. With this as motivation, we will say that an invariant probability measure μ on $P(\mathbb{Z}^\nu)$ is an *invariant equilibrium state with inverse temperature* β *and activity* z if, for some ρ_0,ε_0 such that

$$s((\Phi_N,\hat{\Phi}),(\rho_0,\varepsilon_0)) - \beta\varepsilon_0 + \rho_0 \log z = \sup_{(\rho,\varepsilon)}\{s((\Phi_N,\hat{\Phi}),(\rho,\varepsilon)) - \beta\varepsilon + \rho \log z\} \quad ,$$

μ is an equilibrium state with mean density ρ_0 and mean energy per lattice site ε_0. The above somewhat intricate definition of equilibrium state seems to be the logically natural one; fortunately, it is possible to find a more manageable description of these states:

Theorem B4.2

Let $\Phi = \beta\hat{\Phi} - \log z \cdot \Phi_N$, *and let* $p(\Phi)$ *be defined as in Theorem* B1.2. *Then*

$$p(\Phi) = \sup_{(\rho,\varepsilon)}\{s((\Phi_N,\hat{\Phi}),(\rho,\varepsilon)) - \beta\varepsilon + \rho \log z\} \tag{1}$$

$$= \sup\{s(\mu) - \mu(A_\Phi) : \mu \text{ a translation invariant probability measure on } P(\mathbb{Z}^\nu)\} \quad .$$

A translation invariant probability measure on $P(\mathbb{Z}^\nu)$ *is an invariant equilibrium state with inverse temperature* β *and activity* z *if and only if*

$$p(\Phi) = s(\mu) - \int A_{\Phi} d\mu \quad . \tag{2}$$

Proof. By Theorem B1.2,

$$p(\Phi) = \sup_{x}\{s(\Phi,x) - x\} \quad . \tag{3}$$

On the other hand, standard arguments give

$$s(\Phi,x) = \sup\{s((\Phi_N,\hat{\Phi}),(\rho,\varepsilon)) : \beta\varepsilon - \rho \log z = x\} \tag{4}$$

Inserting (4) in (3) gives the first equality in (1). By Theorem B3.1, for any μ,

$$s(\mu) \le s(\Phi, \int A_{\Phi} d\mu) \; , \quad \text{so}$$

$$s(\mu) - \mu(A_{\Phi}) \le \sup_{x}[s(\Phi,x) - x] = p(\Phi) \quad .$$

Hence

$$\sup_{\mu}\{s(\mu) - \int A_{\Phi} d\mu\} \le p(\Phi) \quad .$$

On the other hand, if x_0 is chosen so that

$$s(\Phi,x_0) - x_0 = p(\Phi) \quad ,$$

and if μ_0 is chosen so that $\int A_{\Phi} d\mu_0 = x_0$; $s(\mu_0) = s(\Phi,x_0)$ (which is possible by Theorem B3.1), we have

$$p(\Phi) = s(\mu_0) - \int A_{\Phi} d\mu_0 \quad ,$$

so the second equality in (1) is proved.

To prove the second assertion of the theorem, assume first that μ is an invariant equilibrium state with inverse temperature β and activity z. By the definition of invariant equilibrium state, if we let

$$\varepsilon_0 = \int A_{\hat{\Phi}} d\mu, \quad \rho_0 = \int A_{\Phi_N} d\mu \quad ,$$

then

$$p(\Phi) = s((\Phi_N,\hat{\Phi}),(\rho_0,\varepsilon_0)) - \beta\varepsilon_0 + \log z \; \rho_0 \quad ,$$

and

$$s((\Phi_N,\hat{\Phi}),(\rho_0,\varepsilon_0)) = s(\mu) \quad .$$

Since

$$\beta\varepsilon_0 - \log z \; \rho_0 = \int A_{\Phi} d\mu, \quad \text{we get}$$

$$p(\Phi) = s(\mu) - \int A_{\Phi} d\mu \quad ,$$

as desired.

Conversely, suppose

$$p(\Phi) = s(\mu) - \int A_{\Phi} d\mu \quad ,$$

and let $\varepsilon_0 = \int A_{\hat{\Phi}} d\mu$; $\rho_0 = \int A_{\Phi_N} d\mu$. Then

$$p(\Phi) = s(\mu) - \beta\varepsilon_0 + \rho_0 \log z, \text{ and}$$

$$s(\mu) \leq s((\Phi_N,\hat{\Phi}),(\varepsilon_0,\rho_0)) \text{ by Theorem B3.1.}$$

On the other hand, by (1),

$$p(\Phi) = \sup_{(\varepsilon,\rho)}\{s((\Phi_N,\hat{\Phi}),(\varepsilon,\rho)) - \beta\varepsilon + \rho \log z\} \quad ,$$

so

$$s((\Phi_N,\hat{\Phi}),(\varepsilon_0,\rho_0)) - \beta\varepsilon_0 + \rho_0 \log z = p(\Phi)$$

and

$s((\Phi_N,\hat{\Phi}),(\varepsilon_0,\rho_0)) = s(\mu)$, so μ is an invariant equilibrium state with inverse temperature β and activity z.

In the light of the preceding theorem, we may suppress $\Phi_N,\hat{\Phi},\beta,z$ from our notation and speak simply of equilibrium states for the potential Φ, i.e., states which satisfy (2).

Proposition B4.3. *The set of invariant equilibrium states for a given potential* Φ *is a non-empty set of probability measures which is convex and closed, hence, compact, in the weak-* topology. The extremal points of this set are exactly the invariant equilibrium states which are extremal points of the set of all invariant probability measures, i.e., are the ergodic equilibrium states.*

The proof of all the assertions follows at once from the remark that the set of invariant equilibrium states for Φ is the set of invariant probability measures where the affine upper semi-continuous function $\mu \to s(\mu) - \int A_{\Phi} d\mu$ takes on its supremum.

B5. TANGENTS TO THE GRAPH OF THE PRESSURE

In this section, we will exploit systematically the fact that the pressure $p(\Phi)$ is a continuous convex function on the Banach space $\mathcal{B}$ of interactions. To do this, we have to develop some techniques concerning differentiation of convex functions. Consider first a convex function h of a real variable x. The line $\{(x,h(x_0) + \alpha(x - x_0)): -\infty < x < \infty\}$ is tangent to the graph of h at x_0 if and only if it lies everywhere below the graph of $h(x)$. The set of α's for which this is the case is a non-empty closed interval; it reduces to a single point if and only if h is differentiable at x_0. We can generalize these remarks to convex functions on a Banach space as follows: Let $h(x)$ be a continuous convex function defined on a non-empty open convex subset of a Banach space X. We will say that a continuous linear functional φ on X is a *tangent functional* to h at x_0 if

$$h(x_0 + y) \geq h(x_0) + \varphi(y) \tag{1}$$

for all y such that $h(x_0 + y)$ is defined (or, equivalently, for all y sufficiently near zero). The set of all tangent functionals to h at x_0 is evidently a weak-* closed convex set of the dual space of X; a simple application of the Hahn-Banach theorem shows that it is non-empty, and the continuity of h implies that it is bounded. Again using the Hahn-Banach theorem, one shows that $h(x)$ has a directional derivative in each direction at x_0 if and only if there is only one tangent functional to h at x_0; in this case, we say that h is differentiable at x_0. From the definition of tangent functional, it follows at once that, if $x_n \to x_0$, if φ_n is a tangent functional to h at x_n, and if φ_n converges to φ in the weak-* topology, then φ is a tangent functional to h at x_0 and, similarly, if h_n is a sequence of convex functions converging pointwise to h, if φ_n is a tangent functional to h_n at x_0, and if φ_n converges in the weak-* topology to φ, then φ is a tangent functional to h at x_0.

If h is a convex function of one variable, then h is differentiable at all but countably many points. One does not have such a sharp statement for functions of severable variables, but it can be proved that, if h is a continuous convex function on a separable Banach space, the set of points where h fails to be differentiable is a set of the first category, i.e., is contained in a countable union of closed nowhere dense sets. (A proof of this statement does not seem to exist in the published literature. Dunford and Schwartz [3], V.9.8, prove the weaker assertion that h is differentiable on a dense set; their method of proof gives in fact our assertion).

The usefulness of the above machinery for the study of equilibrium states derives from a simple remark: If μ is an equilibrium state for the interaction Φ, then $\Psi \to -\int A_\Psi d\mu$ is a tangent functional to p at Φ. Indeed,

$$p(\Phi + \Psi) = \sup_\nu (s(\nu) - \int A_{\Phi+\Psi} d\nu)$$

$$\geq s(\mu) - \int A_\Phi d\mu - \int A_\Psi d\mu = p(\Phi) - \int A_\Psi d\mu \quad .$$

Since two different equilibrium states μ_1 and μ_2 define different tangent functionals, we can conclude immediately that, on the large set of Φ's at which $p(\Phi)$ is differentiable, there is only one invariant equilibrium state. To get a satisfactory correspondence between tangent functionals and equilibrium states, we still need to prove that every tangent functional to p at Φ is of the form $\Psi \to -\int A_\Psi d\mu$ for some equilibrium state μ. To do this, we need a general result about convex functions. We first establish the terminology.

Let h be a continuous convex function defined on an open subset U of a Banach space X and let $x_0 \in U$. A tangent functional φ to h at x_0 is said to be *approximable by unique tangent functionals* if there exists a sequence x_n in X converging to x_0 such that h has a unique tangent functional φ_n at each x_n and such that φ_n converges to φ in the weak-* topology.

Proposition B5.1

Let h *be a continuous convex function on an open set* U *in a separable Banach space* X*, and let* $x_0 \in U$*. The weak-* closed convex hull of the set of tangent functionals at* x_0 *approximable by unique tangent functionals is the set of all tangent functionals at* x_0*.*

(In other words, every tangent functional to h at x_0 is a weak-* limit of convex combinations of tangent functionals which are approximable by unique tangent functionals. We will not prove this proposition; it is Theorem 1 of Lanford and Robinson [7].)

We can now prove:

Theorem B5.2

To each tangent functional to the graph of p *at* Φ*, there corresponds a uniquely determined invariant equilibrium state* μ *for* Φ *such that the tangent functional is given by* $\Psi \to \int A_\Psi d\mu$*. Conversely, each invariant equilibrium state defines a tangent functional by this formula. There is a residual* subset* $\mathcal{D}$ *of* $\mathcal{B}$ *such that, if* $\Phi \in \mathcal{D}$*, there is only one invariant equilibrium state of* Φ*.*

Proof. Everything except the first assertion is already proved. If p has a unique tangent functional at Φ, then, since there is at least one invariant equilibrium state for Φ, nothing remains to be proved. Hence, suppose there are many tangent functionals at Φ. We will first show that every tangent functional at Φ approximable by unique tangent functionals corresponds to an equilibrium state. Thus, let Φ_n be a sequence of potentials converging to Φ such that there is a unique tangent functional to p at Φ_n, and let μ_n be the corresponding sequence of equilibrium states. Assume that $-\int A_\Psi d\mu_n$ converges for all $\Psi \in \mathcal{B}$. This implies that the sequence μ_n converges in the weak-* topology to an invariant probability measure μ on $P(\mathbb{Z}^\nu)$. We want to show that μ is an invariant equilibrium state for Φ, i.e., that

---

* A subset of a topological space is said to be *residual* if its complement is contained in a countable union of closed nowhere dense subsets.

$$p(\Phi) = s(\mu) - \int A_\Phi d\mu \quad .$$

It suffices, in fact, to prove

$$p(\Phi) \leq s(\mu) - \int A_\Phi d\mu$$

since the opposite inequality always holds. But, for each n, $p(\Phi_n) = s(\mu_n) - \int A_\Phi d\mu_n$. By the definition of weak-* convergence,

$$\lim_{n\to\infty} \int A_\Phi d\mu_n = \int A_\Phi d\mu \quad ,$$

and by the continuity of p,

$$\lim_{n\to\infty} p(\Phi_n) = p(\Phi) \quad .$$

Finally, by the upper semi-continuity of s,

$$s(\mu) \geq \lim_{n\to\infty} s(\mu_n), \quad \text{so}$$

$$p(\Phi) \leq s(\mu) - \int A_\Phi d\mu \quad ,$$

so μ is an invariant equilibrium state.

The above argument shows that every tangent functional to p at Φ which is approximable by unique tangent functionals corresponds to an invariant equilibrium state. The same is true, therefore, for any convex combination of tangent functionals approximable by unique tangent functionals, and a simple approximation argument, using again the upper semi-continuity of s, proves it for weak-* limits of these. By Proposition B5.1, there are no other tangent functionals, so the theorem is proved.

So far, we have not said anything about the relation between invariant equilibrium states and limits of finite-volume ensembles. This relation is already implicit in Theorem B3.1 and the results of Chapter I on equivalence of ensembles, but we will rederive directly some elementary facts about infinite-volume limits of the grand-canonical probability distribution.

Proposition B5.3

Let Λ_n be a sequence of finite subsets of $\mathbb{Z}^\nu$ becoming infinitely large in the sense of van Hove. For each n, let H_n be a function on $P(\Lambda_n)$, to be interpreted as the potential energy for external forces acting on the finite system in Λ_n. We assume

$$\lim_{n\to\infty} \frac{\|H_n\|_\infty}{N(\Lambda_n)} = 0 \quad .$$

For each interaction Φ, we define a probability measure μ_n on $P(\Lambda_n)$ by

$$\mu_n(\{X\}) = \frac{\exp[-\Sigma\Phi(X) - H_n(X)]}{\Xi_n(\Phi)} ,$$

$$\Xi_n(\Phi) = \sum_{X\subset\Lambda_n} \exp[-\Sigma\Phi(X) - H_n(X)]$$

Then

(a) *If there is a unique equilibrium state for* Φ, *the measures* μ_n, *averaged over translations, converge to that equilibrium state. (We are using the terminology introduced in Section* B2 *for the passage from measures on* $P(\Lambda_n)$ *to a limiting measure on* $P(\mathbb{Z}^\nu)$)

(b) *Whether or not there is a unique equilibrium state for* Φ, *if the measures* μ_n, *averaged over translations, converge to a measure* μ *on* $P(\mathbb{Z}^\nu)$, *the measure* μ *is an invariant equilibrium state for* Φ.

Proof. We will prove (b) first. Note that

$$\left|\frac{1}{N(\Lambda_n)} \log \Xi_n - \frac{1}{N(\Lambda_n)} \log \Xi_n^0\right| \le \frac{\|H_n\|_\infty}{N(\Lambda_n)} ,$$

where

$$\Xi_n^0(\Phi) = \sum_{X\subset\Lambda} \exp(-\Sigma\Phi(X)) .$$

Since $\lim_{n\to\infty} \frac{1}{N(\Lambda_n)} \log \Xi_n^0(\Phi) = p(\Phi)$, and since

$$\lim_{n\to\infty} \frac{\|H_n\|_\infty}{N(\Lambda_n)} = 0 ,$$

$$\lim_{n\to\infty} \frac{1}{N(\Lambda_n)} \log \Xi_n(\Phi) = p(\Phi) \quad \text{for all} \quad \Phi \in \mathcal{B} .$$

For any n, $\log \Xi_n(\Phi)$ is a differentiable convex function of Φ, and

$$\frac{d}{dt} \frac{1}{N(\Lambda_n)} \log \Xi_n(\Phi + t\Psi)\Big|_{t=0} = -\int d\mu_n^\Phi \frac{\Sigma\Psi}{N(\Lambda_n)} .$$

It is not hard to see that, if μ_n^Φ, averaged over translations, converges to μ, then

$$\lim_{n\to\infty} \int d\mu_n^\Phi \frac{\Sigma\Psi}{N(\Lambda_n)} = \int A_\Psi d\mu$$

Thus, we have a sequence of differentiable convex functions $\frac{1}{N(\Lambda_n)} \log \Xi_n$ converging pointwise to p, and the derivative of $\frac{1}{N(\Lambda_n)} \log \Xi_n$ at Φ converges to $-\int A_\Psi d\mu$. Hence, $\Psi \to -\int A_\Psi d\mu$ is a tangent functional to p at Φ, so μ is an invariant equilibrium state for Φ.

Now suppose that there is only one invariant equilibrium state μ for Φ. Then, by (b), this equilibrium state is the only cluster point for the sequence of averages of μ_n over translations. By compactness, since this sequence has a unique cluster point, it must converge.

C. GIBBS STATES

In the preceding chapter, we defined equilibrium states as invariant probability measures on $P(\mathbb{Z}^\nu)$ maximizing the entropy at fixed energy. In this chapter, we develop another line of attack on the theory of equilibrium states. We introduce a class of probability measures on $P(\mathbb{Z}^\nu)$, which we call Gibbs states, which are defined as solutions of a certain set of linear equations. It then turns out that those Gibbs states which are translation-invariant are precisely the invariant equilibrium states. Thus, equilibrium states are described either as the solutions of a maximization problem or as the translation-invariant solutions of a set of equations. It turns out to be possible to prove under appropriate circumstances (corresponding roughly to systems at low density or high temperature) that there is only one Gibbs state. In particular, there is only one equilibrium state. The results of the preceding chapter than allow us to conclude that all observables converge in probability to constants in the corresponding micro-canonical ensemble.

Gibbs states are useful in other contexts as well. We have already noted that the equilibrium states are the *invariant* Gibbs states. Although the equations defining Gibbs states are formally translation invariant, they may have non-invariant solutions. (This implies, of course, that they have more than one solution.) For a proof that non-invariant Gibbs states exist, see Dobrushin [2] or Ginibre [4]. Thus, Gibbs states provide a natural notion of non-invariant equilibrium state, and it is hoped that the existence of non-invariant Gibbs states has something to do with the existence of crystals. The connection is at the moment a little tenuous, and we will not have much more to say about it.

C1. DEFINITIONS AND GENERAL PROPERTIES

To motivate the definition of Gibbs state, we begin by considering a system contained in a finite box (i.e., a finite subset M of $\mathbb{Z}^\nu$) and interacting by a potential Φ. (We continue to follow the convention introduced in the preceding chapter of absorbing the temperature and activity into the interaction.) To simplify the notation, we write

$$\begin{aligned} U(X) &= \Sigma\Phi(X) \\ W(X,Y) &= U(X \cup Y) - U(X) - U(Y) \qquad (X \cap Y = \phi) \quad . \end{aligned} \tag{1}$$

Thus, $U(X)$ is the potential energy of the configuration X and $W(X,Y)$ is the energy of interaction between X and Y. Let Λ be a subset of M, and split any $Z \in P(M)$ as

$$Z = X \cup Y \quad ,$$

where $X \in P(\Lambda)$ and $Y \in P(M\backslash\Lambda)$.

The grand-canonical probability of finding the configuration $X \cup Y$ in M is given by

$$\frac{\exp[-U(X) - W(X,Y) - U(Y)]}{\Xi_M} \tag{2}$$

Summing this expression over all $X \in P(\Lambda)$ gives the total probability of finding Y in $M\backslash\Lambda$. Dividing the probability of finding X in Λ and Y in $M\backslash\Lambda$ by the probability of finding Y in $M\backslash\Lambda$, we obtain the *conditional probability* of finding X in Λ given that we have Y in $M\backslash\Lambda$. This conditional probability, then, is given by

$$\frac{\exp[-U(X) - W(X,Y)]}{\sum_{X'\subset\Lambda} \exp[-U(X') - W(X',Y)]} \tag{3}$$

(where we have cancelled a common factor of $\exp[-U(Y)]/\Xi_M$.)

The expression (3) has much better properties as M becomes large (with Λ held fixed) than (2) does. To see this, let us assume for simplicity that the potential Φ has finite range R. Then $W(X,Y)$ depends only on the part of Y at a distance no greater than R from Λ. Since Y appears in the expression for the conditional probability only through $W(X,Y)$, we can formally pass to the limit as the box M becomes infinitely large with Λ held fixed. Thus, we expect that the finite-volume analogue of the finite-volume grand canonical probability measure is one such that, for any finite $\Lambda \subset \mathbb{Z}^\nu$, any $X \subset \Lambda$, and any $Y \subset \mathbb{Z}^\nu\ \Lambda$, the conditional probability of finding the configuration X in Λ, given that the configuration outside Λ is Y, is given by formula (3). This is, in heuristic terms, the definition of Gibbs state. To use it as it stands, we would have to invoke the technical definition of "conditional probability" (the elementary notion of conditional probability as the ratio of two probabilities used above in discussing finite systems, cannot be used here since the probability of any single $X \cup Y \subset \mathbb{Z}^\nu$ is zero for most interesting probability measures on $P(\mathbb{Z}^\nu)$.) Rather than introduce this machinery, we will rework the above considerations slightly to give a definition of Gibbs state which, besides avoiding technicalities about conditional probabilities, is especially convenient for our purposes. The reader who knows the definition of conditional probability will readily convince himself that the definition we are about to give is equivalent to the one already given.

We must first of all specify the class of potentials we will consider. It is more or less clear that, for our proposed definition to work, $W(X,Y)$ must make sense even for infinite configurations Y, provided X is finite. We have already remarked that this is certainly the case if the potential Φ has finite range. A convenient larger class of potentials are those such that

$$\sum_{0\in X} |\Phi(X)| < \infty \quad ; \tag{4}$$

for the remainder of this chapter we will assume that (4) holds. The set of potentials satisfying (4) forms a Banach space which is slightly smaller than the Banach space of potentials considered in the preceding chapter; for n-body potentials, however, condition (4) is no more restrictive than the one used before. For disjoint subsets X and Y of $\mathbb{Z}^\nu$, with X finite but Y possibly infinite, we define

$$W(X,Y) = \sum_{\substack{Z \subset X \cup Y \\ Z \cap X \neq \phi \neq Z \cap Y}} \Phi(Z) \tag{5}$$

it is easy to check from (4) that the sum converges and defines a continuous function of Y on $P(\mathbb{Z}^\nu \setminus X)$; moreover, for Y finite, this definition of W agrees with the one already given.

Now let Λ be a finite subset of $\mathbb{Z}^\nu$; let $X \subset \Lambda$ and $Y \subset \mathbb{Z}^\nu \setminus \Lambda$; and define

$$\mu_\Lambda(\{X\}|Y) = \frac{\exp[-U(X) - W(X,Y)]}{\Xi_\Lambda(Y)} \quad , \tag{6}$$

where

$$\Xi_\Lambda(Y) = \sum_{X' \subset \Lambda} \exp[-U(X') - W(X',Y)] \tag{7}$$

For each Y, $\mu_\Lambda(.|Y)$ is a probability measure on $P(\Lambda)$.* If f is a continuous function on $P(\mathbb{Z}^\nu)$, and if $Y \subset \mathbb{Z}^\nu \setminus \Lambda$, we define

$$(\tau_\Lambda f)(Y) = \sum_{X \subset \Lambda} f(X \cup Y)\mu_\Lambda(\{X\}|Y) \tag{8}$$

$\tau_\Lambda f$ is a continuous function on $P(\mathbb{Z}^\nu \setminus \Lambda)$; we think of it as being a continuous function on $P(\mathbb{Z}^\nu)$ which does not depend on the part of the configuration in Λ. Thus, for each Λ, τ_Λ is a bounded linear operation on $C(P(\mathbb{Z}^\nu))$. If M is a finite subset of $\mathbb{Z}^\nu$ which contains Λ, we may also regard τ_Λ as a bounded operator on $C(P(M))$; the statement that the condition probability, with respect to the grand canonical probability measure $\mu^{(M)}$ on $P(M)$ of finding X in Λ given Y in $M \setminus \Lambda$ is equal to $M_\Lambda(\{X\}|Y)$ is easily seen to be equivalent to the statement that

$$\mu^{(M)}(f) = \mu^{(M)}(\tau_\Lambda f)$$

for all $f \in C(P(M))$. Thus, we are led to our precise definition of Gibbs state: A probability measure μ on $P(\mathbb{Z}^\nu)$ is called a *Gibbs state* if, for all finite $\Lambda \subset \mathbb{Z}^\nu$ and all $f \in C(P(\mathbb{Z}^\nu))$,

$$\mu(f) = \mu(\tau_\Lambda f) \quad . \tag{9}$$

---

* At this point, it is a matter of indifference whether we regard $\mu_\Lambda(.|Y)$ as a measure on $P(\Lambda)$ or as a function on $P(\Lambda)$. Because of the sorts of operations one wants to do with μ_Λ, it turns out to look more natural to treat it as a measure in its first argument, and we have done so. We will write both $\mu_\Lambda(E|Y)$ for the measure of a subset E of $P(\Lambda)$ and $\mu_\Lambda(f|Y)$ for the integral of a function f on $P(\Lambda)$; which usage is intended can usually be determined readily from the context.

It is clear from this definition that, for a given interaction, the set of Gibbs states is convex and weak-* closed.

The motivation for the definition should make the following proposition plausible:

Proposition C1.1

Let M_n *be an increasing sequence of finite subsets of* $\mathbb{Z}^\nu$ *such that each finite* M *is contained in some* M_n*. Let* $\mu^{(M_n)}$ *be the grand-canonical probability measure on* $P(M_n)$ *(i.e., the probability measure assigning to each point* X *of* $P(M_n)$ *a probability proportional to* $\exp\{-U(X)\}$*).*

(a) *If* $\mu = \lim_{n\to\infty} \mu^{(M_n)}$ *exists, then* μ *is a Gibbs state*

(b) *If* (M_n) *becomes infinitely large in the sense of van Hove, and if* $\bar{\mu} = \lim_{n\to\infty} \overline{\mu^{(M_n)}}$ *exists, then* $\bar{\mu}$ *is a Gibbs state.*

(We are using the notation introduced in Section B2 for obtaining probability measures on $P(\mathbb{Z}^\nu)$ as limits of probability measures on $P(M_n)$).

Proof. Let f be a continuous function on $P(\mathbb{Z}^\nu)$ and let Λ be a finite subset of $\mathbb{Z}^\nu$. For each M_n, we may regard $P(M_n)$ as contained in $P(\mathbb{Z}^\nu)$, and hence we can regard f as defined on $P(M_n)$. The equation

$$\mu = \lim_{n\to\infty} \mu^{(M_n)}$$

means simply that

$$\mu(f) = \lim_{n\to\infty} \mu^{(M_n)}(f|_{P(M_n)})$$

for all continuous f. If $\Lambda \subset M_n$, we may obtain $\tau_\Lambda f|_{P(M_n)}$ by carrying out the finite-volume version of τ_Λ on $f|_{P(M_n)}$. But we have already remarked that

$$\mu^{(M_n)}(f|_{P(M_n)}) = \mu^{(M_n)}((\tau_\Lambda f|_{P(M_n)}))$$

Thus

$$\mu(f) = \lim_{n\to\infty} \mu^{(M_n)}(f|_{P(M_n)}) = \lim_{n\to\infty} \mu^{(M_n)}(\tau_\Lambda f|_{P(M_n)}) = \mu(\tau_\Lambda f) \quad ,$$

so

$$\mu(f) = \mu(\tau_\Lambda f)$$

for all continuous f and all finite Λ. This proves (a); the proof of (b) is similar.

By strengthening the above proposition a bit, we can convert it in a characterization of Gibbs states. Recalling the definition of μ_Λ and $\mu^{(\Lambda)}$, we see that

$$\mu^{(\Lambda)}(\cdot) = \mu_\Lambda(\cdot|\emptyset)$$

This leads us to consider the measures $\mu_\Lambda(\cdot|Y)$, $Y \in P(\mathbb{Z}^\nu \setminus \Lambda)$, as generalized grand canonical measures for the interaction we are considering. The interpretation of $\mu_\Lambda(\cdot|Y)$ in this context is clear: It is the grand-canonical probability measure for a system in the box Λ, interacting with itself through the potential Φ but also subject to external forces exerted by the specified configuration Y outside Λ. Using the ideas of the proof of Proposition C1.1, one shows easily that, if M_n is any increasing sequence of finite subsets of $\mathbb{Z}^\nu$ which eventually contains any finite Λ, if $Y_n \in P(\mathbb{Z}^\nu \setminus M_n)$ for each n, and if the measures $\mu_{M_n}(\cdot|Y_n)$ converge to a measure μ on $P(\mathbb{Z}^\nu)$, then μ is a Gibbs state. This gives us quite a general procedure for constructing Gibbs states, and we might hope to get all of them in this way. A little reflection shows that we may have to take convex combinations and limits in order to get all Gibbs states; we have, in fact, the following:

Proposition C1.2.

The set of all Gibbs States (for a given interaction Φ) is the weak- closed convex hull of the set of infinite-volume cluster points of finite-volume probability measures of the form* $\mu_M(\cdot|Y)$, $Y \in P(\mathbb{Z}^\nu \setminus M)$.

Proof. Let K denote the weak-* closed convex hull of the set of infinite-volume cluster points of the $\mu_M(\cdot|Y)$. We have already remarked that each such cluster point is a Gibbs state, and that the set of Gibbs states is convex and weak-* closed; hence, K is contained in the set of all Gibbs states. Assume that there is a Gibbs state μ_0 which does not belong to K. By standard separation theorems for convex sets, there is a function $f \in C(P(\mathbb{Z}^\nu))$ such that $\mu_0(f) \notin \{\mu(f): \mu \in K\}$. Since cylinder functions are uniformly dense in $C(P(\mathbb{Z}^\nu))$, we may assume that f is measurable in some finite $\Lambda \subset \mathbb{Z}^\nu$. By scaling and adding a constant, we can assume that

$$\mu_0(f) \le -1; \ \mu(f) \ge 0 \quad \text{for all} \quad \mu \in K \ .$$

To obtain a contradiction, we have only to show that, for arbitrarily large finite subsets M of $\mathbb{Z}^\nu$, there exist configurations $Y \in P(\mathbb{Z}^\nu \setminus M)$ such that $\mu_M(f|Y) \le -1$; then a simple compactness argument will show that there is a cluster point μ of these $\mu_M(\cdot|Y)$'s with $\mu(f) \le -1$, contradicting the fact that $\mu(f) \ge 0$ for all $\mu \in K$. We will show that such configurations Y exist for all $M \supset \Lambda$. Indeed, by the definition of Gibbs state, we have

$$-1 \geq \mu_0(f) = \mu_0(\tau_M f)$$

But by the definition of τ_M,

$$(\tau_M f)(Y) = \mu_M(f|Y)$$

where, in the expression on the right, we have used the fact that f is measurable in $\Lambda \subset M$ to treat f as a function on $P(M)$. But since $\mu_0(\tau_M f) \leq -1$, we must have

$$(\tau_M f)(Y) \leq -1$$

for at least one Y, as asserted.

In the course of proving the above proposition, we have essentially proved the following:

Proposition C1.3

Let μ *be a Gibbs state,* M *a finite subset of* $\mathbb{Z}^\nu$*, and let* μ_M *be the projection of* μ *onto* $P(M)$*. Then* μ_M *is an integral (over* Y*) of the measures* $\mu_M(\cdot|Y)$.

Proof. Let f be a function on $P(M)$. By definition,

$$\mu_M(f) = \mu(f) = \mu(\tau_M f) = \int \mu(dY)\mu_M(f|Y)$$

It is not hard to see that the conclusion of the proposition may be taken as an alternative definition of Gibbs state. Proposition C1.2 implies an interesting relation between the existence of more than one Gibbs state and the occurrence of "long-range order." Suppose that there exist two distinct Gibbs states μ_1 and μ_2. Then there is an $f \in C(P(\mathbb{Z}^\nu))$ which distinguishes between them. Without loss of generality, we can assume that f is measurable in some finite Λ and that $\mu_1(f) \geq 1$; $\mu_2(f) \leq 0$. Then for any M containing Λ, no matter how large, there exist Y_1 and Y_2 in $P(\mathbb{Z}^\nu \setminus M)$ such that

$$\mu_M(f|Y_1) \geq 1; \; \mu_M(f|Y_2) \leq 0 \quad .$$

In other words, the effects of the external configurations Y_1 and Y_2 make themselves felt all the way to the center of M. This phenomenon may be described more graphically when the interaction Φ has finite range R. In this case, $\mu_M(\cdot|Y)$ depends only on the part of Y at a distance $\leq$ R from M. Letting M_R denote the region consisting of M together with all points at a distance $\leq$ R from M, we may regard $\mu_M(\cdot|Y_1)$ and $\mu_M(\cdot|Y_2)$ as describing grand canonical probability measures on M_R in which the parts of the configuration in a shell of thickness R around the boundary are fixed, i.e., grand canonical probability measures with boundary conditions. The existence of more than one Gibbs state thus means that, merely by changing boundary conditions, one can change the grand canonical probability measure in the center of M by an amount which does not go to zero as M becomes infinitely large.

We now turn to the relation between Gibbs states and invariant equilibrium states.

Theorem C1.4

Let μ be a translation invariant probability measure on $P(\mathbb{Z}^\nu)$. Then μ is an invariant equilibrium state if and only if it is a Gibbs state.

Proof. (a) Let μ be a Gibbs state. We want to show that

$$s(\mu) - \mu(A_\Phi) \geq p(\Phi) \quad .$$

Since

$$s(\mu) = \lim_{M\to\infty} \frac{S(\mu,M)}{N(M)}$$

$$\mu(A_\Phi) = \lim_{M\to\infty} \frac{\mu(U_M)}{N(M)}$$

$$p(\Phi) = \lim_{M\to\infty} \frac{\log \Xi(M,\Phi)}{N(M)} \quad ,$$

it will suffice to prove

$$\liminf_{M\to\infty}\left[\frac{S(\mu,M) - \mu(U_M) - \log \Xi(M,\Phi)}{N(M)}\right] \geq 0 \quad .$$

Now $S(\mu,M)$ and $\mu(U_M)$ depend only on the projection μ_M of μ onto $P(M)$. Moreover, $\mu(U_M)$ is an affine function of μ_M and $S(\mu,M)$ is a concave function of μ_M. Thus, since μ_M may be written as an integral over Y of $\mu_M(\cdot|Y)$, it is enough to prove

$$\liminf_{M\to\infty} \inf_{Y\subset\mathbb{Z}^\nu\setminus M} \frac{[S(\mu_M(\cdot|Y)) - \mu_M(U|Y) - \log \Xi(M,\Phi)]}{N(M)} \geq 0 \quad . \tag{10}$$

Now by definition

$$S(\mu_M(\cdot|Y)) = -\sum_{X\subset\Lambda} \mu_M(\{X\}|Y)\log \mu_M(\{X\}|Y) \quad ,$$

so by the definition (6) of $\mu_M(\cdot|Y)$ we get

$$S(\mu_M(\cdot|Y)) = \log\{\Xi_M(Y)\} + \int_{P(M)} \mu_M(dX|Y)[U(X) + W(X,Y)] \quad .$$

Now let

$$V_M = \sup\{|W(X,Y)| : X\subset M,\ Y\subset\mathbb{Z}^\nu\setminus M\}$$

Then

$$\Xi_M(Y) = \sum_{X\subset M} \exp[-U(X) - W(X,Y)] \geq e^{-V_M} \sum_{X\subset M} \exp[-U(X)] = e^{-V_M}\,\Xi(M,\Phi)$$

and

$$\int_{P(M)} W(X,Y)\mu_M(dX|Y) \geq -V_M \quad ,$$

so

$$S(\mu_M(\cdot|Y)) - \mu_M(U_M|Y) - \log \Xi(M,\Phi)$$
$$\geq -2V_M$$

for all Y. Thus, to prove (10), we have only to show that $\frac{V_M}{N_M}$ goes to zero as M becomes infinitely large in the sense of van Hove. This follows readily from the condition (4) on the interaction, so the first half of the theorem is proved.

(b) We want to show that every invariant equilibrium state is a Gibbs state. Suppose first that, for the interaction we are considering there is only one invariant equilibrium state. By Proposition B5.3, this unique invariant equilibrium state is the infinite-volume limit of translation averaged grand-canonical probability measures and hence, by part (b) of Proposition C1.1, is a Gibbs state. Now drop the assumption that the interaction has a unique invariant equilibrium state. By what we have just shown, and a simple approximation argument, it follows that every equilibrium state which is approximable by unique equilibrium states is a Gibbs state (we are using an obvious modification of the terminology introduced in Section B5). Since every equilibrium state is a weak limit of convex combinations of states approximable by unique equilibrium states, the theorem is proved.

C2. DOBRUSHIN'S UNIQUENESS THEOREM

In this section, we will prove a theorem, due to Dobrushin [2], which gives a sufficient condition on the interaction Φ for there to exist only one Gibbs state. We have already alluded to one consequence of the uniqueness of the Gibbs state — it implies that there is only one invariant equilibrium state and hence that all observables converge in probability to constants in the corresponding grand-canonical, canonical, and micro-canonical ensembles.

To see another consequence, we must undo our notational convention of absorbing the temperature and activity into the interaction. Thus, what we have written as U(X) should properly be written as $\beta\hat{U}(X) - \log z\, N(X)$, where $\hat{U}$ is the true interaction energy. Suppose we can show that there exists a unique Gibbs state $\mu_{\beta,z}$ for each (β,z) in some region Δ of the (β,z) plane. Let (β_0,z_0) be in Δ, and let (β_n,z_n) be a sequence of points in Δ converging to (β_0,z_0). We claim that μ_{β_n,z_n} must converge to μ_{β_0,z_0} in the weak-* topology. Indeed, it is easy to see that any weak-* cluster point of (μ_{β_n, z_n}) must be a Gibbs state corresponding to (β_0, z_0), and hence, by uniqueness, must be μ_{β_0, z_0}. But the set of all probability measures is weak-* compact, and a sequence in a compact space with only one cluster point must converge. Thus, the unique Gibbs state varies continuously with β, z. In particular, the energy per lattice site and the density are continuous functions of β, z, so we see that uniqueness of the Gibbs state in a region of the (β, z) plane implies that the system undergoes no first-order phase transitions in that region.

Dobrushin's theorem is valid in a more general context than that of lattice gases, so we generalize our considerations a bit. We start with an index set I and an assignment, for each $x \in I$, of a measurable space Ω_x. In the lattice gas case, we have $I = \mathbb{Z}^\nu$ and $\Omega_x = \{0,1\}$ for all x. In defining Gibbs states, we specified, for each finite $\Lambda \subset I$, a mapping $Y \to \mu_\Lambda(\cdot|Y)$ from $\prod_{x \notin \Lambda} \Omega_x$ to probability measures on $\prod_{x \in \Lambda} \Omega_x$ and we considered measures on $\prod_{x \in I} \Omega_x$ with these functions as conditional probabilities. Here we do something which is in principle less restrictive: We specify, for each $x \in I$ a mapping $Y \to \mu_x(\cdot|Y)$ from $\prod_{y \neq x} \Omega_y$ to probability measures on Ω_x. We assume that each of these mappings is in an appropriate sense measurable. If f is a bounded measurable function on $\prod_x \Omega_x$, we write $f = f(\omega_x, Y)$, where $\omega_x \in \Omega_x$ and $Y \in \prod_{y \neq x} \Omega_y$, and we define

$$\tau_x f(Y) = \int f(\omega_x, Y) \mu_x(d\omega_x|Y) \tag{1}$$

We regard $\tau_x f$ as a function on $\prod_{y \in I} \Omega_y$ which doesn't depend on the Ω_x component of its argument. We are going to obtain a condition on the specified $\mu_x(d\omega_x|Y)$ which guarantees that there is no more than one probability measure μ on $\prod_x \Omega_x$ such that

$$\int f d\mu = \int \tau_x f d\mu \quad \text{for all} \quad x \in I \tag{2}$$

(The theorem says nothing about the existence of such a measure μ; this must be investigated by other methods and is in any case no problem for lattice gases.) To formulate the condition, we define, for each $x,y \in I$, $x \neq y$,

$$\rho_{x,y} = \frac{1}{2} \sup\{\|\mu_x(\cdot|Y) - \mu_x(\cdot|Y')\| : Y, Y' \text{ the same except for the } \Omega_y \text{ component}\} \tag{3}$$

Here, the norm means the measure norm on Ω_x.

If Λ is a finite subset of I, not containing x, we have

$$\|\mu_x(\cdot|Y) - \mu_x(\cdot|Y')\| \leq 2 \sum_{y \in \Lambda} \rho_{x,y} \tag{4}$$

provided $Y = Y'$ outside Λ (here we are regarding Y, Y' as functions $y \to \omega_y, \omega'_y$ defined on $I \setminus \{x\}$). Condition (4) does not automatically hold if Λ is infinite, but we will from now on assume that the μ_x's are such that it does hold for *all*

$\Lambda \subset I\setminus\{x\}$. If $\sum_y \rho_{x,y} < \infty$, this is equivalent to assuming that the mappings

$$Y \to \mu_x(\cdot\,|Y)$$

are uniform limits of cylinder functions. We can now state:

Theorem C2.1

Assume that there exists $\alpha < 1$ *such that*

$$\sum_y \rho_{x,y} \le \alpha \quad \textit{for all} \quad x \in I \ . \tag{5}$$

Then there is at most one probability measure μ on $\prod_x \Omega_x$ *such that*

$$\int f d\mu = \int \tau_x f d\mu \tag{6}$$

for all bounded measurable functions f *and all* $x \in I$.

Note that, for lattice gases with translation-invariant interactions, $\rho_{x,y}$ depends only on $x - y$ so (5) can be rewritten as

$$\sum_{y\in\mathbb{Z}^\nu\setminus\{0\}} \rho_{0,y} < 1$$

Proof. The proof of the theorem will proceed by showing that, for any bounded cylinder function f, there is a sequence $x_1, x_2, x_3 \ldots$ of points of I not necessarily distinct such that

$$\tau_{x_n} \ldots \tau_{x_1} f$$

converges uniformly to a constant $c(f)$ as $n \to \infty$. Once this is proved, it follows that for any probability measure μ satisfying (6).

$$\int f d\mu = c(f)$$

for all cylinder functions f; hence, that any two probability measures satisfying (6) agree on all bounded cylinder functions and therefore must be equal.

For any bounded measurable function f on $\Pi\Omega_x$, and any $x \in I$, we define

$$\delta_x(f) = \sup\{|f(Y) - f(Y')| : Y = Y' \text{ except at } x\} \ . \tag{7}$$

In other words, $\delta_x(f)$ is the maximum change in f that can be produced by changing the x component of its argument. In particular, if f is a cylinder function, then $\delta_x(f) = 0$ for all but finitely many x's. Let $\mathcal{F}$ denote the set of all bounded measurable functions f on $\prod_x \Omega_x$ such that

(a) $\sum_x \delta_x(f) < \infty$

(b) f is a uniform limit of cylinder functions.

Note that all bounded cylinder functions belong to F. Note also that condition (b) guarantees that, for $f \in F$,

$$\sup_X f(X) - \inf_X f(X) \leq \sum_x \delta_x(f)$$

(this is, in fact, the only use we make of (b)). To prove the theorem, it suffices to prove:

(a) For all $x \in I$, $\tau_x F \subset F$

(b) For any $f \in F$, there exists a sequence $x_1, x_2, x_3, \ldots$ in I such that

$$\lim_{n\to\infty} \sum_y \delta_y(\tau_{x_n} \ldots \tau_{x_1} f) = 0 \quad ;$$

Given these two facts we have only to apply the argument outlined in the first paragraph of the proof. Instead of proving (b), it will in fact suffice to prove

(b′) For any $f \in F$, there exist $x_1, \ldots, x_n$ in I such that

$$\sum_y \delta_y(\tau_{x_n} \ldots \tau_{x_1} f) \leq \left(\frac{1+\alpha}{2}\right) \sum_y \delta_y(f) \quad ;$$

(b) then follows by iteration.

The essential step in the proof is the following lemma:

Lemma C2.2

If $f \in F$, *then* $\tau_x f \in F$ *and we have*

$$\begin{aligned} \delta_y(\tau_x f) &= 0 \quad \text{if } y = x \\ &\leq \delta_y(f) + \rho_{x,y}\delta_x(f) \quad \text{if } y \neq x \ . \end{aligned} \tag{8}$$

$$\sum_y \delta_y(\tau_x f) \leq \sum_y \delta_y(f) - (1-\alpha)\delta_x(f) \quad . \tag{9}$$

Proof. (9) follows at once from (8) by summing over y and using

$$\sum_y \rho_{x,y} \leq \alpha \quad .$$

Thus, if f satisfies condition (a) in the definition of F, so does $\tau_x f$. it is easy to check (using (4)) that, if f is a uniform limit of cylinder functions, so is $\tau_x f$. Hence, proving the lemma reduces to proving (8).

Since $\tau_x f$ does not depend on the x component of its argument, we have immediately that $\delta_x(\tau_x f) = 0$. Hence, let $x \neq y$. We may write

$$\tau_x f(\omega_y, \hat{Y}) = \int f(\omega_x, \omega_y, \hat{Y}) \mu_x(d\omega_x | (\omega_y, \hat{Y})) \ , \tag{10}$$

where $\omega_y \in \Omega_y$ and $\hat{Y} \in \prod_{z\neq x,y} \Omega_z$. We recall that, by definition

$$\delta_y(\tau_x f) = \sup_{\omega_y,\omega'_y,\hat{Y}} |\tau_x f(\omega_y,\hat{Y}) - \tau_x f(\omega'_y,\hat{Y})| \quad . \tag{11}$$

Now:

$$\begin{aligned}\tau_x f(\omega_y,\hat{Y}) - \tau_x f(\omega'_y,\hat{Y}) &= \frac{1}{2}\int [f(\omega_x,\omega_y,\hat{Y}) - f(\omega_x,\omega'_y,\hat{Y})] \\ &\times [\mu_x(d\omega_x|\omega_y,\hat{Y}) + \mu_x(d\omega_x|\omega'_y,\hat{Y})] + \frac{1}{2}\int [f(\omega_x,\omega_y,\hat{Y}) \\ &+ f(\omega_x,\omega'_y,\hat{Y})][\mu_x(d\omega_x|\omega_y,\hat{Y}) - \mu_x(d\omega_x|\omega'_y,\hat{Y})] \quad .\end{aligned} \tag{12}$$

We will estimate the two terms on the right separately. The first is easy: $|f(\omega_x,\omega_y,\hat{Y}) - f(\omega_x,\omega'_y,\hat{Y})| \leq \delta_y(f)$, so the absolute value of the first term is no larger than $\delta_y(f)$.

To estimate the second term in (12) we need the following remark: If g is a measurable function such that

$$\sup_\omega\{g(\omega)\} - \inf_\omega\{g(\omega)\} \leq \delta$$

and if $\nu(\omega)$ is a signed measure of total mass zero, then

$$|\int g d\nu| \leq \frac{1}{2}\delta\|\nu\|$$

This is true since, if we put

$$a = \frac{1}{2}\sup_\omega\{g(\omega)\} + \frac{1}{2}\inf_\omega\{g(\omega)\}, \quad \text{then}$$
$$\int g d\nu = \int (g - a)d\nu$$

(since ν has total mass zero), and $|g - a| \leq \delta/2$ everywhere.

Applying this remark with $g(\cdot) = f(\cdot,\omega_y,\hat{Y})$ and $\nu(\cdot) = \mu_x(\cdot|\omega_y,\hat{Y}) - \mu_x(\cdot|\omega'_y,\hat{Y})$ gives

$$\begin{aligned}&|\int f(\omega_x,\omega_y,\hat{Y})[\mu_x(d\omega_x|\omega_y,\hat{Y}) - \mu_x(d\omega_x|\omega'_y,\hat{Y})]| \\ &\leq \frac{1}{2}\cdot\delta_x(f)\|\mu_x(\cdot|\omega_y,\hat{Y}) - \mu_x(\cdot|\omega'_y,\hat{Y})\| \leq \delta_x(f)\rho_{x,y} \quad .\end{aligned}$$

We can estimate

$$|\int f(\omega_x,\omega'_y,Y)[\mu_x(d\omega_x|\omega_y,\hat{Y}) - \mu_x(d\omega_x|\omega'_y,\hat{Y})]|$$

in the same way. Inserting these estimates in the right-hand side of (12) gives

$$|\tau_x f(\omega_y,\hat{Y}) - \tau_x f(\omega'_y,\hat{Y})| \leq \delta_y(f) + \rho_{x,y}\delta_x(f)$$

for all ω_y, ω'_y, $\hat{Y}$, and this is exactly the estimate (8) we want to prove.

To complete the proof of the theorem, we have now only to prove (b′) and this follows in a straigtforward way by iterating the estimates in the lemma. Specifically, we will prove:

(b") Let $f \in F$, and let $x_1,\ldots,x_n$ be distinct points of I; let $\Lambda = \{x_1,\ldots,x_n\}$. Then

$$\sum_y \delta_y(\tau_{x_n}\ldots\tau_{x_1} f) \le \alpha \sum_{y\in\Lambda} \delta_y(f) + \sum_{y\in I\setminus\Lambda} \delta_y(f)$$

(b') follows readily from (b") by taking Λ large enough. To prove (b"), we let $\delta_y^{(j)}$ denote the bound on $\delta_y(\tau_{x_j}\ldots\tau_{x_1} f)$ obtained by iterating (8). Explicitly,

$$\delta_y^{(0)} = \delta_y(f);$$

$$\delta_{x_{j+1}}^{(j+1)} = 0;\ \delta_y^{(j+1)} = \delta_y^{(j)} + \rho_{x_{j+1},y}\,\delta_{x_j}^{(j)} \quad \text{for } y \neq x_{j+1} \quad .$$

Note the following facts about the $\delta_y^{(j)}$.

(a) By (8), $\delta_y^{(j)} \ge \delta_y(\tau_{x_j}\ldots\tau_{x_1} f)$.

(b) By summing the recursive definition of $\delta_y^{(j)}$ over y, we get

$$\sum_y \delta_y^{(j+1)} \le \sum_y \delta_y^{(j)} - (1-\alpha)\delta_{x_{j+1}}^{(j)}$$

(c) Because $x_1,\ldots,x_n$ are all distinct,

$$\delta_{x_{j+1}}^{(j)} \ge \delta_{x_{j+1}}(f)$$

(In general $\delta_y^{(j)} \ge \delta_y^{(j-1)}$ unless $y = x_j$.) Combining (b) and (c) gives

$$\sum_y \delta_y^{(n)} \le \sum_y \delta_y(f) - (1-\alpha)\sum_{j=1}^{n} \delta_{x_j}(f) \quad ;$$

combining this inequality with (a) gives (b) and completes the proof of the theorem.

It is now a simple matter to specialize the above result to get a sufficient condition for the uniqueness of the Gibbs state with a given interaction. In this case $\Omega_x = \{0,1\}$, and, for any two probability measures μ_1 and μ_2 on $\{0,1\}$,

$$\|\mu_1 - \mu_2\| = 2|\mu_1(\{0\}) - \mu_2(\{0\})| \quad .$$

Also

$$\mu_0(\{0\}|Y) = 1/(1 + z\exp[-\beta W(\{0\}|Y)])$$

Thus

$$\rho_{0,y} = \sup_{Y\subset \mathbb{Z}^\nu\setminus\{0,y\}} \{|1/(1 + z\exp[-\beta W(\{0\}|Y)]) - 1/(1 + z\exp[-\beta W\{0\}|Y \cup \{y\})]|\}$$

Now

$$W(\{0\}|Y \cup \{y\}) - W(\{0\}|Y) = \sum_{\substack{X\supset\{0,y\}\\ X\subset Y\cup\{0,y\}}} \hat{\Phi}(X), \quad \text{and}$$

$$\left|\frac{d}{d\lambda}\left[\frac{1}{1+ze^{-\beta\lambda}}\right]\right| \leq \frac{\beta}{4} \quad .$$

so

$$\sum_{y\neq 0} \rho_{0,y} \leq \frac{\beta}{4} \sum_{y\neq 0} \sum_{X\supset\{0,y\}} |\hat{\Phi}(X)| = \frac{\beta}{4} \sum_{0\in X} |\hat{\Phi}(X)|[N(X) - 1] \quad .$$

Thus, for

$$\beta < \frac{4}{\sum_{0\in X} |\hat{\Phi}(X)|[N(X) - 1]}$$

there is a unique Gibbs state. It is also easy to see that, for any given β, there is a unique Gibbs state for each sufficiently small (or sufficiently large) z, provided that

$$\sum_{0\in X} |\hat{\Phi}(X)|[N(X) - 1] < \infty \quad .$$

REFERENCES

[1] Dobrushin, R. L., "The Description of a Random Field by Means of its Conditional Probabilities and Conditions of its Regularity", *Theory of Probability and its Applications* 13, 197-224 (1968).

[2] "The Problem of Uniqueness of a Gibbsian Random Field and the Problem of Phase Transitions", *Functional Analysis and its Applications* 2, 302-312 (1968).

[3] Dunford, N. and J. T. Schwartz, *Linear Operators I*, Interscience, New York (1958).

[4] Ginibre, J., "On Some Recent Work of Dobrushin", *Systèmes à un nombre infini de degrés de liberté*, pp. 163-175, CNRS, Paris (1970).

[5] Griffiths, R. B., and D. Ruelle, "Strict Convexity ("Continuity") of the Pressure in Lattice Systems", *Commun. Math. Physics.* 23, 169-175 (1971).

[6] Lanford, O. E., "Selected Topics in Functional Analysis", *in Statistical Mechanics and Quantum Field Theory: Summer School of Theoretical Physics, Les Houches 1970*, 109-214, Gordon and Breach, New York (1971).

[7] Lanford, O. E., and D. W. Robinson, "Statistical Mechanics of Quantum Spin Systems III", *Commun. Math. Physics* 9, 327-338 (1968).

[8] Robinson, D. W., and D. Ruelle, "Mean Entropy in Classical Statistical Mechanics", *Commun. Math. Physics* 5, 288-300 (1967).

[9] Ruelle, D., *Statistical Mechanics: Rigorous Results*, Benjamin, New York (1969).

[10] __________, "Correlation Functionals", *J. Math. Physics* 6, 201-220 (1965).

[11] __________, "Superstable Interactions in Classical Statistical Mechanics", *Commun. Math. Physics* 18, 127-159 (1970).

LECTURES ON THE COULOMB STABILITY THEOREM

A. Lenard*

1. INTRODUCTION

These lectures deal with the Coulomb Stability Theorem. The theorem says, roughly speaking, that the Hamiltonian operator of non-relativistic quantum mechanics has a lower bound which is proportional to the number of particles. This is a highly non-trivial result because the potential energy of electrically charged particles is unbounded below in a complicated manner. The theorem turns out to be a subtle consequence of a number of disparate facts:

(a) the behavior of certain differential operators and associated inequalities,

(b) some facts of classical potential theory,

(c) the Pauli Exclusion Principle.

The theorem was proved in 1965 by F. J. Dyson and the present writer [1,2]. A detailed and very readable report on this research is available in the lectures of Dyson at the 1966 Brandeis Summer Institute of Theoretical Physics [3] (for some further comments see [4]).

2. THE PROBLEM

We work in the Hilbert space $\mathcal{H}$ whose elements are functions

$$f_{s_1,s_2,\ldots,s_n}(x_1,\ldots,x_n)$$

$$(s_1,s_2,\ldots,s_n = \pm 1,\ x_1,x_2,\ldots,x_n \in \mathbb{R}^3)$$

with norm defined by

$$\|f\|^2 = \sum_{s_1}\cdots\sum_{s_n}\int dx_1\cdots\int dx_n |f|^2 < \infty \quad .$$

Besides the finiteness of the norm, it is required that

$$P_{ij}f = -f$$

for any pair (i,j), $1 \le i < j \le n$, where $P_{ij}f$ is the function arising from f by the substitution $x_i, s_i \leftrightarrow x_j, s_j$ (Pauli Exclusion Principle).

We are concerned with the operator

* Department of Mathematics, Indiana University, Bloomington, Indiana.

$$H = \frac{\hbar^2}{2M} \sum_{1\le i\le n} \Delta_i + e^2 \sum\sum_{1\le i<j\le m} |x_i - x_j|^{-1} - Ze^2 \sum_{1\le i\le n} \sum_{1\le j\le m} |x_i - y_j|^{-1}$$
$$+ Z^2e^2 \sum\sum_{1\le i<j\le m} |y_i - y_j|^{-1} \quad .$$

Here $\hbar$ is Planck's constant, M the mass of the electron, $-e$ the charge of the electron, n the total number of electrons, Ze the charge of the positive point particles fixed at the locations $y_1, y_2, \ldots, y_m$. Δ_i is the Laplacian differential operator with respect to the variable x_i. If $f \in \mathcal{H}$ is in the domain of H and $\|f\| = 1$, then (f,Hf) is the energy of n electrons in the state f in the presence of m nuclei. Our aim is the following

Theorem 1

For f *in the domain of* H *and* $\|f\| = 1$, $(f,Hf) \ge -A(Z)n\varepsilon_0$, *where* $A(Z)$ *is a polynomial of degree* 2 *in* Z, *and* $\varepsilon_0 = Me^4/2\hbar^2$ *is the natural energy unit. The coefficients in* A *are independent of all the physical variables in the problem.*

We make a few remarks, partly of a physical, partly of a mathematical nature.

It was assumed that all positive charges are of the same magnitude Ze. This is only for simplicity; if the positive charges are assumed different, Z_je say, we still obtain the lower bound of the theorem with $Z = \mathrm{Max}_j Z_j$.

The positively charged particles are not treated quantum mechanically, but as fixed sources of an electrostatic field. It is easy to see, however, that a quantum mechanical treatment of the positively charged particles only raises the lower bound for the energy. For discussion of these points see [2], Section 3.

The operator H is defined for $f \in \mathcal{H}$ only if f depends on the variables $x_1, \ldots, x_n$ in a sufficiently regular manner. In a fundamental paper [5] T. Kato has shown that H is defined and is essentially self-adjoint on the domain $\mathcal{D}_0$ whose elements are functions of the form

$$e^{-(|x_1|^2+\ldots+|x_n|^2)}P(x_1,\ldots,x_n) \quad ,$$

where P are polynomials in the rectangular components of the x_j. Thus the true domain $\mathcal{D}_H \subset \mathcal{H}$ of H is the set of f for which sequences $\{f_n\}$ exist with the properties: $f_n \in \mathcal{D}_0$ for all n, $\lim f_n = f$, $\lim Hf_n$ exists; and in this case $Hf = \lim Hf_n$. It is clear that the theorem needs to be proved only for $f \in \mathcal{D}_0$, then by a passage to the limit it is seen to hold for $f \in \mathcal{D}_H$. The advantage of restriction to $\mathcal{D}_0$ is that all $f \in \mathcal{D}_0$ are infinetely differentiable functions (indeed, holomorphic functions, though we never need this property) with fast decrease at infinity. In particular, intregration by parts is justified and we write

$$(f, -\frac{\hbar^2}{2M} \sum_{1\le i\le n} \Delta_i f) = \frac{\hbar^2}{2M}\int dx_1 \ldots \int dx_n \sum_{1\le i\le n} |\mathrm{grad}_i f|^2 \quad .$$

A further remark concerns the "spin variables" $s_1, \dots, s_n$. If we consider fixed values, say $s_1 = s_2 = \dots = s_\mu = +1$, $s_{\mu+1} = \dots = s_m = -1$, the corresponding component of f has the symmetry that $P_{ij}f = -f$ for $1 \le i < j \le \mu$ and $\mu + 1 \le i < j \le n$. Thus we may ignore the spin variables and deal with functions $f(x_1,\dots,x_n)$ only, provided we impose the condition that f is antisymmetric with respect to interchange of two variables, both belonging to one or another of two subsets of the variables $x_1, \dots, x_n$ into which they are partitioned. This version of the Exclusion Principle will be assumed in the following.

3. NTC - INEQUALITIES

In our subject an important role is played by certain inequalities beween three types of integrals:

$$N = \int_\Omega f^2 dx \quad ,$$

$$T = \int_\Omega |\text{grad } f|^2 dx \quad ,$$

and

$$C = \int_\Omega \frac{1}{|x|} f^2 dx \quad .$$

Here Ω is a domain in $\mathbb{R}^3$, generally but not always bounded, and f is a real valued function defined in Ω.

The simplest is $\Omega = \mathbb{R}^3$. The inequality is derived as follows. For any $\alpha > 0$

$$\frac{1}{|x|} \le \frac{\alpha}{4|x|^2} + \frac{1}{\alpha} \quad .$$

If f is of class C^1, then

$$|\text{grad } f|^2 = \frac{1}{|x|}|\text{grad}(|x|^{\frac{1}{2}}f)|^2 + \frac{1}{4|x|^2} f^2 - \frac{1}{2} \text{div}(\frac{x}{|x|^2} f^2) \quad .$$

If, in addition,

$$f(x) = O(|x|^{-\frac{1}{2}}) \text{ (as } x \to \infty) \quad ,$$

then by integrating the above over $\mathbb{R}^3$ we get

$$C \le \frac{\alpha}{4} \int \frac{1}{|x|^2} f^2 dx + \frac{1}{\alpha} N < \alpha\int |\text{grad } f|^2 dx + \frac{1}{\alpha} N \ \alpha T + \frac{1}{\alpha} N \quad .$$

The best choice of α gives

$$C < 2N^{\frac{1}{2}}T^{\frac{1}{2}} \quad .$$

The coefficient 2 is not best possible, but the simplicity of the proof is attractive [6].

We introduce the following terminology. An NTC-inequality is said to hold over a domain $\Omega \subseteq \mathbb{R}^3$ if two positive constants a and b exist such that

$$C \le aN^{\frac{1}{2}}T^{\frac{1}{2}} + bN$$

for all continuous f with piecewise continuous gradient. An NTC-inequality holds uniformly over a family $\{\Omega\}$ of domains if a and b can be chosen independently of Ω in the family.

<u>Theorem 2</u>

Let $\{\Omega\}$ *be the family of all bounded domains such that* $\Omega \supseteq \Omega_\ell = \{x: |x| < \ell\}$. *Then an* NTC-*inequality holds uniformly over this family with* $a = 1$ *and* $b = \frac{3}{2\ell}$. *The constants are best possible.*

We show first that the constants are best possible. Indeed, let $\Omega = \Omega_\ell$ and choose $f(x) = 1$ identically for $x \in \Omega$. Then $N = \frac{4\pi\ell^3}{3}$, $T = 0$, $C = 2\pi\ell^2$ which shows that equality occurs with $b = \frac{3}{2\ell}$. Again, let $\Omega = \Omega_\ell$ and choose $f(x) = e^{-\lambda|x|}$ for some $\lambda > 0$. One then finds, independently of b,

$$\lim_{\lambda\to\infty} \frac{C - bN}{N^{\frac{1}{2}}T^{\frac{1}{2}}} = 1 \quad ,$$

and this shows that no NTC-inequality can hold with any $a < 1$.

Let $\alpha < 0$ be given, and consider the problem of finding the maximum of

$$\int_0^\ell f^2(r)r\,dr - \frac{\alpha}{2}\int_0^\ell f'^2(r)r^2 dr$$

for continuous f on $0 \le r \le \ell$, with piecewise continuous derivative f', such that

$$\int_0^\ell f^2(r)r^2 dr = 1 \quad .$$

This is a classical variational problem. The maximum is the largest eigenvalue λ of the system (Sturm-Liouville problem)

$$\begin{cases} \frac{\alpha}{2}(r^2 f')' + rf = \lambda r^2 f \quad (0 < r < \ell) \\ f'(\ell) = 0 \\ f'(r)r^2 \to 0 \quad \text{as } r \to 0 \quad . \end{cases}$$

The eigenfunction f, belonging to the maximal eigenvalue λ, has no zeros. So if we write $g = f'/f$, we have

$$\begin{cases} \frac{\alpha}{2}(r^2 g)' + \frac{\alpha}{2} r^2 g^2 + r = \lambda r^2 \\ g(\ell) = 0 \\ g(r)r^2 \to 0 \quad \text{as} \quad r \to 0 \end{cases} .$$

Integrating the differential equation between 0 and ℓ we get then

$$\frac{\alpha}{2} \int_0^{\ell} g^2 r^2 dr + \frac{\ell^2}{2} = \frac{\lambda \ell^3}{3} . \tag{$\dagger$}$$

On the other hand, dividing the equation by r^2 and differentiating once more,

$$\frac{\alpha}{2}(g'' + 2gg') + \frac{\alpha}{r} g' = \frac{\alpha g + 1}{r^2} .$$

At points in $0 < r < \ell$ where $g' = 0$ we have

$$\frac{\alpha r^2}{2} g'' = \alpha g + 1 .$$

Thus a maximum cannot occur and so $g(r) \le 0$. At a minimum $g \ge -\frac{1}{\alpha}$.
Thus we have $g^2 \le \frac{1}{\alpha^2}$ and therefore ($\dagger$) shows

$$\frac{\lambda \ell^3}{3} \le \frac{\ell^2}{2} + \frac{\ell^3}{6\alpha}$$

or

$$\lambda \le \frac{3}{2\ell} + \frac{1}{2\alpha}$$

We have now shown that

$$\int_0^{\ell} f^2 r dr - \frac{\alpha}{2}\int_0^{\ell} f'^2 r^2 dr \le \left(\frac{3}{2\ell} + \frac{1}{2\alpha}\right)\int_0^{\ell} f^2 r^2 dr .$$

Remark that for $f = f(x) = f(r,\theta,\varphi,)$ with $r = |x|,\theta,\varphi$ spherical polar coordinates of x, we have $|\text{grad } f|^2 \ge (\frac{\partial f}{\partial r})^2$. Therefore our inequality yields, upon integration with respect to the angles,

$$\int_{\Omega_\ell} f^2 \frac{1}{|x|} dx - \frac{\alpha}{2}\int_{\Omega_\ell} |\text{grad } f|^2 dx \le \left(\frac{3}{2\ell} + \frac{1}{2\alpha}\right)\int_{\Omega_\ell} dx .$$

Using the best possible α this can be written

$$C \le N^{\frac{1}{2}} T^{\frac{1}{2}} + \frac{3}{2\ell} N .$$

This shows that the NTC-inequality holds for the sphere Ω_ℓ, with the best possible constants $a = 1$, $b = \frac{3}{2\ell}$.

It remains to show that it holds, for any other domain $\Omega \supseteq \Omega_\ell$ with the same constants. To see this, let C_ℓ, N_ℓ, T_ℓ denote the respective integrals extended over Ω_ℓ. Obviously, for any f

$$N_\ell \le N$$

$$T_\ell \le T$$

$$C - C_\ell \le \frac{1}{\ell}(N - N_\ell) \quad .$$

Therefore

$$C \le C_\ell + \frac{1}{\ell}(N - N_\ell) \le N_\ell^{\frac12} T_\ell^{\frac12} + \frac{3}{2\ell} N_\ell + \frac{1}{\ell}(N - N_\ell)$$

$$= N_\ell^{\frac12} T_\ell^{\frac12} + \frac{1}{2\ell} N_\ell + \frac{1}{\ell} N \le N^{\frac12} T^{\frac12} + \frac{3}{2\ell} N \quad .$$

This completes the proof of the theorem.

It should not be thought that an NTC-inequality must hold for any domain. Here is a simple counter-example. Let $\rho = (x_2^2 + x_3^2)^{\frac12}$, $\zeta = x_3$, and φ be cylindrical coordinates of $x = (x_1, x_2, x_3)$. Let $\Omega = \{x\colon 0 < \zeta < 1,\ \rho < \zeta^\alpha\}$ where $\alpha > \frac{3}{2}$. If f depends on ζ only, one finds

$$N = \pi \int_0^1 f^2 \zeta^{2\alpha} d\zeta$$

$$T = \pi \int_0^1 f'^2 \zeta^{2\alpha} d\zeta$$

and

$$C = 2\pi \int_0^1 d\zeta f^2 \int_0^{\zeta^\alpha} \frac{\rho d\rho}{\sqrt{\rho^2 + \zeta^2}} = 2\pi \int_0^1 f^2 \sqrt{\zeta^{2\alpha} + \zeta^2}\, d\zeta > 2\pi \int_0^1 f^2 \zeta d\zeta \quad .$$

Suppose that

$$f(\zeta) = \zeta^{\epsilon - 1} \qquad (\epsilon > 0) \quad .$$

Then

$$N = \pi (2\alpha + 2\epsilon - 1)^{-1}$$

$$T = \pi (1 - \epsilon)^2 (2\alpha + 2\epsilon - 3)^{-1}$$

and

$$C > 2\pi (2\epsilon)^{-1} \quad .$$

Thus N and T remain bounded, whereas $C \to \infty$ as $\epsilon \to 0$, and so there is no NTC-inequality for this Ω.

Evidently the cause of the phenomenon is the sufficiently sharp cusp in the boundary of Ω at the origin. If, however, the boundary is more regular, though it passes through the origin, an NTC-inequality will hold. No general result of this sort is known to the writer, but in one useful case something can be proved.

Theorem 3

Let $\{\Omega\}$ *be the family of congruent cubes* Ω *whose edge length is* ℓ.

Then a uniform NTC-*inequality holds for this family with* $a = 8$ *and* $b = \frac{24}{\ell}$.

It is convenient to reformulate the problem in an equivalent way. Take Ω a fixed cube

$$\Omega = \{x: -\frac{\ell}{2} < x_1, x_2, x_3 < \frac{\ell}{2}\}$$

and define

$$C = C(y) = \int_\Omega \frac{f^2(x)\,dx}{|x - y|}$$

where the point y varies over $\mathbb{R}^3$. The uniformity statement now requires that a and b turn out independent of y.

To show this, we first remark that for any given f, $C(y)$ has a maximum for y varying in $\mathbb{R}^3 \setminus \Omega$ when y is some point on the boundary of Ω. (This follows from the fact that $C(y)$ is a harmonic function in $\mathbb{R}^3 \setminus \Omega$.) Thus we may restrict our considerations to y lying inside, or on the boundary of, Ω.

Now let Ω' be the concentric cube of twice the size of Ω

$$\Omega' = \{x: -\ell < x_1, x_2, x_3 < \ell\} \quad .$$

Extend the function f, defined over Ω, to a function f defined over Ω', by repeated reflections in the faces of Ω. If we now denote by N', T', C', the corresponding integrals over Ω', we have evidently

$$N' = 8N$$

$$T' = 8T$$

and

$$C' > C \quad .$$

On the other hand, Theorem 2 is applicable to the cube Ω', since y is at least a distance $\ell/2$ from the boundary of Ω'. Thus

$$C < C' \le N'^{\frac{1}{2}} T'^{\frac{1}{2}} + \frac{3}{2}\left(\frac{\ell}{2}\right)^{-1} N' = 8N^{\frac{1}{2}} T^{\frac{1}{2}} + \frac{24}{\ell} N \quad .$$

This proves the theorem.

The constants a and b are surely far from the best possible, though we have not proved this.

4. MORE INEQUALITIES

In this section we combine NTC-inequalities with other simple inequalities to obtain results that will be needed later.

Let f be a smooth function defined over a domain $\Omega \subseteq \mathbb{R}^3$ and assume that the Newtonian potential

$$\varphi(x) = \int_\Omega \frac{f(y)}{|y - x|}\,dy$$

exists and that its energy is finite

$$\int_{\mathbb{R}^3} |\operatorname{grad} \varphi(x)|^2 dx < \infty \quad .$$

The vector $\operatorname{grad} \varphi(x)$ is continuous across the boundary of Ω, and we may integrate by parts over $\mathbb{R}^3$ even though

$$\Delta\varphi(x) = \begin{cases} f(x) & x \in \Omega \\ 0 & \text{otherwise} \end{cases}$$

is possible discontinuous at the boundary of Ω.

Let g be a function over $\mathbb{R}^3$ with $\int |\operatorname{grad} g|^2 dx < \infty$. Then we have

$$4\pi(\int_\Omega fg dx)^2 = \frac{1}{4\pi}(g\Delta\varphi dx)^2 = \frac{1}{4\pi}(\int \operatorname{grad} g \cdot \operatorname{grad}\varphi dx)^2$$

$$\leq \frac{1}{4\pi}\int |\operatorname{grad} g|^2 dx \cdot \int |\operatorname{grad} \varphi|^2 dx \tag{1}$$

$$= \int |\operatorname{grad} g|^2 dx \iint_{\Omega\Omega} \frac{f(x)f(y)}{|x - y|} dxdy \quad .$$

Theorem 4

Let Ω be a domain such that for the family $\{\Omega\}$ consisting of all domains congruent to Ω a uniform NTC-inequality holds with constants a,b. Then for any f defined over Ω, g defined over $\mathbb{R}^3$, continuous with piecewise continuous gradient,

$$4\pi(\int f^2 g dx)^2 \leq (\int_{\mathbb{R}^3} |\operatorname{grad} g|^2 dx) \times [a(\int_\Omega \operatorname{grad} f^2 dx)^{\frac{1}{2}}(\int_\Omega f^2 dx)^{\frac{3}{2}} + b(\int f^2 dx)^2] \quad .$$

To prove this, we use Eq.(1) above with f^2 replacing f there. Then

$$\int_\Omega \frac{f^2(x)dx}{x - y} \leq a(\int_\Omega |\operatorname{grad} f|^2 dx)^{\frac{1}{2}}(\int_\Omega f^2 dx)^{\frac{1}{2}} + b\int_\Omega f^2 dx \quad ,$$

independently of y by assumption. This completes the proof.

In the remainder of this section we put Theorems 2 and 4 to use in obtaining certain estimates useful to our main problem.

Let Ω be a cube of edge length ℓ, and let $y_1, y_2, \ldots, y_m$ $(m \geq 2)$ be points inside, or on the boundary of, Ω. Let f be a continuous function over Ω with piecewise continuous gradient. As usual we write

$$N = \int_\Omega f^2 dx$$

$$T = \int_\Omega |\operatorname{grad} f|^2 dx \quad .$$

Instead of C we consider now the more complicated sum

$$W = \sum_{j=1}^m C(y_j) = \sum_{j=1}^m \int \frac{f^2}{|x - y_j|} dx \quad .$$

We shall obtain an estimate of W in terms of N,T and

$$w = \sum\sum_{1\leq i<j\leq m} \frac{1}{|y_i - y_j|} \quad .$$

Let first f be extended into a cube Ω', concentric with Ω and twice the linear size, by repeated reflections in the faces of Ω. Obviously $W < W'$, $8N = N'$, $8T = T'$ where the dash denotes the corresponding integrals over Ω'. Let now $r_1, r_2, \ldots, r_m > 0$ be such that the spheres $S_j = \{x: |x - y_j| < r_j\}$ do not intersect but are all $\subset \Omega'$. This is accomplished by choosing r_j to be the smallest positive number among

$$\frac{1}{2}|y_j - y_1|,\ \frac{1}{2}|y_j - y_2|,\ \ldots,\ \frac{1}{2}|y_j - y_m| \text{ and } \frac{\ell}{2}\ .$$

We note

$$\sum_{j=1}^{m} \frac{1}{2r_j} \le 2\sqrt{3}\, w < 4w\ .$$

Define

$$g(x) = \sum_{j=1}^{m} \operatorname{Min}\left(\frac{1}{|x - y_j|}, \frac{1}{r_j}\right)$$

and

$$h(x) = \begin{cases} \dfrac{1}{|x - y_j|} - \dfrac{1}{r_j} & (x \in S_j) \\ 0 & (\text{otherwise}) \end{cases}$$

We have then

$$w' = \int_{\Omega'} f^2 (g + h)\,dx\ .$$

Since for $x \in S_j$ we have $g(x) > \frac{1}{r_j}$,

$$\int_{\Omega'} f^2 h\,dx = \sum_{j=1}^{m} \int_{S_j} f^2 \left(\frac{1}{|x - y_j|} - \frac{1}{r_j}\right) dx \le \sum_{j=1}^{m} \int_{S_j} \left(\frac{\alpha}{2}|\operatorname{grad} f|^2 + \frac{1}{2\alpha} f^2 + \frac{1}{2r_j} f^2\right) dx$$

$$\le \frac{\alpha}{2} T' + \frac{1}{2\alpha} N' + \frac{1}{2}\int_{\Omega'} f^2 g\,dx\ .$$

Therefore

$$W' \le T'^{\frac{1}{2}} N'^{\frac{1}{2}} + \frac{3}{2}\int_{\Omega} f^2 g\,dx\ .$$

The last integral is estimated by Theorem 4. Here we need an elementary computation of potential theory

$$\frac{1}{8\pi}\int |\operatorname{grad} g|^2 dx = w + \sum_{j=1}^{m} \frac{1}{2r_j}$$

which, as we have seen, is less than $5w$. Thus

$$W' < T'^{\frac{1}{2}} N'^{\frac{1}{2}} + \frac{3}{2}(10w)^{\frac{1}{2}} \left(a_1 T'^{\frac{1}{2}} N'^{\frac{3}{2}} + \frac{b_1}{2\ell} N'^2\right)^{\frac{1}{2}}$$

where a_1 and b_1 are the constants which occur in the NTC-inequality valid uniformly for cubes of unit edge length (Theorem 3).

Our final result is

$$W < 8T^{\frac{1}{2}}N^{\frac{1}{2}} + 12(10w)^{\frac{1}{2}}(a_1 T^{\frac{1}{2}}N^{\frac{3}{2}} + \frac{b_1}{2\ell} N^2)^{\frac{1}{2}} \quad .$$

The significant feature of this inequality is the fact that the bound depends only on the *square root* of T and w.

5. THE ELECTROSTATIC INEQUALITY

A basic tool is an inequality whose source is closely related to an ancient fact know to Isaac Newton: that non-overlapping spherically symmetric charge distributions exert forces on each other as if their total charge was concentrated in their center.

Let $S_1, S_2, \ldots, S_n$ be spheres with respective centers $x_1, x_2, \ldots, x_n \in \mathbb{R}^3$ and radii $r_1, r_2, \ldots, r_n$. Suppose there is a uniform surface layer of change on each S_i; let its surface density be $e_i/(4\pi r_i^2)$, where the e_i may have either sign. This produces a continuous Newtonian potential in $\mathbb{R}^3$.

$$\phi(x) = \sum_{i=1}^{n} \frac{e_i}{4\pi r_i^2} \int_{S_i} \frac{d\sigma_y}{|y - x|} \quad ,$$

where $d\sigma$ is the usual surface area element. The gradient of ϕ is discontinuous across the S_i, but the total electrostatic energy is finite

$$\frac{1}{8\pi} \int_{\mathbb{R}^3} |\mathrm{grad}\, \phi(x)|^2 dx = \frac{1}{2} \sum_{i=1}^{n} \sum_{j=1}^{n} \frac{e_i}{4r_i^2} \frac{e_j}{4r_j^2} \int_{S_i} d\sigma_x \int_{S_j} d\sigma_y \frac{1}{|x - y|} < \infty \quad .$$

If two spheres S_i, S_j do not overlap then simply

$$\int_{S_i} d\sigma_x \int_{S_j} d\sigma_y \frac{1}{|x - y|} = \frac{(4\pi r_i^2)(4\pi r_j^2)}{|x_i - x_j|}$$

If they do overlap the expression is more complicated and we write

$$\int_{S_i} d\sigma_x \int_{S_j} d\sigma_y \frac{1}{|x - y|} = (4\pi r_i^2)(4\pi r_j^2)\left(\frac{1}{|x_i - x_j|} - \Delta(|x_i - x_j|, r_i, r_j)\right) \quad .$$

The function Δ is found explicitly to be

$$\Delta(s,a,b) = \begin{cases} \frac{1}{s} - \mathrm{Min}(\frac{1}{a},\frac{1}{b}) & (0 < s \le |a - b|) \\ \frac{(a + b - s)^2}{4abs} & (|a - b| \le s \le a + b) \\ 0 & (a + b \le s) \quad . \end{cases}$$

It is a monotone decreasing, convex, function of s in the interval $0 < s \le a + b$, beyond which it vanishes. It satisfies the inequalities

$$\frac{1}{s} - \frac{1}{a} < \Delta(s,a,a) < \frac{1}{s} \quad .$$

We note that

$$\lim_{s\to 0}\left(\frac{1}{s} - \Delta(s,a,a)\right) = \frac{1}{a} \quad .$$

We are lead to an expression for the total electrostatic energy in the form

$$\frac{1}{8\pi}\int |\text{grad } \phi|^2 dx = \sum_{i=1}^{n} \frac{e_i^2}{2r_i} + \sum\sum_{1\le i<j\le n} \frac{e_i e_j}{|x_i - x_j|} - \sum\sum_{1\le i<j\le n} e_i e_j \Delta(|x_i - x_j|, r_i, r_j) \quad .$$

Theorem 5

$e_1, \ldots, e_n$ *being arbitrary real and* $r_1, \ldots, r_n$ *arbitrary positive numbers,* $x_1, \ldots, x_n$ *arbitrary points in* $\mathbb{R}^3$, *we have*

$$\sum\sum_{1\le i<j\le n} \frac{e_i e_j}{|x_i - x_j|} > -\sum_{1\le i\le n} \frac{e_i^2}{2r_i} + \sum\sum_{1\le i<j\le n} e_i e_j \Delta(|x_i - x_j|, r_i, r_j) \quad .$$

Proof. The difference of the two sides is just the integral $\frac{1}{8\pi}\int |\text{grad } \phi|^2 dx$.

6. CUBICAL PARTITION OF SPACE

In this section we perform a purely geometrical construction in $\mathbb{R}^3$. Its essence is the partitioning of almost all (in the sense of Lebesgue measure) of $\mathbb{R}^{3n}$ into a disjoint union of sets U of the form

$$U = C_1 \times C_2 \times \ldots \times C_n,$$

where each C_i is a cube in $\mathbb{R}^3$ and their distances are closely related to their sizes.

In $\mathbb{R}^3$ we consider cubes of the form

$$C = \{x = (x_1, x_2, x_3,): a_i \le x_i < a_i + L \quad (i = 1,2,3)\},$$

where $L = 2^\nu$ for some integer ν and a_i are integral multiples of L ("binary cubes"). Binary cubes are either disjoint or one is contained in another. Those of a given size form a partition of $\mathbb{R}^3$.

The same notions exist in $\mathbb{R}^6$. Let C be a binary cube in $\mathbb{R}^6$. We say that C is *separated* if

$$\alpha L \le |x - y|$$

for every $(x,y) \in C$, where $x,y \in \mathbb{R}^3$, L is the edge length of C, and α is a positive constant fixed beforehand. If $x \ne y$ in $\mathbb{R}^3$ then evidently there is a *maximal separated* binary cube C of $\mathbb{R}^6$ which contains (x,y). Thus the subset

$$\{(x,y): x,y \in \mathbb{R}^3, x \ne y\} \subset \mathbb{R}^6$$

is partitioned into maximal separated binary cubes. Let $x \neq y \in \mathbb{R}^3$ and let C be the corresponding maximal separated binary cube of $\mathbb{R}^6$. Let $C' \supset C$ be the next larger binary cube. Since it is not separated it contains (x',y') such that

$$|x' - y'| < 2\alpha L \quad .$$

on the other hand

$$|x - x'|, \ |y - y'| < 2\sqrt{3}\, L \quad ,$$

and so by the triangle inequality

$$|x - y| < (2\alpha + 4\sqrt{3})L = \alpha' L \quad .$$

Now we consider $\mathbb{R}^{3n}$, or rather the subset

$$\{\xi = (x_1,x_2,\ldots,x_n): x_i \neq x_j \quad \text{for} \quad 1 \leq i < j \leq n\} \quad .$$

Associated with any pair (i,j), $1 \leq i < j \leq n$, we have a partition P_{ij} of this set into disjoint subsets. A particular subset belonging to P_{ij} is defined by

$$\{\xi: x_k \in \mathbb{R}^3 \quad \text{for} \quad k \neq i, j; \ (x_i,x_j) \in C\} \quad ,$$

where C is a maximal separated binary cube in $\mathbb{R}^6$ constructed above. Let P be the coarsest common refinement of the $P_{ij} (1 \leq i < j \leq n)$.

A typical set U belonging in the partition P has the form $U = C_1 \times C_2 \times \ldots \times C_n$ where $C_1, \ldots, C_n$ are binary cubes in $\mathbb{R}^3$. Let $\xi = (x_1,x_2,\ldots,x_n) \in \mathbb{R}^{3n}$ be such that $x_i \neq x_j$ for $1 \leq i < j \leq n$, and let L_{ij} be the edge length of the binary cube in $\mathbb{R}^6$ which contains (x_i,x_j) and which belongs to the partition P_{ij}. As we have seen

$$\alpha L_{ij} \leq |x_i - x_j| < \alpha' L_{ij}$$

with $\alpha' = 2\alpha + 4\sqrt{3}$. Now

$$L_i = \operatorname*{Min}_{j \neq i} L_{ij} \quad .$$

Let

$$R_i(\xi) = \operatorname*{Min}_{j \neq i} |x_i - x_j| \quad .$$

Then

$$\begin{cases} \alpha L_i \leq R_i(\xi) < \alpha' L_i \\ i = 1, 2, \ldots, n \\ \xi \in U \end{cases} \qquad (*)$$

We have thus proved the following.

<u>Theorem 6</u>

The subset $(\mathbb{R}^{3n})_0 \subset \mathbb{R}^{3n}$ *of those* $\xi = (x_1,\ldots,x_n)$ *for which* $x_i \neq x_j$ *for* $1 \leq i < j \leq n$, *can be partitioned into sets of the form* $U = C_1 \times C_2 \times \ldots \times C_n$, *where* C_i *is a cube in* $\mathbb{R}^3$ *of edge length* L_i *and where* (*) *holds.*

Note that $\mathbb{R}^{3n} \setminus (\mathbb{R}^{3n})_0$ has measure zero and therefore integrals over $\mathbb{R}^{3n}$ can be written

$$\int dx_1 \ldots \int dx_n \ldots = \sum_{\mathcal{U}} \int_{C_1} dx_1 \ldots \int_{C_n} dx_n \ldots$$

where the sum runs over all $\mathcal{U}$ of the partition $\mathcal{P}$.

7. LOWER BOUND FOR THE ENERGY IN TERMS OF K_1

Suppose f is a smooth function over $\mathbb{R}^{3n}$, in the domain of essential self-adjointness of the Hamiltonian H (cf. Section 2). We want a lower bound for

$$E = \frac{\hbar^2}{2M} T + e^2 W$$

where

$$T = \int dx^{3n} \sum_{i=1} |\text{grad}_i f|^2$$

and

$$W = \int dx^{3n} f^2 \sum_{1 \le i < j \le n+m} \frac{Z_i Z_j}{|x_i - x_j|} .$$

Here we have written $Z_1 = \ldots = Z_n = -1$, $Z_{n+1} = \ldots = Z_{n+m} = Z$; and $x_{n+1}, \ldots, x_{n+m}$ are the fixed points of $\mathbb{R}^3$ at which, the positive charges Ze are located. Write

$$E = \sum_{\mathcal{U}} E(\mathcal{U}) = \frac{\hbar^2}{2M} \sum_{\mathcal{U}} T(\mathcal{U}) + e^2 \sum_{\mathcal{U}} W(\mathcal{U}) ,$$

where $\mathcal{U} = C_1 \times C_2 \times \ldots \times C_n$ runs over the sets of the partitions $\mathcal{P}$ of Theorem 6.

$$T(\mathcal{U}) = \int_{C_1} d^3x_1 \ldots \int_{C_n} d^3x_n \sum_{i=1} |\text{grad}_i f|^2 ,$$

and $W(\mathcal{U})$ is defined similarly.

With any configuration $(x_{n+1}, \ldots, x_{n+m})$ of the positive charges there exists another one, with m possibly diminished, and such that all $x_j (n + 1 \le j \le m)$ lie in or on the boundaries of the C_i, and such that $W(\mathcal{U})$ does not exceed the value for the first configuration. The reason is that for any fixed $\mathcal{U}$ the quantity $W(\mathcal{U})$, viewed as a function of an $x_j (n + 1 \le j \le m)$, is a harmonic function (a Newtonian potential) in the domain $\{x_j \in \mathbb{R}^3 : x_j \ne x_k \text{ for } k \ne j \text{ and } x_j \notin \overline{C}_1, \ldots, \overline{C}_n\}$ where $\overline{C}_i$ denotes the closure of C_i. A harmonic function is either not minimized (in this case it decreases monotonely as $x_j \to \infty$ in some manner), or tends to a minimum as x_j tends for some point on the boundary.

We assume then that all positive charges lie in the $\overline{C}_i$.

According to the electrostatic inequality

$$W(\mathcal{U}) > \int_{\mathcal{U}} d^{3n}x f^2 [- \sum_{1 \le i \le n+m} \frac{Z_i^2}{2r_i} + \sum_{1 \le i < j \le n+m} Z_i Z_j \Delta(|x_i - x_j|, r_i, r_j)] . \tag{2}$$

Here the r_i are as yet arbitrary positive numbers. We now make the following choices

$$\alpha = 8\sqrt{3}$$

and

$$r_i = 4\sqrt{3}\, L_i$$

where L_i is the edge-length of C_i. Then for $x_i \in C_i$, $x_j \in C_j$ we have $|x_i - x_j| \geq \alpha L_i, \alpha L_j$ by Theorem 6. Therefore

$$|x_i - x_j| \geq \frac{\alpha}{2}(L_i + L_j) = r_i + r_j \quad .$$

We recall the properties of the function $\Delta = \Delta(s,a,b)$, in particular that it vanishes for $s \geq a + b$. Thus all terms in the double sum of Eq.(2) which have x_i, x_j in different cubes vanish.

One is therefore led to examine the quantity

$$\frac{\hbar^2}{2M} T - \frac{e^2}{2r}(1 + mZ^2)N + e^2\int_C d^3x f^2[-Z\sum_{1\leq i\leq m}\Delta(|x - y_i|,r,r) + Z^2\sum\sum_{1\leq i<j\leq m}\Delta(|y_i - y_j|,r,r)] \; .$$

Here C is a cube in $\mathbb{R}^3$, the y_i are in $\overline{C}$, $r = 4\sqrt{3}\, L$ and L is the edge length of $\overline{C}$. Making use of

$$\frac{1}{s} - \frac{1}{r} < \Delta(s,r,r) < \frac{1}{s} \quad ,$$

we see that the above quantity is larger than

$$\frac{\hbar^2}{2M} T - Ze^2 W + Z^2e^2 wN - \frac{e^2}{2r}(1 + m^2Z^2)N \quad . \tag{3}$$

Here the letters T, N, W, and w are used in the same sense as in the second half of Section 2. For $m \geq 2$

$$\frac{1}{2}\, Z^2e^2 w \geq \frac{1}{2}\, Z^2e^2 \frac{1}{\sqrt{3}L}\, \frac{m(m-1)}{2} = \frac{Z^2e^2}{r}\, m(m-1) \geq \frac{Z^2e^2m^2}{2r} \quad .$$

Therefore Eq.(3) exceeds

$$\frac{\hbar^2}{2M} T - Ze^2 W + \frac{1}{2}\, Z^2e^2 wN - \frac{e^2}{2r} N$$

At this point make use of the upper bound on W derived in Section 4. The crucial fact is that the expression so obtained is *bounded below as a function of* w *and* T. One finds explicitly the best lower bound

$$-A_1(Z)N\, \epsilon_0 - A_2 N \frac{e^2}{L}$$

where

$$\epsilon_0 = \frac{Me^4}{2\hbar^2_0} \quad ,$$

$$A_1(Z) = (8Z + 720a_1)^2 \quad ,$$

$$A_2 = \frac{1}{4\sqrt{3}} + 360b_1 \quad .$$

The numerical constants a_1 and b_1 come in through the NTC-inequality for unit cubes, see Section 4.

We may summarize our results in terms of the notation at the beginning of this section as follows. Let $\mathcal{U}$ be any set in the partition $\mathcal{P}$. Then

$$E(\mathcal{U}) > - A_1(Z)n\,\epsilon_0\,N(\mathcal{U}) - A_2 e^2 N(\mathcal{U}) \sum_{i=1}^{n} \frac{1}{L_i} ,$$

where L_i are the edge lengths of the C_i defining $\mathcal{U}$, and $N(\mathcal{U}) = \int_{\mathcal{U}} d^{3n}x f^2$.

This result comes close to proving Theorem 1, in particular it incorporates correctly the dependence on the Coulomb potential energy. On the other hand, no use was made of any symmetry property of the wavefunction f. That will have to be done before the second term in our inequality can also be brought into the desired form.

For any $\xi = (x_1, x_2, \ldots, x_n) \in \mathbb{R}^{3n}$ define

$$R_i(\xi) = \operatorname*{Min}_{j \neq i} |x_i - x_j| \quad .$$

Let

$$K_1 = \sum_{i=1}^{n} \int_{R^{3n}} \frac{f^2(\xi)}{R_i(\xi)}\, d^{3n}x \quad .$$

Note that for $\xi \in \mathcal{U}$ one has $R_j(\xi) < \alpha' L_i$ where $\alpha' = 2\alpha + 4\sqrt{3} = 20\sqrt{3}$. Therefore, our inequality yields upon summing over $\mathcal{U}$

$$E > -A_1(Z)n\,\epsilon_0 - A_2 \alpha' e^2 K_1$$

for any f normalized according to

$$N = \|f\|^2 = \int_{\mathbb{R}^{3n}} f^2 d^{3n}x = 1 \quad .$$

8. AN INEQUALITY FOR K_1, K_2 AND T

Our last result shows the role of a quantity K_1 in obtaining a lower bound for the energy independently of the positions of the positive charges. It is now our task to obtain another inequality between T and K_1 by means of which it will be possible to eliminate K_1 altogether. Unfortunately, there is a small technical difficulty which makes the proof somewhat longer at this point than would be the case if electrons had no spin. For in that case $f = f(x_1, x_2, \ldots, x_n)$ would be a function totally antisymmetric with respect to interchange of any pairs of variables. As it is (cf. discussion in Section 2) we must regard f antisymmetric with respect to interchange of two variables only if they are in the same group, say among $x_1, x_2, \ldots, x_\nu$ or among $x_{\nu+1}, \ldots, x_n$, where ν is some integer, $0 \leq \nu \leq n$. This more limited symmetry of f leads to a detour where it is first necessary to derive an inequality between T, K_1 and an analogous quantity K_2 (not using the symmetry of f), then afterward deriving an inequality between K_2 and T making use of the symmetry.

In this section it is then not assumed that f has any symmetry.

We have written $R_i(\xi)$ for the smallest among $|x_j - x_i|(j \neq i)$. We denote the next smallest by $R_i'(\xi)$. Analogously to the definition of K_1, we define

$$K_2 = \int d^{3n}x f^2(\xi) \sum_{i=1}^{n} \frac{1}{R_i'(\xi)} \quad .$$

Evidently $K_1 > K_2$. We now show that in fact $K_1 > \frac{3}{2}K_2$, and that the difference is

$$K_1 - \frac{3}{2} K_2 = O(n^{\frac{1}{2}}T^{\frac{1}{2}}) \quad ,$$

where

$$T = \int d^{3n}x \sum_{i=1}^{n} |\mathrm{grad}_i f|^2 \quad .$$

This estimate is proved as follows

$$K_1 = \sum_{i,j,k} \int d^{3n}x \frac{f^2}{|x_i - x_j|} \quad ,$$

where the sum is over triplets (i,j,k), $1 \leq i, j, k \leq n$, $(i,j,k$ different), and the integration over the domain

$$\{\xi: |x_i - x_j| < |x_i - x_k| < |x_i - x_s| \quad \text{for} \quad s \neq i,j,k\}$$

In this domain $R_i = |x_i - x_j|$, $R_i' = |x_i - x_k|$. Doing the integration over x_j first, we use the NTC-inequality (Theorem 2) for the sphere

$$x_j: |x_j - x_i| < |x_k - x_i|\}$$

Thus for any $\alpha > 0$

$$K_1 \leq \sum_{i,j,k} \int d^{3n}x \left(\frac{3}{2}\frac{f^2}{|x_i - x_k|} + \frac{\alpha}{2}|\mathrm{grad}_j f|^2 + \frac{f^2}{2\alpha}\right.$$

$$= \frac{3}{2} K_2 + \frac{n}{2\alpha} + \frac{\alpha}{2} \sum_{i=1}^{n} \int_{\mathbb{R}^{3n}} d^{3n}x \sum_{j \in J_i} |\mathrm{grad}_j f|^2 \quad ,$$

where $J_i = J_i(\xi)$ is the subset of $\{1,2,\ldots,n\}$ defined by

$$J_i = \{j: |x_j - x_i| = R_i\} \quad ,$$

(i.e., x_j is nearest neighbor of x_i). Now, it is geometrically clear that there cannot be more than some fixed number p of such points x_i to whom the same x_j is nearest neighbor (among the whole set $x_1,\ldots,x_n$). Thus

$$K_1 \leq \frac{3}{2} K_2 + \frac{n}{2\alpha} + \frac{p\alpha}{2} T \quad ,$$

and by choosing the best α

$$K_1 \leq \frac{3}{2} K_2 + (pnT)^{\frac{1}{2}} \quad .$$

An estimate of p is obtained this way. In a triangle the largest side is opposite the largest angle. Therefore if in the triangle ABC we have AB, AC < BC then the angle at A is $> \frac{\pi}{3}$. Thus in a circular cone with vertex at A and opening angle $\frac{\pi}{3}$ there cannot be two points B and C such that A is nearest neighbor to both B and C. Suppose now that $B_1, B_2, \ldots, B_\nu$ are such that A is the nearest neighbor of each (among $A, B_1, B_2, \ldots, B_\nu$). Let C_ω be the circular cone at opening $\frac{\pi}{3}$, vertex A, and axis ω. It contains $\mu(\omega) \leq 1$ among the points B_1, $\ldots$, B_ν. Obviously,

$$\int \mu(\omega) d\omega = \nu\Omega \quad ,$$

where $d\omega$ is the solid angle element swept out by the axis of C_ω and Ω is the solid angle in the interior of C_ω. One has $\Omega = 2\pi(1 - \cos \frac{\pi}{6}) = \pi(2 - \sqrt{3})$. Therefore,

$$\nu \leq \frac{4\pi}{\Omega} = 4(2 + \sqrt{3}) < 15 \quad .$$

Thus we may take $p^{\frac{1}{2}} = 4$.

9. ANTISYMMETRIC FUNCTIONS

Theorem 7

Let f *be a function with continuous second derivatives in the open ball* $\Omega_R = \{x \in \mathbb{R}^3 : |x| < R\}$ *and continuous in the closed ball* $\overline{\Omega}_R$, *and let* $\int_{\Omega_R} f dx = 0$. *Then*

$$\int_{\Omega_R} |\text{grad } f|^2 dx > \frac{4}{R^2} \int_{\Omega_R} f^2 dx$$

Proof. The problem is to minimize the integral

$$\int_{\Omega_R} |\text{grad } f|^2 dx \quad ,$$

subject to

$$\int_{\Omega_R} f^2 dx = 1 \quad ,$$

and

$$\int_{\Omega_R} f dx = 0 \quad .$$

This is an elementary variational problem whose solution can be explicitly determined. The value of the minimum is the smallest *non-vanishing* eigenvalue λ belonging to system

$$\begin{cases} -\Delta f(x) = \lambda f(x) & (x \in \Omega_R) \\ \dfrac{\partial f(x)}{\partial n} = 0 & (x \in \partial\Omega_R) \end{cases} \quad .$$

One finds $\lambda > 4$. This proves the theorem.

A corrolary of this theorem is the following result.

Theorem 8

Let $f(x_1,\ldots,x_n)$ *be antisymmetrical, and smooth on* $\Omega_R^n = \Omega_R \times \ldots \times \Omega_R$. *Then*

$$\int_{\Omega_R^n} \sum_{i=1}^n |\text{grad}_i f|^2 d^{3n}x \geq \frac{4(n-1)}{R^2}\int_{\Omega_R^n} f^2 d^{3n}x \quad .$$

Proof. For $i = 1, 2, \ldots, n$ set

$$g_i = \int_{\Omega_R} f d^3x_i \quad .$$

Then g_i is independent of x_i and antisymmetric in the remaining variables. Write

$$f = \frac{1}{v}(g_1 - g_2 + g_3 - \ldots + (-1)^{n-1}g_n) + h$$

where v is the volume of Ω_R. Then h is antisymmetric too. Moreover,

$$\int_{\Omega_R} g_i d^3x_j = 0 \quad , \qquad (j \neq i)$$

and therefore

$$\int_{\Omega_R} h d^3x_i = 0 \quad . \qquad (\text{all } i)$$

It follows that $g_1, g_2, \ldots, g_n$ and h are orthogonal on Ω_R^n and so

$$\int_{\Omega_R^n} f^2 d^{3n}x = \int_{\Omega_R^n} (h^2 + \frac{1}{v^2}\sum_{i=1}^n g_i^2) d^{3n}x \quad .$$

One verifies easily that these same functions are also orthogonal with respect to the inner product

$$((u,v)) = \int_{\Omega_R^n} \sum_{i=1}^n \text{grad}_i u \cdot \text{grad}_i v \quad .$$

Therefore

$$\int_{\Omega_R^n} \sum_{i=1}^n |\text{grad}_i f|^2 d^{3n}x = ((f,f)) = ((h,h)) + \frac{1}{v^2}\sum_{i=1}^n ((g_i,g_i)) \quad .$$

It follows from Theorem 7 that

$$((h,h)) \geq \frac{4n}{R^2}\int_{\Omega_R^n} h^2 d^{3n}x \quad ,$$

and

$$((g_i,g_i)) \geq \frac{4(n-1)}{R^2}\int_{\Omega_R^n} g_i^2 d^{3n}x \quad ,$$

the factor $n - 1$ arising because g_i depends only on $n - 1$ variables. Putting things together, the proof is thus completed.

We note that for large n Theorem 8 is a very poor inequality. For this reason it is remarkable that the antisymmetry of the wave-function is used in the proof of the main theorem only in this weak sense. For some further remarks see reference [3].

10. THE ROLE OF ANTISYMMETRY: AN INEQUALITY FOR T AND K_2

Let f be a normalized function over $\mathbb{R}^{3n}$, and T be defined as in Section 8. We assume that f is smooth enough so that $T < \infty$. We also assume that f is antisymmetric with respect to interchange of x_i with x_j, provided i and j belong to the same part of $\{1,2,\ldots,n\}$, the latter set having been partitioned in some manner into two (possibly empty) parts.

Let $I \subseteq \{1,2,\ldots,n\}$, and

$$T_I(y) = \int d^{3n}x \sum_{i \in I} |\mathrm{grad}_i f|^2$$

the integral being extended over the domain

$$\{(x_1,\ldots,x_n) \in \mathbb{R}^{3n}: |x_i - y| < r \text{ if } i \in I,\ |x_i - y| \geq r \text{ if } i \notin I\} \quad .$$

We can write then

$$T_I(y) = \int_{\mathbb{R}^{3n}} d^{3n}x \prod_{i \in I} \chi_y(x_i) \prod_{j \notin I} [1 - \chi_y(x_j)] \sum_{k \in I} |\mathrm{grad}_k f|^2$$

where

$$\chi_y(x) = \begin{cases} 1 & |x - y| < r \\ 0 & |x - y| \geq r \quad . \end{cases}$$

clearly

$$\sum_I T_I(y) = \int d^{3n}x \sum_{i=1}^{n} |\mathrm{grad}_i f|^2 \chi_y(x_i) \quad .$$

But

$$\int d^3y \chi_y(x) = \frac{4\pi r^3}{3} \quad .$$

Therefore

$$\frac{4\pi r^3}{3} T = \int d^3y \sum_I T_I(y) \geq \int d^3y \sum_{\#I \geq 3} T_I(y)$$

where $\#I$ is the cardinal number of I.

Consider now $T_I(y)$ with $\#I \geq 3$. At least 3 variables x_i are confined to the sphere $\Omega_r = \{x \in \mathbb{R}^3: |x - y| < r\}$, the rest are outside. Note that among those that are inside Ω_r, there are at least two with respect to whose interchange f is antisymmetric. Thus Theorem 8 is applicable. Let ν_1 and ν_2 denote the number of variables confined to Ω_r, belonging respectively to the two parts of $\{1,2,\ldots,n\}$. Clearly,

$$(\nu_1 - 1) + (\nu_2 - 1) = \nu_1 + \nu_2 - 2 + \#I - 2 \geq \frac{1}{3}(\#I) \quad .$$

Therefore Theorem 8 yields

$$T_I(y) \geq \frac{\#I}{r^2}\int d^{3n}x f^2$$

where the integration is over the same domain as that for $T_I(y)$. We have thus shown

$$\frac{4\pi r^3}{3} T \geq \frac{1}{r^2} \sum_{\#I \geq 3} (\#I) \int dy \int d^{3n}x f^2 = \frac{1}{r^2}\int_{\mathbb{R}^{3n}} d^{3n}x f^2 \sum_{\nu=3}^{n} \nu\omega_\nu$$

where $\omega_\nu = \omega_\nu(x_1, x_2, \ldots, x_n)$ is the volume in $\mathbb{R}^3$ of the set

$$\{y\colon \#\{j\colon |y - x_j| < r\} = \nu\} \quad .$$

(# always stands for cardinal number of a set.) A little thought shows

$$\sum_{\nu=3}^{n} \nu\omega_\nu = \sum_{j=1}^{n} \eta_j \quad ,$$

where η_j is the volume in $\mathbb{R}^3$ of the set

$$\{y\colon |y - x_j| < r,\ \#\{k\colon |y - x_k| < r\} \geq 3\} \quad . \tag{4}$$

Now let R'_j be the second nearest neighbor distance of x_j among the other points $x_1, \ldots, x_n$. The set

$$\{y\colon |y - x_j| < r - R_j\}$$

is empty if $R'_j \geq r$, and if $R'_j < r$ it is a sphere of volume

$$\frac{4\pi(r - R'_j)^3}{3}$$

centered on x_j. If y is in this set and x_k, x_ℓ are the first and second nearest neighbors of x_j we have

$$|y - x_j| < r - R'_j < r$$

$$|y - x_k| \leq |y - x_j| + |x_j - x_k| < (r - R'_j) + R'_j = r$$

$$|y - x_\ell| < r \quad \text{(similarly)} \quad .$$

Thus y belongs to the set defined by Eq (4) above. Thus we have

$$\eta_j > \frac{4\pi(r - R'_j)^3}{3}$$

We have now shown

$$\frac{4\pi r^3}{3} T > \frac{1}{r^2}\int_{\mathbb{R}^{3n}} d^{3n}x f^2 \sum_{j=1}^{n} \frac{4\pi}{3}(r - R'_j)^3_+$$

where $(t^3)_+$ means t^3 if $t > 0$, and 0 if $t \leq 0$.

An elementary integration gives

$$\frac{1}{3}\int_0^s \frac{dr}{r^2}\left(1 - \frac{R}{r}\right)^3_+ = \frac{1}{4sR}\left(1 - \frac{R}{2}\right)^3\left(1 - \frac{R}{s}\right)_+ \geq \frac{1}{4sR} - \frac{1}{s^2} \quad .$$

It follows then that for any $s > 0$

$$T > \frac{K_2}{4s} - \frac{n}{s^2}$$

where K_2 is defined as in Section 8. The best choice of s yields

$$K_2 < 8(nT)^{\frac{1}{2}}$$

This is the desired inequality.

11. COMPLETION OF THE PROOF

We have obtained the inequalities

$$K_1 < \frac{3}{2} K_2 + 4(nT)^{\frac{1}{2}}$$

and

$$K_2 < 8(nT)^{\frac{1}{2}}$$

in Sections 8 and 10 respectively. They imply

$$K_1 < 16(nT)^{\frac{1}{2}} \quad .$$

Thus if A_3 is a constant, there is another constant A_4 such that

$$\frac{\hbar^2}{2M} T - A_3 e^2 K_1 \geq - A_4 n \epsilon_0 \quad .$$

The last step is to unite this with the result obtained at the end of Section 7. To do this one must divide up the kinetic energy $\hbar^2 T/2M$ into two parts, say $\hbar^2 T/4M$ each, and use one part in Section 7, one in the last inequality written above. What this amounts to is an extra factor 2 that comes from replacing M by $2M$ in the definition of ϵ_0. Our final inequality is then

$$E > -A(Z) n \epsilon_0$$

where $A(Z)$ is a quadratic expression with universal numerical coefficients.

This completes the proof of Theorem 1.

If the good will of the attentive reader has not been abused by this long-winded argument, and if he is acquainted with references [1] and [2], he will appreciate one final comment. There is a subtle improvement in these lectures over [1] and [2] which is partly hidden by the notation there. The point is that in our previous work we conducted the argument in such a manner that it was necessary to assume a uniform bound for the absolute value of all charges *positive and negative*. Thus the result of our previous papers gives only a fourth-degree dependence on the parameter Z (see Section 2 of [2]). The present improvement results not in a radical change in method but rather is a more careful way the various arguments are put together.

The challenge to shorten the proof in a really essential way still stands.

REFERENCES

[1] Dyson, F.J., and Lenard, A., *Journ. Math. Phys.*, 8, 423 (1967).

[2] Lenard, A., and Dyson, F.J., *Journ. Math. Phys.*, 9, 698 (1968).

[3] *Statistical Physics, Phase Transitions and Superfluidity*, Brandeis University Summer Institute in Theoretical Physics 1966, Gordon and Breach Publishers, New York (1968).

[4] *Proceedings of the Fifth Annual Eastern Theoretical Physics Conference*, Brown University 1966, Benjamin Inc., New York (1967).

[5] Kato, T., *Trans. Am. Math. Soc.*, 70, 195 (1951).

[6] Courant, R., and Hilbert, D., *Methods of Mathematical Physics*, Interscience Publishers Inc., New York (1953), Vol I, Chapter VI, Section 5.

LECTURES ON THE THERMODYNAMIC LIMIT FOR COULOMB SYSTEMS

E. H. Lieb* and J. L. Lebowitz**

1. INTRODUCTION

These lecture notes present an outline of the proof of the existence of the thermodynamic limit for Coulomb systems. A brief statement of the main results has appeared previously, [9], and the full work will appear shortly, [11]. What we have tried to do in these notes is to present the ideas and methods used in constructing this proof while leaving out most of the details of the analysis. In some places, such as section 3, we treat only the simplest kind of Coulomb system: two species of charged particles (one positive and one negative) whose only interaction is through the Coulomb potential. In other places we simply state various lemmas and theorems without proof.

The basic pre-requirement for the existence of a thermodynamic limit for Coulomb systems is the Dyson-Lenard Theorem, [1], which gives a lower bound to the energy of a system of charged particles. It is therefore very fortunate that the proof of this theorem is presented in a particularly nice form, in Professor Lenard's lectures which are included in this volume.

Statement of the Problem

Statistical Mechanics as developed by Gibbs and others rests on the hypothesis that equilibrium properties of matter can be completely described in terms of a phase-space average, or partition function, $Z = \mathrm{Tr}\{\exp(-\beta H)\}$, with H the Hamiltonian and β the reciprocal temperature. It was realized early that there were grave difficulties in justifying this assumption in terms of basic microscopic dynamics. These questions, which involve the time evolution of macroscopic systems, have still not been satisfactorily resolved, but the great success of equilibrium statistical mechanics in offering qualitative and quantitative equilibrium explanations for such varied phenomena as superconductivity, specific heats of crystals, chemical equilibrium constants, etc., have left little doubt about the essential correctness of the partition function method. However, since Z cannot be evaluated explicitly for any reasonable physical Hamiltonian H, comparison with experiment always involves some

---

* Department of Mathematics, Massachusetts Institute of Technology, Cambridge Massachusetts. Supported in part by National Science Foundation grant GP26526.

** Department of Physics, Belfer Graduate School of Science, Yeshiva University New York, New York. Supported in part by AFOSR #F44620-71-C-0013.

uncontrolled approximations. Hence, the following problem deserves attention: Is it true that the thermal properties of matter obtained from an exact evaluation of the partition function would be extensive and otherwise have the same form as those postulated in the science of thermodynamics? In particular, does the thermodynamic, or bulk, limit exist for the Helmholtz free energy/unit volume derived from the partition function, and if so, does it have the appropriate convexity, i.e., stability properties?

To be more precise: Let $\{\Lambda_j\}$ be a sequence of bounded open sets (domains) in $\mathbb{R}^d$ with Λ_j becoming infinitely large as $j \to \infty$ in some 'reasonable way' which will be specified later. (We shall be concerned primarily with $d = 3$ but many of our results are valid for all d.) The volume (Lebesgue measure) of Λ_j will be denoted by $V(\Lambda_j)$ and $V(\Lambda_j) \to \infty$ as $j \to \infty$. Consider now a sequence of systems consisting of s species of particles in the domains $\{\Lambda_j\}$. Let $\underset{\sim}{N}_j = (N_j^1,\ldots,N_j^s)$ be the particle number vector specifying the system in Λ_j, i.e., N_j^i, is a non-negative integer and is the number of particles of species i contained in Λ_j. The canonical partition function of the <u>jth</u> system at reciprocal temperature β is then given by

$$Z(\beta,\underset{\sim}{N}_j;\ \Lambda_j) = \sum_{\alpha=0}^{\infty} \exp[-\beta E_\alpha(\underset{\sim}{N}_j;\ \Lambda_j)] \equiv \exp[V(\Lambda_j)g(\beta,\underset{\sim}{\rho}_j;\ \Lambda_j)] \quad , \tag{1.1}$$

where $E_\alpha(\underset{\sim}{N}_j;\ \Lambda_j)$ are the energy levels of the <u>jth</u> system, $\underset{\sim}{\rho}_j \equiv N_j/V(\Lambda_j)$ is the particle density vector, and $-kTg(\beta,\underset{\sim}{\rho}_j;\ \Lambda_j)$ is the Helmholtz free energy/unit volume of the <u>jth</u> system. According to statistical mechanics, knowledge of g determines all the equilibrium properties of this system. The question to be studied is the following: Given a sequence of particle density vectors $\{\underset{\sim}{\rho}_j\}$ which approach a limit $\underset{\sim}{\rho}$ as $j \to \infty$, does $g(\beta,\underset{\sim}{\rho}_j;\ \Lambda_j)$ approach a limit, $g(\beta,\underset{\sim}{\rho})$, as $j \to \infty$ and is this limit independent in some sense of the particular sequence of domains $\{\Lambda_j\}$ and density vectors $\{\underset{\sim}{\rho}_j\}$ used in going to the limit? If so, does the limiting free energy density have, as a function of $\underset{\sim}{\rho}$ and β the convexity properties required for thermodynamic stability, i.e., is $g(\beta,\underset{\sim}{\rho})$ convex in β and concave in $\underset{\sim}{\rho}$? (With regard to β, we see from (1.1) that each $g(\beta,\underset{\sim}{\rho}_j;\ \Lambda_j)$ is convex in β. Therefore, if this limit $g(\beta,\underset{\sim}{\rho})$ exists it will automatically be convex in β. Consequently we can set $\beta = 1$ and omit mention of β, and shall do so henceforth.)

The proof of the above for the free energy obtained from the canonical ensemble and the proof that the 'same' results are obtained, in the thermodynamic limit, from the microcanonical and grand canonical ensembles as well, has come to be recognized (by some people) as one of the basic goals of statistical mechanics and is referred to as proving the existence of the thermodynamic limit.

<u>Background: Tempering and the Coulomb Potential</u>

Various authors have evolved a technique for proving the existence of the thermodynamic limit for systems whose Hamiltonians satisfy certain conditions. (The different names associated with this development are: Van Hove, Lee and Yang, van Kampen, Wills, Mazur and van der Linden, Griffiths, and in particular Ruelle and Fisher. The reader is referred to Fisher [3] and Ruelle [17] for an exposition and references. For a synopsis and more references see also Lebowitz [8] and Griffiths [5].) In particular it was necessary to assume that the interaction between the particles constituting the microscopic units of macroscopic matter were short range or 'tempered'. This means that there exists a fixed distance $r_0 \geq 0$ and constants $C \geq 0$ and $\varepsilon > 0$ such that the inter-domain interaction potential energy between N_1 particles in a domain Λ_1, N_2 particles in a domain Λ_2, ..., and N_K particles in a domain Λ_K, has a bound in terms of the minimum distances r_{ij} between Λ_i and Λ_j,

$$I(N_1,\ldots,N_K) \equiv U(N_1,\ldots,N_K) - \sum_{i=1}^{K} U(N_i) \leq C \sum_{i\neq j}^{K} r_{ij}^{-(d+\varepsilon)} N_i N_j \tag{1.2}$$

whenever $r_{ij} \geq r_0$ *for all* $i \neq j$. We have written here $U(N) \equiv U(\underset{\sim}{x}_1,\ldots,\underset{\sim}{x}_N)$ for the total potential energy of N particles at positions $\underset{\sim}{x}_\ell \in \mathbb{R}^d$. (We shall generally not indicate that the particles belong to different species when this is not essential and shall denote $\Sigma_1^s N^\alpha$ by N).

The requirement of tempering unfortunately excludes the Coulomb potential which is the true potential relevant for real matter. That a nice thermodynamic limit exists for systems with Coulomb forces is a fact of common experience, but the proof that it does so is a much more subtle matter than for short range forces. It is screening, brought about by the long range nature of the Coulomb force itself, that causes the Coulomb force to behave as if it were short range. This has the consequence, as we shall prove in these notes, that when the sequence of systems are *overall neutral* then the approach of $g(\underset{\sim}{\rho}_j; \Lambda_j)$ to its limit $g(\underset{\sim}{\rho})$ and the properties of $g(\underset{\sim}{\rho})$ are the same as those obtained for systems with tempered interactions (except that the ρ^i, $i = 1, \ldots, s$ are constrained by the neutrality requirement). In particular $g(\underset{\sim}{\rho})$ is the same for different 'shapes' of the domains in $\{\Lambda_j\}$. This shape independence disappears when the constraint of charge neutrality is lifted and systems with a 'non-negligible' amount of net charge are considered. The true long range nature of the Coulomb force now becomes manifest, leading in some cases to a shape dependent limit of the free energy density and in other cases (when the excess charge is large) to an infinite limit (cf. section IV).

Background: H-stability and the Dyson-Lenard Theorem

The basic condition on the N body Hamiltonian

$$H(N) = \sum_{\ell=1}^{N} (p_\ell^2/2m_\ell) + U(\underset{\sim}{x}_1,\ldots,\underset{\sim}{x}_N) \quad , \tag{1.3}$$

where m_ℓ is the mass and $\underset{\sim}{p}_\ell$ the momentum (momentum operator in quantum mechanics) of the ℓth particle, required for the existence of thermodynamics is that there exist a constant $B < \infty$, such that for all N

$$E_0(N) \geq -BN \quad . \tag{1.4}$$

Here $E_0(N)$ is the ground state energy of the N particle system in infinite space, $\underset{\sim}{x}_i \in \mathbb{R}^d$, defined by

$$E_0(N) = \operatorname*{Inf}_{\Psi}[(\Psi,H(N)\Psi)/(\Psi,\Psi)] \tag{1.5}$$

with the $\Psi(\underset{\sim}{x}_1,\ldots,\underset{\sim}{x}_N)$ elements of a properly constructed Hilbert space in which $H(N)$ is a self-adjoint operator. The functions $\Psi(\underset{\sim}{x}_1,\ldots,\underset{\sim}{x}_N)$ have to satisfy the proper symmetry relations whenever the coordinates of two particles belonging to the same species are interchanged: $\Psi \to \Psi$ or $\Psi \to -\Psi$ for bosons or fermions respectively. (Since the spin does not appear directly in the Hamiltonian we can, and do, treat particles of the same type having different values of their spins in the z-direction as belonging to different species.)

We shall refer to condition (1.4) as H-stability. Heuristically, H-stability insures against collapse of the system. Mathematically it provides an upper bound to the sequence $\{g(\underset{\sim}{\rho}_j;\Lambda_j)\}$ and this bound plays an essential role in the proof. It should be emphasized however that H-stability does not in itself imply a thermodynamic limit. As an example, it is trivial to prove H-stability for charged particles all of one sign, and it is equally obvious that the thermodynamic limit does not exist in that case.

To satisfy (1.4) it is clearly sufficient that the potential energy $U(N)$ by itself have a lower bound of the same form:

$$\operatorname{Inf} U(\underset{\sim}{x}_1,\ldots,\underset{\sim}{x}_N) \geq -NB \quad , \quad \text{all } \underset{\sim}{x}_1, \ldots, \underset{\sim}{x}_N \quad . \tag{1.6}$$

Indeed for a classical system (1.6) is also necessary for (1.4). There are a large variety of interaction potentials for which the existence of the lower bound (1.6) can be verified explicitly. The simplest of these is the case when $U(N)$ can be written as the sum of a positive term and a term consisting of a sum of pair potentials $v(\underset{\sim}{x}_i - \underset{\sim}{x}_j)$ which is bounded below and has the asymptotic behaviors,

$$\lim_{|\underset{\sim}{r}|\to 0+} |\underset{\sim}{r}|^{d+\delta} v(\underset{\sim}{r}) \to +\infty \qquad \text{and} \qquad \lim_{|\underset{\sim}{r}|\to\infty} |\underset{\sim}{r}|^{d+\varepsilon} v(\underset{\sim}{r}) \geq 0 \tag{1.7}$$

for some $\delta > 0$ and $\varepsilon > 0$. (This result is due to Morrey [12] who appears to have been the first to consider bounds of the form (1.6) for non-Coulomb potentials.) More general types of potentials satisfying (1.6) have been considered by other authors [3], [17].

It is clear, however, that (1.6) will not be satisfied by a system of point charges with charges q_i of different signs, $i = 1, \ldots, N$. The interparticle Coulomb potential has the form, for $d = 3$,

$$U_c(\underset{\sim}{x}_1,\ldots,\underset{\sim}{x}_N) = \tfrac{1}{2} \sum_{i \neq j} q_i q_j |\underset{\sim}{x}_i - \underset{\sim}{x}_j|^{-1} \quad , \tag{1.8}$$

and the potential energy of even a single pair of oppositely charged particles has no lower bound. Interestingly though, if the particles have hard cores, i.e., U(N) contains in addition to its Coulomb part (1.8) a term which is $+\infty$ if $|x_i - x_j| < R$, then Onsager [14] showed that (1.6) is satisfied. Onsager's proof is so simple that we shall present it here (in a form communicated to us by Penrose*). Since the particles cannot approach each other any closer than a distance R, the effect of the Coulomb interaction between the particles will be the same [13] if the charge of each particle is distributed in *any* spherically symmetric way within a ball of radius ½ R centered on the position of that particle, e.g., a uniform charge density. Now, as is well known from electrostatics [7]

$$\tfrac{1}{2} \sum_{i\neq j} q_i q_j |x_i - x_j|^{-1} = \tfrac{1}{2}\int_{\mathbb{R}^d} E^2(x)\,dx$$

$$- \sum_{i=1} \text{(self energy of the } \underline{i\text{th}} \text{ particle)} \geq -NB \qquad , \tag{1.9}$$

where $E(x)$ is the electrostatic field, and B is the maximum self energy of any of the balls.

Onsager's results were generalized somewhat by Fisher and Ruelle [4]. This work, however, still left open the question of whether a system of point Coulomb charges, which may be taken as the building blocks of real matter, has a lower bound of the form (1.4). Now when dealing with a quantum system of charges, the non-existence of a lower bound to $-|x_i - x_j|^{-1}$ might appear not as serious as in the classical case since we expect that the Heisenberg uncertainty principle, which prevents particles from having their positions 'close to each other' without also having a large kinetic energy, will insure the existence of a lower bound to the ground state energy. This is indeed the case for any finite system, (-13.5 electron volts for a system composed of one electron and one proton), and generally $E_0(N) > -\infty$, for any N, [6]. We need, however, a bound proportional to N and this, it turns out, the uncertainty principle alone cannot provide. The required result was proven by Dyson and Lenard [1,10], who showed that (1.4) holds for a system of point Coulomb charges when *all* species with negative and/or positive charges are fermions. This is happily the case in nature where the electrons are fermions. (When neither of the charges are fermions, Dyson [2] found an *upper* bound to the ground state energy that is proportional $-N^{7/5}$; hence such a system will not be thermodynamically stable.)

We note here that the Dyson-Lenard lower bound is valid whenever the masses of the fermion particles are finite (the masses only affect the numerical value of B which is of no interest here). Hence it remains valid if the kinetic energy term in the Hamiltonian is multiplied by some δ, $0 < \delta < 1$, e.g., $\delta = \tfrac{1}{2}$.

---

* See also Penrose's comments [16] on using electromagnetic energy considerations to establish the thermodynamic limit for charged and magnetic systems.

Basic Inequalities and Outline of the Notes

Let us consider a system of $\underset{\sim}{N} = (N^1,\ldots,N^s)$ particles in a domain Λ with a Hamiltonian $H(N;\Lambda)$

$$H(N;\Lambda) = -\tfrac{1}{2}\hbar^2 \sum_{i=1}^{N} (m_1)^{-1}\Delta_i + U_c(\underset{\sim}{x}_1,\ldots,\underset{\sim}{x}_N) + U_T(\underset{\sim}{x}_1,\ldots,\underset{\sim}{x}_N) \quad . \tag{1.10}$$

Here $N \equiv \sum_{\alpha=1}^{s} N^\alpha$, and $\underset{\sim}{x}_i \in \Lambda$ is the coordinate of a particle of species one for $1 \le i \le N^1$, and of a particle of species two for $N^1 < i \le N^1 + N^2$, etc. $U_c(\underset{\sim}{x}_1,\ldots,\underset{\sim}{x}_N)$ is the Coulomb potential defined in (1.8), so that $m_i = m_1$, $q_i = e_1$ for $1 \le i \le N^1$, etc., with m_α and e_α, $\alpha = 1, \ldots, s$, the mass and charge of a particle of the αth species. $U_T(\underset{\sim}{x}_1,\ldots,\underset{\sim}{x}_N)$ is a tempered and stable potential satisfying (1.2) and (1.6) (which is also translationally and rotationally invariant). It is not altogether useless to include tempered potentials along with the true Coulomb potentials because one might wish to consider model systems in which ionized molecules are the elementary particles. *Although we shall omit U_T in most of these notes, it should be understood that all the stated theorem are valid for the full Hamiltonian* (1.10). $H(N;\Lambda)$ is a self-adjoint operator, defined via the Friedrichs extension. (In the physicists language this corresponds to using a Hilbert space in which the wave functions vanish on the boundary of Λ.) When the statistics of the particles satisfy the conditions of the Dyson-Lenard theorem then $H(N;\Lambda)$ will satisfy the inequality (note the factor $\tfrac{1}{2}$)

$$H(N;\Lambda) \ge -\frac{\hbar^2}{4} \sum_{i=1}^{N} (m_i)^{-1}\Delta_i - N\Phi \tag{1.11}$$

with Φ some constant, $\Phi < \infty$. The canonical partition function $Z(\underset{\sim}{N};\Lambda)$ and the corresponding $g(\underset{\sim}{\rho};\Lambda)$ will have the bounds

$$Z(N;\Lambda) \le \exp[N\Phi] \prod_{\alpha=1}^{s} Z_{0,\alpha}(N^\alpha;\Lambda) \quad , \tag{1.12a}$$

$$g(\underset{\sim}{\rho};\Lambda) \le \rho\Phi + \sum_{\alpha=1}^{s} g_{0,\alpha}(\rho^\alpha;\Lambda) \quad , \tag{1.12b}$$

where $\rho \equiv \sum_{\alpha=1}^{s} \rho^\alpha$ and $Z_{0,\alpha}$ (resp. $g_{0,\alpha}$) is the partition function (resp. -free energy/unit volume) of an ideal gas (fermion or boson according to the statistic of species α) of particles with masses $m'_\alpha = 2m_\alpha$. The inequality (1.13) readily yields a uniform bound on any sequence $g(\underset{\sim}{\rho}_j;\Lambda_j)$ whenever the $\underset{\sim}{\rho}_j$ are in a compact subset of $\mathbb{R}^s$, (with $\rho_j^i \ge 0$).

We now give a sketch of the method used in our proof. As usual, one first proves the existence of the limit for a standard sequence of domains. The limit for an arbitrary domain is then easily arrived at by packing that domain with the standard ones. The basic inequality that is needed is that if a domain Λ contains K disjoint sub-domains $\Lambda_1, \Lambda_2, \ldots, \Lambda_K$ and *if the inter-domain interaction be neglected* then

$$Z\left(\sum_{i=1}^{K} \underset{\sim}{N}_i;\ \Lambda\right) \geq \prod_{i=1}^{K} Z(\underset{\sim}{N}_i;\ \Lambda_i) \tag{1.13a}$$

or

$$g\left(\sum_{i=1}^{K} f_i \underset{\sim}{\rho}_i;\ \Lambda\right) \geq \sum_{i=1}^{K} f_i g(\underset{\sim}{\rho}_i;\ \Lambda_i) \quad , \tag{1.13b}$$

where $f_i \equiv V(\Lambda_i)/V(\Lambda)$ is the fraction of the volume of Λ occupied by Λ_i. If the distance between every pair of sub-domains is not less than r_0, one can use (1.2) to obtain a useful bound on the tempered part of the omitted inter-domain interaction energy, $I_T(N_1,\ldots,N_K)$.

The normal choice for the standard domains are cubes Γ_j, with Γ_{j+1} being composed of 2^d copies of Γ_j, together with 'corridors'. One chooses $\underset{\sim}{N}_{j+1} = 2^d \underset{\sim}{N}_j$. Neglecting I_T one would have from (1.13b) that $g(\underset{\sim}{\rho}_{j+1};\ \Gamma_{j+1}) \geq g(\underset{\sim}{\rho}_j;\ \Gamma_j)$ which, since $g(\underset{\sim}{\rho}_\ell;\ \Gamma_\ell)$ is bounded above implies the existence of a limit. To justify neglect of I_T one makes the corridors increase in thickness with increasing j; although V_j^c, the corridor volume, approaches ∞ one makes $V_j^c/V_j \to 0$ in order that $\Sigma_1^{2^d} f_i \to 1$ as $j \to \infty$. The positive ε of (1.2) allows one to accomplish these desiderata.

Obviously, such a strategy will fail with Coulomb forces, but fortunately there is another way to bound the inter-domain energy. The essential point is that it is not necessary to bound this energy for all possible states of the systems in the sub-domains; it is only necessary to bound the 'average' interaction between domains which is much easier. This is expressed mathematically by using the Peierls-Bogoliubov inequality (Jensen's inequality in the classical case) to show that for $\underset{\sim}{N} = \Sigma_1^K \underset{\sim}{N}_i$

$$Z(\underset{\sim}{N};\ \Lambda) \geq \exp[-\langle I(N_1,\ldots,N_K)\rangle] \prod_{i=1}^{K} Z(\underset{\sim}{N}_i;\ \Lambda_i) \quad , \tag{1.14}$$

where $\langle I\rangle$ is the *average* inter-domain energy in an ensemble where each sub-domain is *independent*. Where $\langle I\rangle$ vanishes, (1.14) reduces to (1.13a) and, in general, there is a corresponding equation for $g(\underset{\sim}{\rho};\ \Lambda)$ as in (1.13b).

To prove (1.14) consider the case $K = 2$ and let $\{\Psi_j\}$, $j = 1, 2, \ldots$, be a set of functions consisting of all properly symmetrized and normalized functions of the form

$$\Psi_j \equiv \Psi_{n,m} \equiv \Phi_n(\underset{\sim}{x}_1,\ldots,\underset{\sim}{x}_{N_1};\ \Lambda_1)\, \chi_m(\underset{\sim}{x}_{N_1+1},\ldots,\underset{\sim}{x}_{N_1+N_2};\ \Lambda_2) \tag{1.15}$$

where the $\{\Phi_n\}$ and $\{\chi_m\}$ are a *complete* orthonormal set of eigen functions in the Hilbert spaces of $H(N_1;\ \Lambda_1)$ and $H(N_2;\ \Lambda_2)$. The $\{\Psi_j\}$ are clearly an orthonormal set (possible incomplete) in the Hilbert space of $H(N_1 + N_2;\ \Lambda)$. Hence

$$\begin{aligned} Z(\underset{\sim}{N}_1 + \underset{\sim}{N}_2;\ \Lambda) &\geq \sum_j (\Psi_j, \{\exp[-H(N_1 + N_2;\ \Lambda)]\}\Psi_j) \\ &= \sum_j (\Psi_j, \{\exp[-H(N_1;\ \Lambda_1) - H(N_2;\ \Lambda_2) - I(N_1,N_2)]\}\Psi_j) \end{aligned} \tag{1.16}$$

where the last equality follows from the fact that the support of the $\{\Phi_n\}$ is in Λ_1 and the support of the $\{X_m\}$ is in Λ_2. The convexity of the exponential function implies (Peierls-Bogoliubov inequality) for any pair of self-adjoint operators $\underset{\sim}{A}$ and $\underset{\sim}{B}$ and *any* set of normalized vectors $\{f_\ell\}$ in the domain of $\underset{\sim}{A}$ and $\underset{\sim}{B}$ that

$$\sum_\ell (f_\ell,[\exp(\underset{\sim}{A} + \underset{\sim}{B})]f_\ell) \geq \sum_\ell \exp\{(f_\ell,[\underset{\sim}{A} + \underset{\sim}{B}]f_\ell)\}$$
$$\geq \{\sum_\ell \exp(f_\ell,\underset{\sim}{A}f_\ell)\} \exp[\langle \underset{\sim}{B}\rangle_{\underset{\sim}{A}}] \qquad (1.17)$$

where

$$\langle \underset{\sim}{B}\rangle_{\underset{\sim}{A}} \equiv \{\sum_\ell [\exp(f_\ell,\underset{\sim}{A}f_\ell)](f_\ell,\underset{\sim}{B}f_\ell)\}/\{\sum_\ell \exp(f_\ell,\underset{\sim}{A}f_\ell)\} \quad . \qquad (1.18)$$

Applying (1.17) to (1.16) and remembering that $\{\Phi_n\}$ and $\{X_m\}$ are complete in the Hilbert spaces of $H(N_1;\Lambda_1)$ and $H(N_2;\Lambda_2)$ respectively yields the desired inequalities,

$$Z(\underset{\sim}{N}_1 + \underset{\sim}{N}_2;\Lambda) \geq Z(\underset{\sim}{N}_1;\Lambda_1)Z(\underset{\sim}{N}_2;\Lambda_2 : I_1)$$
$$\geq Z(\underset{\sim}{N}_1;\Lambda_1)Z(\underset{\sim}{N}_2;\Lambda_2)\exp[-\langle I(N_1,N_2)\rangle] \quad . \qquad (1.19)$$

Here $Z(\underset{\sim}{N}_2;\Lambda_2 : I_1)$ is the partition function of N_2 particles in Λ_2 with a Hamiltonian

$$H(N_2;\Lambda_2 : I_1) = H(N_2;\Lambda_2) + I_1(\underset{\sim}{x}_{N_1+1},\ldots,\underset{\sim}{x}_{N_1+N_2}) \quad , \qquad (1.20)$$

with

$$I_1 \equiv \mathrm{Tr}_1\{I(N_1,N_2)\exp[-H(N_1;\Lambda_1]\}/Z(\underset{\sim}{N}_1;\Lambda_1) \quad , \qquad (1.21)$$

the subscript 1 indicating that the trace is taken with respect to the $\{\Phi_m\}$. Hence I_1 is the value of the *inter-domain* interaction energy for a given configuration of the N_2 particles in Λ_2 *averaged* over the canonical ensemble of N_1 particles in Λ_1 which are *unaffected* by the presence of the particles in Λ_2. Similarly,

$$\langle I(N_1,N_2)\rangle \equiv \frac{\mathrm{Tr}_{1,2}\{I(N_1,N_2)\exp[-H(N_1;\Lambda_1)]\exp[-H(N_2;\Lambda_2)]\}}{Z(\underset{\sim}{N}_1;\Lambda_1)Z(\underset{\sim}{N}_2;\Lambda_2)} \quad , \qquad (1.22)$$

the trace now being taken over a complete set of functions corresponding to a system consisting of a specified set of N_1 particles in Λ_1 and another set of N_2 particles in Λ_2. The quantity $\langle I(N_1,N_2)\rangle$ thus corresponds to the average of the interaction between the particles in Λ_1 and those in Λ_2 when the states and the probabilities of different states in each box are *completely unaffected* by the presence of the other box. This corresponds to taking the average of $I(N_1,N_2)$ with a density matrix which is a direct product of the unperturbed density matrices in Λ_1 and Λ_2.

We now make the observation, which is one of the crucial steps in our proof, that if Λ_1 and/or Λ_2 are balls then, because of the rotational (and translational) symmetry of the Hamiltonian the unperturbed density matrices (corresponding to no interaction between Λ_1 and Λ_2) are spherically symmetric about the centers of Λ_1 and/or Λ_2. This implies in particular that the average unperturbed charge density in Λ_1 and/or Λ_2 is spherically symmetric and hence by Newton's theorem the Coulomb contribution to $\langle I(N_1,N_2)\rangle$ in (1.19) is the same as would be obtained if all the charges in the ball domain were concentrated at its center and would *vanish* when the ball is over-all neutral.

This clearly generalizes to the Coulomb part of $\langle I(N_1,\ldots,N_K)\rangle$ in (1.14) and this leads us to choose balls, rather than cubes, for our standard domains. There is of course a price to be paid for this since balls do not pack into each other as nicely as cubes do and necessitates our packing the standard ball domain B_j not only with balls of type B_{j-1} but with balls of types $B_0, B_1, \ldots, B_{j-2}$ as well. The geometrical problem involved in this is handled in Section 2.

We mention here that the use of (1.14) permits us to prove the existence of the thermodynamic limit for systems containing electric or magnetic dipoles which interact with long range, i.e., non tempered, potentials (falling off only as $|x_i-x_j|^{-3}$). The average interaction between domains will vanish since the expectation value of the dipole moment of any particle will be zero in the absence of an external electric or magnetic field, Griffiths (1968)*.

For such systems it is essential, however, that the particles have hard cores since otherwise they will not satisfy the H-stability condition. Indeed, $E_0(N)$ will not be bounded below. For this reason we cannot include spin-spin couplings between the elementary charges in our analysis. These couplings are intrinsically of a relativistic nature and present entirely new problems (Dyson, private communication).

Needless to say, we do not deal with the strong (nuclear) and weak interactions. As pointed out by Dyson [2], the magnitude of the nuclear forces is so large that they would give completely different binding energies for molecules and for crystals if they played any role in the thermal properties of ordinary matter. We are also neglecting, of course, gravitational forces which certainly are important for large aggregates of matter and thus might be thought important in the 'thermodynamic limit'. To quote Onsager [15], however, "The common concept of a homogeneous phase implies dimensions that are large compared to the molecules and small compared to the moon." When we speak of the thermodynamic limit which is mathematically the infinite system limit, we have in mind its physical application to systems containing, $10^{22} \sim 10^{28}$ particles, i.e. systems which are large enough for surface effects to be

---

* Griffiths' proof for dipoles does not use (1.19) but relies on the complete symmetry between 'up and down' orientations of the dipoles. Using such symmetry, Griffiths (unpublished) was able to prove the existence of the thermodynamic limit for a system of charged particles in which the positive and negative particles are identical under charge conjugation, e.g., positrons and electrons. When such an additional symmetry is present the rotational invariance of the Hamiltonian becomes unimportant and it is not necessary to use balls as we do. Unfortunately such symmetries are not present in real systems.

negligible and yet small enough for internal gravitational effects also to be completely negligible. (An external gravitational field will of course have some effect but does not present any fundamental problem.)

2. PACKING A DOMAIN WITH BALLS

In this section we address ourselves to a geometric construction which is fundamental to our proof of the existence of the thermodynamic limit, namely the possibility of packing a ball or a cube by smaller balls such that the packing is both complete and rapid.

We shall always use the word *domain* to mean a bounded, open set in $\mathbb{R}^d$. If Λ is a domain and $B = \{B_i\}$ is a denumerable family of domains such that $B_i \subset \Lambda$ for all i, we shall say that B is *packed* in Λ if the $\{B_i\}$ are all disjoint. The packing is *complete* if $\Sigma_i V(B_i) = V(\Lambda)$ where $V(\Lambda)$ is the volume (Lebesgue measure) of Λ.

Definition. For a domain $\Lambda \subset \mathbb{R}^d$ and a real number h we define

$$\Lambda_h = \{\underset{\sim}{r}: \underset{\sim}{r} \in \Lambda, d(\underset{\sim}{r}; \sim\Lambda) < h\} \qquad \text{for } h > 0$$

$$= \{\underset{\sim}{r}: \underset{\sim}{r} \in \sim\Lambda, d(\underset{\sim}{r}; \Lambda) \leq -h\} \qquad \text{for } h \leq 0 \quad , \tag{2.1}$$

where $d(;)$ is the distance function and $\sim$ denotes complement. We also define $V(h; \Lambda)$ to be the volume of Λ_h.

We shall frequently make use of the fact (Lemma 2 of section 8 in Fisher [3] that the number N_y, of cubes of side $2y$ that can be packed in a Λ satisfies the inequality

$$N_y \geq (2y)^{-d}[V(\Lambda) - V(2y\sqrt{d}; \Lambda)] \quad . \tag{2.2}$$

Definition. Let σ_d be the volume of a ball of unit radius in $\mathbb{R}^d$. $g_d \equiv 2^{-d}\sigma_d$ is the fraction of the volume of a cube of side $2y$ filled by a ball of radius y when the ball is packed in the cube. We also define $\alpha_d \equiv (2^d - 1)2\sqrt{d}$.

Clearly, for a ball B of radius $r \geq 2y\sqrt{d} \geq 0$

$$V(2y\sqrt{d}; B) \leq V(-2y\sqrt{d}; B) \leq \alpha_d \sigma_d r^{(d-1)} y \quad . \tag{2.3}$$

The main theorem we wish to prove is that we can find a sequence of balls of decreasing radius, of which the jth type has radius $r_j = \delta^j$ (with $\delta < 1$), such that we can completely pack a unit d-dimensional ball ($r = 1$) with these and, moreover, we can do this rapidly.

Theorem 2.1

Let p *be a positive integer and, for all positive integers* j, *define radii* $r_j = (1 + p)^{-j}$ *and integers* $n_j = p^{j-1}(1 + p)^{j(d-1)}$. *Then if* $1 + p \geq \alpha_d + g_d^{-1}$

it is possible to pack $\bigcup_{j=1}^{\infty}$ (n_j *balls of radius* r_j) *in a unit* d*-dimensional ball. The packing is complete since* $\sum_{j=1}^{\infty} n_j r_j^d = 1$.

Proof. We shall give an explicit construction for accomplishing the packing stated in the theorem by using (2.2) and (2.3). First cover the unit ball by a cubic array of cubes of side $2r_1$. We shall show that there are n_1 of these cubes which are contained in the unit ball. We can place a ball of radius r_1 at the center of each of these cubes. We then cover the unit ball by a cubic array of cubes of side $2r_2$ and show that there are n_2 of these cubes which are contained in the unit ball and which do not intersect the first n_1 balls. The argument is repeated inductively. Thus, we have to show that after placing all balls up to and including those of radius r_j we can pack n_{j+1} in a cubic array into Ω_j, which is the interior of the unfilled portion of the unit ball. (We must prove this for $j \geq 0$, with $r_0 \equiv 0$.) For $j \geq 0$,

$$V(\Omega_j) = \sigma_d - \sigma_d \sum_{k=0}^{j} n_k r_k{}^d = \sigma_d \left(\frac{p}{p+1}\right)^j \quad .$$

Clearly, $V(2\sqrt{d}r_{j+1};\ \Omega_j)$ is bounded above by M_j which is the sum of the $V(-2\sqrt{d}r_{j+1};B)$ for each ball of $\bigcup_{\ell=0}^{j}$ (n_ℓ balls of radius r_ℓ) separately, plus $V(2\sqrt{d}r_{j+1};\ B)$ for the unit ball. Thus, by (2.3), if $2\sqrt{d}r_{j+1} < r_j$ (which is true when p satisfies the hypothesis)

$$V(2\sqrt{d}r_{j+1};\ \Omega_j) \leq M_j \leq \alpha_d \sigma_d r_{j+1}\{1 + \sum_{k=0}^{j} n_k r_k^{d-1}\}$$

$$= (p^j + p - 2)(p - 1)^{-1}(1 + p)^{-(j+1)} \alpha_d \sigma_d \equiv \hat{M}_j \quad . \tag{2.4}$$

Using (2.2) it is sufficient to show that

$$(2r_{j+1})^d n_{j+1} \leq [V_j - \hat{M}_j] \leq [V(\Omega_j) - V(2\sqrt{d}r_{j+1};\ \Omega_j)] \quad .$$

Inserting the relevant quantities, we require that

$$1 \leq g_d[p + 1 - \alpha_d \frac{1 + p^{-j}(p - 2)}{p - 1}]$$

for all $j \geq 0$. By the hypothesis $p \geq 2$. Then $p^{-j}(p - 2) \leq (p - 2)$ and hence it is sufficient that

$$1 \leq g_d[p + 1 - \alpha_d] \quad ,$$

which agrees with the hypothesis.

The minimum ratio of successive radii, $1 + p$, required by this construction is 27 for $d = 3$. We note that the fraction of volume of the unit ball occupied by all the balls of radius r_j is

$$f_j = n_j r_j^d = p^{-1}\gamma^j \quad , \tag{2.5}$$

where

$$\gamma = p(1 + p)^{-1} < 1 \quad . \tag{2.6}$$

Moreover, the fraction of volume left unfilled after the balls of type j have been packed is γ^j. This implies that the packing is "exponentially fast".

It can be shown that Theorem 2.1 is also true if "unit d-dimensional ball" is replaced by "d-dimensional cube of volume σ_d".

3. THERMODYNAMIC LIMIT FOR SPHERICAL AND GENERAL DOMAINS

In this section we shall prove the existence of the thermodynamic limit for a two component system of charges e_1 and e_2 in a standard sequence of balls. We shall assume that $e_1 > 0$, $e_2 < 0$ and $|e_1/e_2|$ is rational and that the particles interact via the Coulomb potential alone. To do so we shall define a sequence of standard balls $\{B_j\}$ of increasing radii $\{R_j\}$, $j = 0, 1, \ldots$.

Definition. Let $1 + p$ satisfy the condition of Theorem 2.1 and be *even*. (The fact that $1 + p$ is even will not be used until later.) Choose an $R_0 > 0$. The balls, $B_0, B_1, \ldots$, forming the *standard sequence*, are chosen to have radii

$$R_j = R_0(1 + p)^j \quad . \tag{3.1}$$

The volume of B_j will be denoted by V_j.

The packing described in Theorem 2.1 will be referred to as the *standard packing* of the ball B_K with balls $\{B_j\}$, $j = 0, 1, \ldots, K - 1$.

Filling of Balls with Particles

In the following we shall fill the standard balls with particles in various ways. However, we shall always observe the following convention: Each ball will have charge neutrality. We take q particles of type 1 and ℓ particles of type 2 such that $qe_1 + \ell e_2 = 0$ and such that q and ℓ have no common divisor as the fundamental unit, and this will be referred to simply as a multiplet. Densities and (multiplet) numbers will be in terms of this unit.

We define

$$g_j(\rho) = (V_j)^{-1} \ell n Z\,(N = \rho V_j;\, B_j) \quad , \tag{3.2}$$

where N is the number of multiplets and where we have set $\beta = 1$ for convenience.

Since N must be an integer, an obvious restriction is thereby placed on ρ. However, following Fisher [3] we can define g for all ρ by linear interpolation as follows:

Definition. Let $f(N)$ be a function from the integers to the reals. If $n = N + \eta$, with N an integer and $0 \le \eta \le 1$, we extend $f(\cdot)$ to the reals by $f'(n) \equiv f(N) + \eta[f(N + 1) - f(N)]$.

The usefulness of this definition is made manifest by the following lemma.

Lemma 3.1

Let $\mathbb{Z}^+$ *be the non-negative integers,* $\mathbb{R}^+$ *the non-negative reals, and* $\mathbb{R}$ *the reals. Let* $f, h_1, h_2, \ldots, h_M$ *be functions from* $\mathbb{Z}^+$ *to* $\mathbb{R}$ *and let* $f', h_1', \ldots, h_M'$ *be the extended functions from* $\mathbb{R}^+$ *to* $\mathbb{R}$ *as in the above definition. Let* $N_j \in \mathbb{Z}^+$ *and* $n_j \in \mathbb{R}^+$. *If* $f(\Sigma_1^M N_j) \geq \Sigma_1^M h_j(N_j)$ *for all* $\{N_j\}$ *then* $f'(\Sigma_1^M n_j) \geq \Sigma_1^M h_j'(n_j)$ *for all* $\{n_j\}$.

The proof follows by induction on M. The case $M = 1$ is obvious and $M = 2$ is proved in Fisher [3], footnote 25.

Let us now consider a standard packing of B_K and place N_j multiplets in all balls of type j, $j = 0, 1, \ldots, K - 1$. The total number of multiplets in B_K is then

$$N = \Sigma_0^{K-1} N_j n_{K-j} \quad , \tag{3.3}$$

so that

$$\rho = N/V_K = p^{-1} \Sigma_0^{K-1} \rho_j \gamma^{K-j} \quad . \tag{3.4}$$

Our fundamental inequality on the partition function of a subdivided domain, together with the vanishing of the average Coulomb interaction for neutral balls implies:

Theorem 3.1

Let $\rho_0, \ldots, \rho_{K-1}$ *be non-negative reals and let* $\rho = p^{-1} \Sigma_0^{K-1} \rho_j \gamma^{K-j}$. *Then*

$$g_K(\rho) \geq \frac{1}{p} \sum_{j=0}^{K-1} \gamma^{K-j} g_j(\rho_j) \quad . \tag{3.5}$$

Limit of $g_k(\rho)$ as $k \to \infty$

Our next task is to use Theorem 3.1 to establish the thermodynamic limit of $g_k(\rho)$ for the standard sequence of balls. To accomplish this we define, for each $j \geq 0$, a standard density sequence (depending on ρ) as follows:

$$\rho_j = \rho \ \text{ for } \ j > 0; \ \rho_j = \rho(1 - \gamma)^{-1} \ \text{ for } \ j = 0 \quad .$$

It is understood that when $\rho = 0$, $g_j(0) = 0$.

With ρ held fixed, let us denote $g_j(\rho_j)$ simply by g_j. Then, from Theorem 3.1,

$$g_k = \frac{1}{p} \sum_{j=0}^{k-1} \gamma^{k-j} g_j + c_k$$

for $k > 0$, where c_k is a non-negative real number.

The solution of (3.6), valid for $k > 0$, is easily found to be

$$g_k = \gamma c_k + (1 - \gamma) \sum_{j=1}^{k} c_j + (1 - \gamma) g_0 \quad . \tag{3.7}$$

Equation (3.7) establishes a limit for g_k because: (a) g_0 is finite; (b) As each $c_j \geq 0$, and as we know that g_k has an upper bound by H-stability, the sum involving the c's must converge. This implies that $c_k \to 0$ and hence (3.7) must have a limit. We shall call this limit $g(\rho)$.

Further examination of (3.7) leads to a lower bound for g which is proportional to ρ for sufficiently small ρ.

Our analysis of (3.6) thus yields

Theorem 3.2

Let ρ *be a fixed multiplet density. Then* $g(\rho) = \lim_{k\to\infty} g_k(\rho)$ *exists and is finite. Furthermore, there exists a* $\rho_1 > 0$ *such that for* ρ *in the closed-open interval* $[0,\rho_1)$, $g(\rho)$ *is bounded below by* $\alpha\rho$ *with* α *finite and independent of* ρ.

Convexity of the Free Energy

With the limit $g(\rho)$ in hand we can next establish convexity. It is here that we use the fact that $1 + p$ was chosen to be even. This permits us to place densities ρ'_j corresponding to a final density ρ' (with $\rho'_0 = \rho'(1-\gamma)^{-1}$, $\rho'_j = \rho'$, $j > 0$) in half the balls of each type and densities ρ''_j in the other half. Taking the limit $j \to \infty$ yields

$$g(\tfrac{1}{2}\rho' + \tfrac{1}{2}\rho'') \geq \tfrac{1}{2} g(\rho') + \tfrac{1}{2} g(\rho'') \quad . \tag{3.8}$$

We can now follow the standard arguments used for non-Coulomb systems to establish the concavity and hence continuity of $g(\rho)$. Similarly, the approach of $g_j(\rho)$ to $g(\rho)$ can be shown (by means of Dini's theorem) to be uniform on any closed interval $[0,\rho']$, $\rho' < \infty$.

Neutral Multicomponent Systems with Coulomb and Tempered Interactions in General Domains

Thus far we have established the limit and the convexity of the free energy/unit volume for an overall neutral system composed of two species of charged particles interacting with Coulomb forces only and confined to the standard sequence of balls. This permitted us to deal with a neutral multiplet as though it were a single particle.

We shall now state the general theorem on the properties of the free energy/unit volume for an overall neutral system composed of s species of particles with charges $e_1, \ldots, e_s$. We suppose these charges to be rational fractions of each other so that, in appropriate units, the e_i may be taken to be integers. In nature all elementary charges are in fact integral multiples of the electron charge. The e_i may not be all of one sign, but we do allow some of them to be zero. We shall represent particle numbers by a vector $\underset{\sim}{N} = (N^1,\ldots,N^s)$, so that charge neutrality is represented by $\underset{\sim}{N} \cdot \underset{\sim}{E} = 0$ with $\underset{\sim}{E} = (e_1, \ldots, e_s)$. In a like manner we shall represent particle densities by a vector $\underset{\sim}{\rho}$.

The particles comprising our system may have, in addition to their Coulomb interactions, other kinds of interaction potentials as long as those interactions are tempered and the full Hamiltonian is H-stable (this will always be true when the additional interactions are themselves stable). When these tempered interactions include hard cores, there will be some convex domain in $\mathbb{R}^s$ in which a vector $\underset{\sim}{\rho}$ must lie in order for the density to be less than the close packing density. We shall denote the fact that $\underset{\sim}{\rho}$ is in this domain by writing $|\underset{\sim}{\rho}| < \rho_c$.

We consider a general sequence of domains $\{\Lambda_j\}$ tending to infinity in a reasonable way. To define reasonable we introduce the following conditions on a sequence of domains in $\mathbb{R}^d$:

(a) A sequence of domains $\{\Lambda_j\}$ *tends to infinity in the sense of Van Hove* if $V(\Lambda_j) \to \infty$ and $V(h; \Lambda_j)/V(\Lambda_j) \to 0$ as $j \to \infty$ for each fixed h. (For definitions see (2.1)).

(b) A sequence of domains $\{\Lambda_j\}$ satisfies the *ball condition* if there exists a $\delta > 0$ such that

$$V(\Lambda_j)/V(\hat{B}_j) \geq \delta \quad , \tag{3.9}$$

where $\hat{B}_j$ is the ball of smallest radius containing Λ_j.

(c) A sequence of domains $\{\Lambda_j\}$ *tends to infinity in the sense of Fisher* if $V(\Lambda_j) \to \infty$ and if there exists a continuous function $\pi: \mathbb{R}^1 \to \mathbb{R}^1$, with $\pi(0) = 0$ such that

$$V(\alpha[V(\Lambda_j)]^{1/d}; \Lambda_j)/V(\Lambda_j) \leq \pi(\alpha) \tag{3.10}$$

for all α and all j.

Obviously, condition (c) implies (a). It also implies condition (b) as shown in Fisher [3]. On the other hand, neither condition (a) nor (b) implies the other, nor do conditions (a) and (b) together imply (c).

Definition. A regular sequence of domains, $\{\Lambda_j\}$, in $\mathbb{R}^d$ is one satisfying conditions (a) and (b) if only strongly tempered potentials (in addition to the Coulomb potential) are present. If weakly tempered potentials are also present then the stronger condition (c) must be satisfied.

Our final result for neutral systems, which we state here without proof, is

Theorem 3.3

Let $\{\Lambda_j\}$ be a regular sequence of domains. Let $\{\underset{\sim}{N}_j\}$ be a sequence of non-negative, integer valued particle number vectors satisfying the neutrality condition, $\underset{\sim}{N}_j \cdot \underset{\sim}{E} = 0$, and let $\underset{\sim}{\rho}_j = V(\Lambda_j)^{-1}\underset{\sim}{N}_j$. If $\lim_{j\to\infty} \underset{\sim}{\rho}_j = \underset{\sim}{\rho}$, with $|\underset{\sim}{\rho}| < \rho_c$ then

(a) $\lim_{j\to\infty} g(\underset{\sim}{\rho}_j;\ \Lambda_j) = g(\underset{\sim}{\rho})$ *exists and is independent of the sequence of domains or particle numbers.*

(b) $g(\underset{\sim}{\rho})$ *is continuous and concave in the convex domain* $D = \{\underset{\sim}{\rho}:\ |\underset{\sim}{\rho}| < \rho_c\} \cap \{\underset{\sim}{\rho}:\ \underset{\sim}{\rho} \cdot \underset{\sim}{E} = 0\}$ *and* $g(0) = 0$.

(c) *Let* K *be a compact subset of* D. *Suppose that for each* $\underset{\sim}{\rho} \in K$ *we have a sequence* $\{\underset{\sim}{N}_j(\underset{\sim}{\rho})\}$ *and the corresponding sequence* $\{\underset{\sim}{\rho}_j(\underset{\sim}{\rho})\}$ *with the additional hypothesis that* $\underset{\sim}{\rho}_j(\underset{\sim}{\rho}) \to \underset{\sim}{\rho}$ *uniformly on* K. *Then* $g_j(\underset{\sim}{\rho}_j(\underset{\sim}{\rho})) \to g(\underset{\sim}{\rho})$ *uniformly on* K.

4. SYSTEMS WITH NET CHARGE

In the last section we showed that a sequence of systems of charged particles has a thermodynamic limit when the finite systems in the sequence have no net charge, that is $\underset{\sim}{N}_j \cdot \underset{\sim}{E} = 0$. The free energy density in this limit, $-g(\underset{\sim}{\rho})$, is independent of the shape of the domains Λ_j and depends only on the limit of the particle density vector $\underset{\sim}{N}_j/V(\Lambda_j)$.

It is intuitively clear that this condition of strict charge neutrality, $\underset{\sim}{N}_j \cdot \underset{\sim}{E} = 0$, is unnecessarily restrictive. We expect that a 'small' amount of uncompensated charge will have no effect on the free energy density in the thermodynamic limit while a 'large' amount of uncompensated charge will lead to a divergent free energy density in that limit. The dividing line between 'small' and 'large' should be when the excess charge Q_j, in a domain Λ_j, increases in proportion to the 'surface area' of Λ_j as $j \to \infty$. In this case we expect the thermodynamic limit of the free energy density to exist but that its value depends also on the limiting *shape* of the domains Λ_j.

These expectations come from macroscopic electrostatic theory [7] which shows that the lowest energy configuration for any net charge Q confined to a domain Λ is obtained when Q is concentrated at the boundary of Λ. This configuration of the charge is described in electrostatics by a two dimensional charge density $\sigma(\underset{\sim}{x})$, $\underset{\sim}{x} \in S_\Lambda$, where S_Λ is the surface of Λ. (We shall only consider three dimensional systems here, that is $\Lambda_j \subset \mathbb{R}^3$.) This surface charge density will be such as to make the electrostatic potential constant in the interior of Λ, i.e., there will be no electric field in Λ. The electrostatic energy of this surface layer is equal to $\frac{1}{2} Q^2/C(\Lambda)$ where $C(\Lambda)$ is the *capacitance* of Λ.

For a given domain shape, $C(\Lambda)$ is proportional to $[V(\Lambda)]^{1/3}$ and the electrostatic energy per unit volume will thus be proportional to $[Q/V^{2/3}]^2$, the square of the 'average surface charge density'. Hence for a sequence of domains $\{\Lambda_j\}$ with volumes $\{V_j\}$ and capacitances $\{C_j\}$ each containing a net charge Q_j such that as $j \to \infty$, $V_j \to \infty$, $C_j/V_j^{1/3} \to c$ and $Q_j/V_j^{2/3} \to \sigma$, the minimum electrostatic energy per unit volume $\tilde{e}_j$ will also approach a limit

$$\tilde{e} = \lim_j \tilde{e}_j = \frac{1}{2}\frac{\sigma^2}{c} \quad . \tag{4.1}$$

Note that (4.1) refers solely to the *macroscopic* electrostatic energy per unit volume of the charge Q in the domain Λ or on the surface S_Λ. We shall now state a theorem which shows that in the thermodynamic limit the difference between the free energy densities of a neutral system, obtained in section 3 and of a system containing some extra charged particles is given precisely by (4.1). For technical reasons the theorem is proved only for a sequence of domains whose shapes approach ellipsoids in the sense defined below. This is more restrictive than is desirable or (probably) necessary as will be clear from the derivation of the theorem.

Definition. Let E be an open ellipsoid of unit volume and capacity c_E. A sequence of domains $\{\Lambda_j\}$, $j = 1, 2, \ldots$, will be called *asymptotically similar to* E if $V(\Lambda_j) \to \infty$ and if there exist ellipsoids $\{E'_j\}$ and $\{E''_j\}$ similar to E such $E'_j \subset \Lambda_j \subset E''_j$ and $V(E''_j)/V(E'_j) \to 1$ as $j \to \infty$. The capacity of Λ_j will clearly lie between the capacities of E'_j and E''_j. These latter capacities are $c_E[V(E'_j)]^{1/3}$ and $c_E[V(E''_j)]^{1/3}$ respectively.

Theorem 4.1

Let $\{\Lambda_j\}$ *be a sequence of domains asymptotically similar to an ellipsoid* E, *and let* $\{\underset{\sim}{N}_j\}$, *and* $\{\underset{\sim}{n}_j\}$ *be sequences of integer particle number vectors such that* $\underset{\sim}{N}_j \cdot \underset{\sim}{E} = 0$, $\underset{\sim}{n}_j \cdot \underset{\sim}{E} = Q_j$, *and*

$$\lim_{j\to\infty} \underset{\sim}{N}_j/V(\Lambda_j) = \underset{\sim}{\rho} \ , \quad \lim_{j\to\infty} \underset{\sim}{n}_j/V(\Lambda_j) = 0 \ , \quad \lim_{j\to\infty} Q_j[V(\Lambda_j)]^{-2/3} = \sigma \quad .$$

Then if $|\underset{\sim}{\rho}| < \rho_c$,

$$\lim_{j\to\infty} g([\underset{\sim}{N}_j + \underset{\sim}{n}_j]/V(\Lambda_j);\ \Lambda_j) = g(\underset{\sim}{\rho}) - \frac{1}{2}\frac{\sigma^2}{c_E} \quad .$$

Remarks. (a) Since $E'_j \subset \Lambda_j \subset E''_j$, it follows from the basic inequality that $Z(\underset{\sim}{N}_j + \underset{\sim}{n}_j;\ E''_j) \geq Z(\underset{\sim}{N}_j + \underset{\sim}{n}_j;\ \Lambda_j) \geq Z(\underset{\sim}{N}_j + \underset{\sim}{n}_j;\ E'_j)$. Moreover, since $V(E''_j)/V(E'_j) \to 1$ as $j \to \infty$ it is sufficient to prove the theorem for a sequence of ellipsoids $\{E_j\}$ similar to E, whose volumes are the same as that of the $\{\Lambda_j\}$. With each E_j we associate a pair of homothetic ellipsoids, E_j^- and E_j^+ similar to E_j such that $E_j^- \subset E_j \subset E_j^+$ and $V(E_j^+)/V(E_j^-) \to 1$ as $J \to \infty$. The volumes and capacities of E_j^-, E_j, and E_j^+ will be denoted by $(L_j^-)^3$, L_j^3, $(L_j^+)^3$ and C_j^-, C_j, C_j^+ respectively. Clearly $C_j^\pm = c_E L_j^\pm$ and $C_j = c_E L_j$. The interiors of the ellipsoidal shells $E_j^+\backslash E_j$ and $E_j\backslash E_j^-$ will be called D_j^+ and D_j^- respectively.

(b) The reason for the introduction of ellipsoidal domains, is their well known electrostatic property [7] that a *uniform* three-dimensional charge density τ in an ellipsoidal shell such as D_j^+ (defined above) has a self energy $\frac{1}{2}\tau^2 V(D_j^+)^2/C'_j$ and produces a *constant* potential $\tau V(D_j^+)/\bar{C}_j$ in the interior of E_j, with

$c_E L_j \le \bar{C}_j \le C'_j \le c_E L_j^+$. This fact will enable us to obtain bounds on the partition functions for the domains $\{E_j\}$ in a simple manner. Identical methods would work also for any other sequence of domains for which there are shell domains surrounding each Λ_j with the above mentioned properties of the shells $D_j^{\pm}$.

(c) The proof of Theorem (4.1) will proceed by establishing bounds on the free energy of these systems. For this we shall need the free energies of two kinds of neutral systems: the first kind consists of $\underset{\sim}{N}_j$ particles in E_j^-; the second kind is a system in E_j^+ which contains an additional species of particles so that it has altogether $s + 1$ species. The new species, which, following Aristotle, we call *hyle* will be labeled by the index zero. Its charge e_0 will be ± 1 (in units in which all e_i, $i = 1, \ldots, s$, are integers). The sign of e_0 will be chosen as the opposite of the sign (which we shall take to be independent of j) of the excess charge Q_j, that is $e_0 Q_j < 0$. The new neutral system will have an $s + 1$ component particle number vector $\underset{\sim}{N}_j + \underset{\sim}{n}_j + \underset{\sim}{n}_j^0 = (n_j^0, N_j^1 + n_j^1, \ldots, N_j^s + n_j^s)$ with $n_j^0 = |Q_j|$, $n_j^0 e_0 = -Q_j$ so that the system is overall neutral. The hyle particles will only have Coulomb interactions and will be fermions in order to comply with the Dyson-Lenard theorem.

Lower Bound on the Partition Functions of Charged Ellipsoids

We consider a packing of E_j^- with balls and we distribute the $\underset{\sim}{N}_j$ particles, $\underset{\sim}{N}_j \cdot \underset{\sim}{E} = 0$, among the balls such that each ball is neutral and call the resulting partition function $Z(\underset{\sim}{N}_j;\ B(E_j^-))$. The remaining $\underset{\sim}{n}_j$ particles we place in D_j^-. It then follows from our basic inequality and the fact that each ball is neutral that

$$Z(\underset{\sim}{N}_j + \underset{\sim}{n}_j;\ E_j) \ge Z(\underset{\sim}{N}_j;\ B(E_j^-))Z(\underset{\sim}{n}_j;\ D_j^-) \quad . \tag{4.2}$$

It can be shown, using Theorem 3.1, that the packing for each j can be chosen so that upon taking the logarithm of (4.2) and dividing by $V(E_j)$ one obtains

$$\liminf_{j\to\infty}\{g(\underset{\sim}{N}_j + \underset{\sim}{n}_j]/V(E_j);\ E_j) - [V(E_j)]^{-1}\ell nZ(\underset{\sim}{n}_j;\ D_j^-)\} \ge g(\underset{\sim}{\rho}) \quad . \tag{4.3}$$

Since $\underset{\sim}{n}_j/V(E_j) \to 0$, the only contribution from $\ell nZ(\underset{\sim}{n}_j;\ D_j^-)/V(E_j)$ which survives when $j \to \infty$ is the Coulomb self energy of the charges in D_j^-. We now use the following general inequality for the partition function of a system of N particles in a domain Λ, with a Hamiltonian H;

$$\ell nZ(N;\ \Lambda) \ge -J^{-1} \sum_{\alpha=1}^{J} (\psi_\alpha, H\psi_\alpha) \quad , \tag{4.4}$$

where $\{\psi_\alpha\}$, $\alpha = 1, \ldots, J$ is *any* properly symmetrized and normalized set of functions of the N particle coordinates $\underset{\sim}{x}_i$, $i = 1, \ldots, N$ and spins, which vanish unless $\underset{\sim}{x}_i \in \Lambda$. Applying (4.4) to $Z(\underset{\sim}{n}_j : D_j^-)$ with a choice of ψ_α which corresponds to the $\underset{\sim}{n}_j$ particles being situated in little balls centered on the vertices of a cubical lattice covering D_j^-, we obtain a lower bound on this self energy corresponding to a uniform distribution of the charge Q_j in D_j^-,

$$\lim_{j\to\infty} \sup[V(E_j)]^{-1} \ell n Z(\underset{\sim}{n}_j;\, D_j^-) \geq -\frac{1}{2}\frac{\sigma^2}{c_E} \quad . \tag{4.5}$$

This yields

$$\lim_{j\to\infty} \inf g([\underset{\sim}{N}_j + \underset{\sim}{n}_j]/V(E_j);\, E_j) \geq g(\underset{\sim}{\rho}) - \frac{1}{2}\frac{\sigma^2}{c_E} \quad . \tag{4.6}$$

Upper Bound on the Partition Functions of Charged Ellipsoids

Let $Z(\underset{\sim}{N}_j + \underset{\sim}{n}_j + \underset{\sim}{n}_j^0;\, E_j^+)$ be the partition function of a system in the domain E_j^+ having $s + 1$ species with $n_j^0 = |Q_j|$ hyle particles of charge $e_0 = -Q_j/|Q_j|$, as in remark (c) after Theorem 4.1. The masses m_0 of the hyle particles may be chosen arbitrarily. We then have

$$Z(\underset{\sim}{N}_j + \underset{\sim}{n}_j^0;\, E_j^+) \geq Z(\underset{\sim}{N} + \underset{\sim}{n}_j\!:\, E_j) Z(n_j^0;\, D_j^+\!:\, W_j) \quad . \tag{4.7}$$

Here $Z(n_j^0;\, D_j^+\!:\, W_j)$ is the partition function of n_j^0 particles of species zero whose Hamiltonian consists of a kinetic energy term, a Coulomb pair interaction term, and an *external one-body electrostatic potential* $w_j(\underset{\sim}{x}_i)$, $i = 1, \ldots, n_j^0$, produced by the (canonical ensemble) average charge density of the $\underset{\sim}{N}_j + \underset{\sim}{n}_j$ particles in E_j. Taking logarithms in (4.7) and dividing by $V(E_j^+)$ gives the upper bound

$$\lim_{j\to\infty} \sup\{g([\underset{\sim}{N}_j + \underset{\sim}{n}_j]/V(E_j^+);\, E_j^+) + [V(E_j^+)]^{-1} \ell n Z(n_j^0;\, D_j^+\!:\, W_j)\} \leq g(\underset{\sim}{\rho}) \quad . \tag{4.8}$$

Here, too, the only contribution from $\ell n Z(n_j^0;\, D_j^+\!:\, W_j)$ which survives in the limit is the Coulomb energy which now consists of two parts: the self energy of the charges in D_j^+ and the mutual electrostatic energy between the charge $-Q_j$ in D_j^+ and Q_j in E_j. Now if the charge $-Q_j$ were smeared out uniformly in D_j^+ then, because of the properties of the ellipsoidal shells mentioned in remark (b), the sum of these two energies would be $\frac{1}{2} Q_j^2/C_j' - Q_j^2/\bar{C}_j$ with C_j' and $\bar{C}_j$ both approaching C_j as $j \to \infty$. It can be shown indeed by using inequality (4.4) with a suitable choice of $\{\psi_\alpha\}$, that

$$\lim_{j\to\infty} \inf[V(E_j)]^{-1} \ell n Z(n_j^0;\, D_j^+\!:\, W_j) \geq \frac{1}{2}\frac{\sigma^2}{c_E} \quad . \tag{4.9}$$

Combining this with (4.8) and (4.6) yields Theorem 4.1.

When the magnitude of the charge contained in Λ_j, $Q_j = \underset{\sim}{M}_j \cdot \underset{\sim}{E}$, (where $\underset{\sim}{M}_j$ is an integer particle number vector), increases faster than $V(\Lambda_j)^{2/3}$, i.e., $|Q_j| V(\Lambda_j)^{-2/3} \to \infty$, then it is possible to show that $g(\underset{\sim}{M}_j/V(\Lambda_j);\, \Lambda_j) \to \infty$ for *any regular sequence of domains* $\{\Lambda_j\}$.

5. GRAND CANONICAL ENSEMBLE

The grand canonical partition function for a system of s species in a domain Λ_j with chemical potentials μ_i, $i = 1, \ldots, s$, is defined as

$$\Xi(\underset{\sim}{\mu};\ \Lambda_j) = \sum_{N^1=0}^{\infty} \cdots \sum_{N^s=0}^{\infty} \exp[\underset{\sim}{\mu} \cdot \underset{\sim}{N}]Z(\underset{\sim}{N};\ \Lambda_j) \quad , \tag{5.1}$$

where $\underset{\sim}{\mu} = (\mu_1,\ldots,\mu_s)$, and we have set $\beta = 1$. The grand canonical pressure is defined as

$$\pi(\underset{\sim}{\mu};\ \Lambda_j) = V(\Lambda_j)^{-1}\ell n\Xi(\underset{\sim}{\mu};\ \Lambda_j) \quad . \tag{5.2}$$

We also define the *neutral* grand canonical partition function Ξ', by restricting the summations in the right side of (5.1) to neutral systems for which $\underset{\sim}{N} \cdot \underset{\sim}{E} = 0$. The function Ξ' will clearly depend only on that part of the vector $\underset{\sim}{\mu}$ which is perpendicular to $\underset{\sim}{E}$, i.e., on $\underset{\sim}{\mu}' = \underset{\sim}{\mu} - (\underset{\sim}{\mu} \cdot \underset{\sim}{E})\underset{\sim}{E}/(\underset{\sim}{E} \cdot \underset{\sim}{E})$, and will thus be a function of only $s - 1$ independent variables,

$$\Xi'(\underset{\sim}{\mu}';\ \Lambda_j) = \sum_{(\underset{\sim}{N}\cdot\underset{\sim}{E}=0)}\!\!\cdots\sum \exp[\underset{\sim}{\mu}' \cdot \underset{\sim}{N}]Z(\underset{\sim}{N};\ \Lambda_j) \quad . \tag{5.3}$$

Similarly,

$$\pi'(\underset{\sim}{\mu}';\ \Lambda_j) = V(\Lambda_j)^{-1}\ell n\Xi'(\underset{\sim}{\mu}';\ \Lambda_j) \tag{5.4}$$

is the neutral grand canonical pressure.

As in section 4, we shall confine our attention here to domains $\Lambda_j \subset \mathbb{R}^3$.

Remark. As is well known, if $\mu > 0$, the grand canonical partition function Bose gas is infinite for large j (Bose-Einstein condensation). One can prove, [17], that if the particles interact with a tempered super-stable potential then this pressure does exist for all μ, while for a tempered potential which is only stable the pressure exists only for small values of μ (depending on β), i.e., $\mu < f(\beta)$.

For Coulomb systems to be H-stable the Dyson-Lenard theorem requires that all charged bosons have charges of the same sign. We can show that if the only bosons present are charged ones then $\lim_{j\to\infty} \pi(\underset{\sim}{\mu};\ \Lambda_j)$ exists for all values of the μ_i, $(-\infty \le \mu_i < \infty,\ i = 1, \ldots, s)$; see Lemma 5.3. If, however, our systems contains some species of neutral bosons, say $e_1 = e_2 = \ldots = e_\ell = 0$, $\ell \le s - 2$, then the corresponding μ_i, $i = 1, \ldots, \ell$ will have to be appropriately small unless the tempered potentials involving these uncharged particles satisfy some super-stability condition. Since the part of the proof which involves the uncharged components does not differ from the standard ones we shall assume from now on that all the species are charged with $e_1, \ldots, e_a > 0$ and $e_{a+1}, \ldots, e_s < 0$. We shall assume that species $a + 1, \ldots, s$ are fermions and that some or all of species $1, \ldots, a$ may be bosons.

We shall now state the main theorem of this section.

Theorem 5.1

For any regular sequence of domains $\{\Lambda_j\}$, $\pi(\underset{\sim}{\mu}) = \lim_{j\to\infty} \pi(\underset{\sim}{\mu};\ \Lambda_j) = \lim_{j\to\infty} \pi'(\underset{\sim}{\mu}';\ \Lambda_j) = \pi'(\underset{\sim}{\mu}')$ *exists and is related to the Helmholtz free energy density by*

$$\pi'(\underset{\sim}{\mu}') = \sup_{|\underset{\sim}{\rho}|<\rho_c} [\underset{\sim}{\rho} \cdot \underset{\sim}{\mu}' + g(\underset{\sim}{\rho})] \quad , \tag{5.5}$$

the supremum being taken only over values of $\underset{\sim}{\rho}$ *for which* $\underset{\sim}{\rho} \cdot \underset{\sim}{E} = 0$.

Proof. The proof that $\lim \pi'(\mu'; \Lambda_j)$ exists and is given by (5.5) is analogous to Fisher's [3] proof of a similar result for one component systems interacting only with tempered potentials with the additional result that the μ_i are arbitrary even if some of the components are bosons. (The reason for this is that if the boson density is large then the fermion density must also be large to insure charge neutrality. See Lemma 5.3.) The new element entering Theorem 5.1 is the equality of $\pi(\underset{\sim}{\mu}; \Lambda_j)$ and $\pi'(\underset{\sim}{\mu}'; \Lambda_j)$ in the thermodynamic limit. This means in essence that the terms in the grand partition function for which $\underset{\sim}{N}_j \cdot \underset{\sim}{E} \neq 0$ do not contribute to the pressure in this limit and hence $\lim_{j\to\infty} \pi(\underset{\sim}{\mu}; \Lambda_j)$ depends only on $s - 1$ variables. Now since $\pi(\underset{\sim}{\mu}; \Lambda_j) \geq \pi'(\underset{\sim}{\mu}'; \Lambda_j)$, Theorem 5.1 will be established if we can prove that $\pi(\underset{\sim}{\mu}; \Lambda_j) \leq \pi(\underset{\sim}{\mu}'; \Lambda_j) + \delta_j$ with $\delta_j \to 0$ as $j \to \infty$. This is accomplished with the help of the following three lemmas which we shall give here without proof (assuming for simplicity that there are no hard cores).

Lemma 5.1

Let $\underset{\sim}{M} = (M^1,\ldots,M^s)$ *be an integer particle number vector such that* $\underset{\sim}{M} \cdot \underset{\sim}{E} = Q$. *It is then possible to decompose* $\underset{\sim}{M}$ *into a "neutral" part* $\underset{\sim}{N}$ *and a "charged" part* $\underset{\sim}{n}$, $\underset{\sim}{M} = \underset{\sim}{N} + \underset{\sim}{n}$ *such that:* (a) $\underset{\sim}{N}$ *and* $\underset{\sim}{n}$ *are both integer particle number vectors;* (b) $\underset{\sim}{N} \cdot \underset{\sim}{E} = 0$, $\underset{\sim}{n} \cdot \underset{\sim}{E} = Q$; (c) *it is impossible to decompose* $\underset{\sim}{n}$ *into a non-zero neutral part and a charged part;* (d) $|\underset{\sim}{n}| \equiv \sum_{i=1}^{s} n^i \leq \lambda|Q|$ *with* λ *a constant.*

Lemma 5.2

Let $\{\Lambda_j\}$ *be a regular sequence of domains with* $V(\Lambda_j) = V_j$ *and let* K *be a compact subset of* $\{\rho : |\rho| < \rho_c\}$. *Let* $\underset{\sim}{\mu}$ *be a fixed chemical potential. Then there exists a sequence of numbers* $\{\varepsilon_j\}$ *(depending on* K *and* $\underset{\sim}{\mu}$*) tending to zero as* $j \to \infty$, *such that*

$$\underset{\sim}{\mu} \cdot \underset{\sim}{n} V_j^{-1} + g(\underset{\sim}{M} V_j^{-1}; \Lambda_j) - g(\underset{\sim}{N}; \Lambda_j) \leq \varepsilon_j \quad , \tag{5.6}$$

whenever $\underset{\sim}{M} V_j^{-1} \in K$ *and* $\underset{\sim}{M} = \underset{\sim}{N} + \underset{\sim}{n}$ *as in Lemma* 5.1.

Lemma 5.3

Let $\{\Lambda_j\}$ *be a sequence of regular domains with volumes* $\{V_j\}$. *Then there exists some fixed, strictly positive constants* k *and* α *independent of* j *such that*

$$Z(\underset{\sim}{M};\Lambda_j) \le \{\prod_{i=1}^{a} Z^{+}_{0,i}(M_i;\Lambda_j)\}\{\prod_{i=a+1}^{s}(M_i!)^{-1}[V_j]^{M_i}\}$$
$$\cdot \exp\{k\sum_{i=1}^{s} M_i - \alpha V_j^{-\frac{1}{3}}(\underset{\sim}{\mu}\cdot\underset{\sim}{E})^2\}$$

for j *sufficiently large. Here,* $Z^{+}_{0,i}(M_i;\Lambda_j)$ *is the ideal Bose gas partition function of* M_i *bosons of species* i *in the domain* Λ_j.

The proof of Theorem 5.1 now proceeds as follows: Using Lemma 5.3 we establish that

$$\Xi(\underset{\sim}{\mu};\Lambda_j) \le 2\sum_{M^1=0}^{[V_j/v_0]}\cdots\sum_{M^s=0}^{[V_j/v_0]} \exp[\underset{\sim}{\mu}\cdot\underset{\sim}{M}]Z(\underset{\sim}{M};\Lambda_j) \tag{5.7}$$

for j sufficiently large, where v_0 is some fixed small volume. The inequality (5.7) is easily obtained for non-Coulomb systems when the interactions among the bosons is superstable. The physical content of Lemma 5.3 is that the Coulomb energy is as efficacious as a superstable interaction in this respect; the Coulomb energy discourages a large excess of bosons over fermions. The number of terms in (5.7) is at most $(1 + V_j/v_0)^s$. If we now write $\underset{\sim}{M} = \underset{\sim}{N} + \underset{\sim}{n}$ as in Lemma 5.1 and use (5.6) we readily find that,

$$\Xi'(\underset{\sim}{\mu};\Lambda_j) \le \Xi(\underset{\sim}{\mu};\Lambda_j) \le 2\sum_{(\underset{\sim}{N}\cdot\underset{\sim}{E}=0)}\cdots\cdots\sum e^{\underset{\sim}{\mu}\cdot\underset{\sim}{N}} Z(\underset{\sim}{N};\Lambda_j)$$
$$\left\{\sum_{n^1=0}^{[V_j/v_0]}\cdots\sum_{n^s=0}^{[V_j/v_0]} e^{\underset{\sim}{\mu}\cdot\underset{\sim}{n}}\frac{Z(\underset{\sim}{N}+\underset{\sim}{n};\Lambda_j)}{Z(\underset{\sim}{N};\Lambda_j)}\right\}$$
$$\le 2(2 + V_j/v_0)^s[\exp(\varepsilon_j V_j)]\Xi'(\underset{\sim}{\mu};\Lambda_j) \quad , \tag{5.8}$$

so that

$$\pi'(\underset{\sim}{\mu};\Lambda_j) \le \pi(\underset{\sim}{\mu};\Lambda_j) \le \pi'(\underset{\sim}{\mu};\Lambda_j) + \delta_j \quad , \tag{5.9}$$

and $\delta_j \to 0$ as $j \to \infty$. Equation (5.9) proves the equivalence of $\pi'(\underset{\sim}{\mu};\Lambda_j)$ and $\pi(\underset{\sim}{\mu};\Lambda_j)$ in the thermodynamic limit. The proof of the existence of $\pi'(\underset{\sim}{\mu})$ and (5.5) is identical to that for systems with tempered potentials.

Remark. Theorem 5.1 shows in a striking way the special nature of the Coulomb potential. In the absence of the Coulomb potential, but for any tempered potential, one can, by properly choosing the various chemical potentials μ_i, induce essentially any desired ratio of the densities ρ_i of the various species. For Coulomb potentials, on the other hand, only neutral densities are permitted in the thermodynamic limit. To be more specific, it can be readily shown that $\langle Q\rangle_j V_j^{-1} \to 0$ as $j \to \infty$, where $\langle Q\rangle_j$ is the expectation value of the charge in Λ_j, for an arbitrary choice of the chemical potentials μ_i, i = 1, ..., s.

An interesting question arises about the behavior of the charge fluctuations $\langle[Q - \langle Q\rangle_j]^2\rangle_j V_j^{-1}$ as $j \to \infty$. It seems certain on the basis of our previous results that this will approach zero (probably as $V_j^{-2/3}$) when $j \to \infty$, but we have not established this rigorously.

6. THE MICROCANONICAL ENSEMBLE FOR NEUTRAL SYSTEMS

In the foregoing pages we discussed the existence and properties of the canonical and grand canonical free energies per unit volume. The microcanonical ensemble is an ensemble of even more physical and historical importance. From it the requisite thermodynamic properties of the canonical and grand-canonical ensembles may be deduced directly on general grounds, but the converse is not true. The microcanonical partition, function $\Omega(E,\underset{\sim}{N};\,\Lambda)$, is a function of energy, E, the domain, Λ, and the particle number vector $\underset{\sim}{N}$. There are many ways to define Ω, but in any case one defines an entropy/unit volume, σ, as a function of density, $\underset{\sim}{\rho}$, and energy/unit volume, ε, by

$$\sigma(\varepsilon,\underset{\sim}{\rho};\,\Lambda) \equiv V^{-1}\ell n\Omega(\varepsilon V,\underset{\sim}{\rho}V;\,\Lambda) \quad , \tag{6.1}$$

where $V = V(\Lambda)$. In addition to showing that σ has a thermodynamic limit which is concave in $(\varepsilon,\underset{\sim}{\rho})$, one also has to show that the various definitions of Ω yield the same limiting σ function. (See Ruelle [17] and references quoted therein.)

Instead of following the usual route of first defining σ and then its inverse function $\varepsilon(\sigma,\underset{\sim}{\rho};\,\Lambda)$, we define ε directly to suit our purposes. We then show that it has all the requisite thermodynamic properties for *neutral systems in general domains* as we did in section 3 for the canonical free energy. *It can also be shown that our definitions of ε and σ (which is defined to be the inverse of our ε function) agree with the usual definitions in the thermodynamic limit.* The "equivalence" of the microcanonical ensemble to the canonical and grand-canonical ensembles in this limit is a consequence of the general arguments already developed for non-Coulomb systems (cf. Ruelle [17]).

The Microcanonical Energy Function ε

Definition. Consider a quantum system in a domain Λ (of volume V) with particle density $\underset{\sim}{\rho}$. Let $E_1 \le E_2 \le \ldots$ be the eigenvalues of the Hamiltonian arranged in increasing order (including multiplicity). Let $\sigma \in \mathbb{R}^1$ and let $\ell \ge 1$ be the smallest integer $\ge \exp(\sigma V)$. Then the *energy function* is defined by

$$\varepsilon(\sigma,\underset{\sim}{\rho};\,\Lambda) \equiv (V\ell)^{-1}\sum_{i=1}^{\ell} E_i \quad . \tag{6.2}$$

Remarks. (a) H-stability provides the lower bound

$$\varepsilon(\sigma,\underset{\sim}{\rho};\,\Lambda) \ge |\underset{\sim}{\rho}|\phi \quad , \tag{6.3}$$

for some constant, ϕ.

(b) The range of $V\varepsilon(\sigma,\underset{\sim}{\rho};\Lambda)$ is $[E_1,\infty)$ since the Hamiltonian is unbounded above.

(c) It is clear from the definitions that ε is non-decreasing in σ. Hence, the energy function has a pseudo-inverse called the *entropy function* which will be denoted by $\sigma(\varepsilon,\underset{\sim}{\rho};\Lambda)$. It is given explicitly by

$$\sigma(\varepsilon,\underset{\sim}{\rho},\Lambda) = \sup\{\sigma\colon \varepsilon(\sigma,\underset{\sim}{\rho};\Lambda) \le \varepsilon\} \quad . \tag{6.4}$$

Implicit in Eq. (6.2) is the notion that each E_i is defined for all $\underset{\sim}{\rho}$ by linear interpolation. Thus, the definition, (6.4), of σ is not the same as one would obtain if one defined σ for non-integral particle numbers by linear interpolation of σ. In other words, we have given priority to the energy function. It is also to be noted that while the domain of ε (in σ) is $(-\infty,\infty)$, the domain of σ (in ε) is $[E_1/V,\infty)$.

We now use the *minimax principle* which states that if $\{\psi_i\}$, $i = 1, \ldots, \ell$ is a set of ℓ orthonormal functions (called variational functions) in the domain of the Hamiltonian, H, and that if we form the ℓ-square Hermitian matrix A whose elements are $A_{ij} = (\psi_i, H\psi_j)$, and label the eigenvalues of A as $\lambda_1 \le \lambda_2 \le \ldots \le \lambda_\ell$ then $\lambda_i \ge E_i$ for $i = 1, \ldots, \ell$. In particular, for integral particle numbers,

$$\varepsilon(\sigma,\underset{\sim}{\rho};\Lambda) \le (V\ell)^{-1}\mathrm{Tr}A \quad , \tag{6.5}$$

where $\exp(\sigma V) = \ell$. This formula shows the advantage of our definition of ε because all we need to know are the diagonal elements of A.

To apply this principle, let $\Lambda \supset \Lambda_1 \cup \Lambda_2$, with Λ_1 and Λ_2 disjoint, and let $\underset{\sim}{N} = \underset{\sim}{N}_1 + \underset{\sim}{N}_2$ be the respective particle number in the various domains. If $\{\psi_i^1, E_i^1\}$, $i = 1, \ldots, n_1$ (resp. $\{\psi_i^2, E_i^2\}$, $i = 1, \ldots, n_2$) are the first n_1 (resp. n_2) eigenfunctions and eigenvalues in Λ_1 (resp. Λ_2), we can form the set of $n_1 n_2$ variational functions in Λ by $\psi_{ij} = \psi_i^1 \otimes \psi_j^2$. To evaluate the right hand side of (6.5) we need consider only $A_{ij,ij}$ and this is given by

$$A_{ij,ij} = E_i^1 + E_j^2 + U_{ij} \quad , \tag{6.6}$$

where U_{ij} is the expectation value of the inter-domain part of the potential energy. Obviously, (6.6) generalizes in a trivial way when Λ contains more than two disjoint subdomains.

The average interaction, U_{ij}, consists of a non-Coulomb, but tempered part and a Coulomb part. The former can be easily bounded and we shall ignore it in these notes. Bounding the Coulomb part U_{ij}^C is slightly more complicated.

Suppose that Λ_1 in the previous discussion is a ball, B. Each index i denoting the eigenfunctions and eigenvalues of the Hamiltonian in B can best be written as a pair (α,m) where α denotes the principal quantum numbers, including the angular momentum, $L(\alpha)$ (irreducible representation of the rotation group), and m denotes the magnetic quantum number (row of the representation). The energy E_i depends only on α and not on m. Suppose further that n_1 is such that for every

α all the levels (α,m) with $-L(\alpha) \le m \le L(\alpha)$ appear in the list $1, \ldots, n_1$ if any one (α,m') does. In that case we shall say that n_1 is perfect. When we do the sum $\sum_{m=-L}^{L} U^C_{(\alpha,m),j}$, which is part of the sum in (6.5), we have to evaluate an average charge density in Λ_1 which involves integrals over all but one of the N_1 particle coordinates in B, such as

$$I_\alpha(\underset{\sim}{r}) = \sum_{m=-L}^{L} \int_{B^{N_1-1}} |\psi^1_{(\alpha,m)}(\underset{\sim}{r},\underset{\sim}{r}_2,\ldots,\underset{\sim}{r}_{N_1})|^2 d\underset{\sim}{r}_2 \ldots d\underset{\sim}{r}_{N_1} \quad .$$

Clearly I_α depends only on the distance of r from the center of B. If, in addition, we postulate that $N_1 \cdot E = 0$, i.e., that Λ_1 contains a neutral mixture of particles, then the average Coulomb potential outside of Λ_1 will vanish by Newton's theorem. That is

$$\sum_{i=1}^{n_1} U^C_{ij} = 0 \quad \text{for all } j \quad , \tag{6.7}$$

regardless of the shape of Λ_2 and of its constituent particles. If n_1 is not perfect, it lies between two perfect numbers μ and ν, $\mu < n_1 < \nu$, $\nu - \mu = 2L(\alpha) + 1 \equiv t$, where α is the last principle quantum number appearing in the first n_1 levels. The sum $\sum_{i=1}^{\mu} U^C_{ij} = 0$ and can be ignored. We are then left with

$$\widetilde{U} = \sum_{i=\mu+1}^{n_1} U^C_i \quad \text{where} \quad U^C_i \equiv \sum_{j=1}^{n_2} U^C_{ij} \quad .$$

The key fact is that we can relabel the last t levels in ν such that $\widetilde{U} \le 0$. This is so because $\sum_{i=\mu+1}^{\nu_1} U^C_i = 0$.

Writing, for $i = 1, 2$, $x_i = V_i/V$ and $\exp(\sigma_i V_i) = n_i$, then if $\exp(\sigma V) = n_1 n_2$ we have $\sigma = x_1\sigma_1 + x_2\sigma_2$. If we now denote the energy function of Λ_i by ε_i and if Λ_1 is a ball, then the preceding discussion shows that

$$\varepsilon(x_1\sigma_1 + x_2\sigma_2, x_1\underset{\sim}{\rho}_1 + x_2\underset{\sim}{\rho}_2; \Lambda) \le x_1\varepsilon_1(\sigma_1,\underset{\sim}{\rho}_1; \Lambda_1) + x_2\varepsilon_2(\sigma_2,\underset{\sim}{\rho}_2; \Lambda_2) \quad . \tag{6.8}$$

It can be shown that (6.8) is true even when n_1, n_2, $\underset{\sim}{N}_1$ and $\underset{\sim}{N}_2$ are not integral and that it generalizes to more than two subdomains provided all but one of them is a ball. Thus, we have established precisely the analogue of the inequalities on the g function of section 3. Therefore, the same analysis as that given in section 3 will lead to the same conclusions for the energy function.

Our results are summarized in the following theorem:

Theorem 6.1

(a) *Let* $\{\Lambda_j, \underset{\sim}{N}_j\}$ *be a sequence of regular domains and integer valued particle number vectors satisfying the neutrality condition* $\underset{\sim}{N}_j \cdot \underset{\sim}{E} = 0$ *and such that* $\underset{\sim}{\rho}_j = V(\Lambda_j)^{-1}\underset{\sim}{N}_j$ *satisfies* $|\underset{\sim}{\rho}| < \rho_c$. *Let a sequence of entropies* $\{\sigma_j\}$ *also be given and suppose that* $\underset{\sim}{\rho}_j \to \underset{\sim}{\rho}$ *with* $|\underset{\sim}{\rho}| < \rho_c$ *and* $\sigma_j \to \sigma$. *Then the energy functions* $\varepsilon(\sigma_j, \underset{\sim}{\rho}_j; \Lambda_j)$ *converge to a function* $\varepsilon(\sigma,\rho)$ *which is independent of the particular sequence.*

(b) $\varepsilon(\sigma,\boldsymbol{\rho})$ *is continuous and convex in* $(\sigma,\boldsymbol{\rho})$ *in the domain*

$$D = \{(\sigma,\boldsymbol{\rho}) : |\boldsymbol{\rho}| < \rho_c, \boldsymbol{\rho}\cdot\boldsymbol{E} = 0, -\infty < \sigma < \infty\} \quad .$$

It is also non-decreasing in σ.

(c) $\varepsilon(\sigma,\boldsymbol{0}) = 0$.

(d) *Let* K *be a compact subset of* D. *Suppose that for each* $(\sigma,\boldsymbol{\rho}) \in K$ *we have a sequence* $\{\sigma_j(\sigma,\boldsymbol{\rho}),\boldsymbol{\rho}_j(\sigma,\boldsymbol{\rho})\}$ *which approaches* $(\sigma,\boldsymbol{\rho})$ *uniformly on* K. *Then* $\varepsilon(\sigma_j,\boldsymbol{\rho}_j;\Lambda_j)$ *approaches* $\varepsilon(\sigma,\boldsymbol{\rho})$ *uniformly on* K.

(e) *The entropy function,* $\sigma(\varepsilon,\boldsymbol{\rho},\Lambda)$, *also approaches a limit* $\sigma(\varepsilon,\boldsymbol{\rho})$ *uniformly on compacts.*

(f) $\sigma(\varepsilon,\boldsymbol{\rho})$ *is continuous, and concave in* $(\varepsilon,\boldsymbol{\rho})$ *in the domain* $D = \{(\varepsilon,\boldsymbol{\rho}) : |\boldsymbol{\rho}| < \rho_c, \boldsymbol{\rho}\cdot\boldsymbol{E} = 0, \varepsilon > \varepsilon_1(\boldsymbol{\rho})$, *where* $\varepsilon_1(\boldsymbol{\rho}) = \lim_{j\to\infty} E_1(\boldsymbol{\rho}_j;\Lambda_j)V(\Lambda_j)^{-1}$ *and* $E_1(\boldsymbol{\rho};\Lambda)$ *is the lowest eigenvalue of the Hamiltonian in* Λ. *It is also non-decreasing in* ε *and its range is not bounded above.*

(g) $\sigma(\varepsilon,\boldsymbol{\rho})$ *and* $\varepsilon(\sigma,\boldsymbol{\rho})$ *are inverse functions.*

REFERENCES

[1] Dyson, F. J. and A. Lenard, *J. Math. Phys.* 8, 423 (1967).

[2] Dyson, F. J., *J. Math. Phys.* 8, 1538 (1967).

[3] Fisher, M. E., *Arch. Ratl. Mech. Anal.* 17, 377 (1964).

[4] Fisher, M. E. and D. Ruelle, *J. Math. Phys.* 7, 260 (1966).

[5] Griffiths, R. B., "Rigorous Results and Theorems", to appear in *Phase Transitions and Critical Points*, C. Domb and M. S. Green, editors, Academic Press.

[6] Kato, T., *Perturbation Theory for Linear Operators*, Springer, (1966).

[7] Kellog, O. D., *Foundations of Potential Theory*, Dover Publication, (1953).

[8] Lebowitz, J. L., *Ann. Rev. Phys. Chem.* 19, 389 (1968).

[9] Lebowitz, J. L. and E. Lieb, *Phys. Rev. Lett.* 22, 631 (1969).

[10] Lenard, A. and F. J. Dyson, *J. Math. Phys.* 9, 698 (1968).

[11] Lieb, E. H. and J. L. Lebowitz, "The Constitution of Matter", to appear in *Advances in Math.* December 1972.

[12] Morrey, C. B., *Comm. Pure. Appl. Math.* 8, 279 (1955).

[13] Newton, I., *Philosophia Naturalis Principia Mathematica* (1687): translated by A. Motte, revised by F. Cajori, University of California Press, Berkeley, California, (1934), Book 1, p. 193, propositions 71, 76.

[14] Onsager, L., *J. Phys. Chem.* 43, 189 (1939).

[15] Onsager, L., in *The Neurosciences*, p. 75, G. C. Quarton, T. Melinchick and F. O. Schmitt, editors, Rockefeller University Press, (1967).

[16] Penrose, O. and E. R. Smith, *Comm. Math. Phys.* 26, 53 (1972).

[17] Ruelle, D., *Statistical Mechanics*, W. A. Benjamin, Inc., New York, (1969).

SOME PROBLEMS CONNECTED WITH THE DESCRIPTION OF COEXISTING PHASES AT LOW TEMPERATURES IN THE ISING MODEL

G. Gallavotti,* A. Martin-Löf,** and S. Miracle-Solé***

INTRODUCTION

In the following we are going to give an account of some recent results describing a system in which two phases can co-exist starting from the basic assumptions of statistical mechanics. We will consider the 2-dimensional Ising model with nearest neighbour interaction and without external magnetic field, this being the "simplest" model which can be shown to undergo a phase transition at low temperature. This transition is manifested e.g. by the instability of the average magnetization and other averages with respect to perturbations in the boundary conditions or in the external field. For example, if the boundary spins are all +1 (-1) then the average magnetization of an infinite system will be $+m^*$ $(-m^*)$, $m^* > 0$ being the spontaneous magnetization. We are going to use the terminology of Dobrushin [3] and Lanford-Ruelle [8] (applicable to much more general spin systems) and call Gibbs-state or equilibrium state of a finite system a probability distribution for its configurations defined by the Boltzmann factor and by fixing some boundary condition. Such a probability distribution is characterized by specifying the correlation functions $\langle\sigma_{x_1}\ldots\sigma_{x_n}\rangle$ for all finite families $\{\sigma_{x_1},\ldots,\sigma_{x_n}\}$ of spins on the lattice. A Gibbs-state of an infinite system is then defined as a probability distribution, or family of correlation functions, which are limits of correlation functions of an increasing family of systems with some boundary conditions. The occurrence of a phase transition in the above sense then reveals itself by the existence of more than one Gibbs-state for the infinite system; the correlation functions can have different limits depending on the boundary conditions. The set of possible Gibbs-states for the infinite system can be seen to form a convex compact set of probability distributions in a suitable topology, and an arbitrary Gibbs-state can be represented as a convex linear combination ("mixture") of extremal (in the sense of convexity theory) Gibbs-states. Such extremal states are often identified with "pure phases". They have correlations which decay over large distances: $\langle\sigma_{x_1}\ldots\sigma_{x_n}\sigma_{y_1}\ldots\sigma_{y_m}\rangle \to \langle\sigma_{x_1}\ldots\sigma_{x_n}\rangle\langle\sigma_{y_1}\ldots\sigma_{y_m}\rangle$ as $d(x_1,\ldots,x_n;y_1,\ldots,y_m) \to \infty$ at least in the Cesaro sense.

---

* Istituto di Matematica dell' Universita di Roma nell' ambito del GNAFA.

** Department of Mathematics, Royal Institute of Technology, Stockholm.

*** Facultad de Ciencias, Universidad de Zaragoza.

This somewhat abstract definition of "pure phase" is elaborated in Section 2 for the model we consider. It is shown that at low temperature there are only two extremal translationally invariant Gibbs-states for the infinite system namely those obtained by taking as boundary conditions: all spins at the boundary equal to +1 or -1. These states have average magnetization $\pm m^*$ respectively and have decaying correlations. An arbitrary translationally invariant Gibbs-state is then a "mixture" of these two states in the sense that its correlation functions are given by $\langle \sigma_{x_1} \ldots \sigma_{x_n} \rangle = \alpha \langle \sigma_{x_1} \ldots \sigma_{x_n} \rangle_+ + (1-\alpha) \langle \sigma_{x_1} \ldots \sigma_{x_n} \rangle_-$ for some α with $0 \le \alpha \le 1$. From the proof of the above relation it is furthermore seen that it is a consequence of the fact that in a large finite system with some given boundary condition a typical configuration can be described as a mixture (in the ordinary sense) of large regions, in which the state is described by either of the two extremal probability distributions; the average proportions of the + regions and the - regions being approximately α and $(1-\alpha)$ respectively. "Large" in the above picture means that the arcs of the region near the boundaries separating + and - regions, where boundary effects are noticeable, is negligible. This means that the correlation functions $\langle \sigma_{x_1} \ldots \sigma_{x_n} \rangle$ above can be interpreted as describing the state of a family $\{\sigma_{x_1+a}, \ldots, \sigma_{x_n+a}\}$ with a chosen "at random" in the box containing the system, because with probability α the family will fall well inside a + region and with probability $1-\alpha$ well inside a - region.

In Section 3,5, and 6 we study a situation in which the separation of the two phases can be investigated and the surface tension coming from their "surface" of separation can be exhibited. A simple boundary condition producing a separation of the two phases is the one we consider: The lattice is periodic in the horizontal direction (i.e. it is a vertical cylinder) and it has the boundary condition + on the top edge and - on the bottom edge. We show in Section 3 and 5 that if one considers the canonical ensemble with a fixed magnetization $m = \alpha m^* + (1-\alpha)(-m^*)$ and these boundary conditions, then a typical configuration will consist of one + region on top of one - region. In these regions the average magnetizations are very nearly $+m^*$ and $-m^*$ respectively, and the proportions of the areas are very nearly α and $(1-\alpha)$. Furthermore the border going around the cylinder, which separates the two regions, has a length which does not exceed the circumference of the cylinder very much. In Section 3 and 6 we show that there is a surface tension associated with this border, which does not depend on α, and which therefore can be called the surface tension between two coexisting phases. (To be distinguished from the surface tension between the fixed spins at the top e.g. and the + phase below.)

In Section 4 we develop a "cluster theory" of the two pure phases, which is a basic tool in the proofs of Section 5 and 6. In Section 7 we discuss some problems related to those treated in the preceding sections. The results of Section 2 are due to Gallavotti, Miracle-Solé [7]. The results and proofs concerning the phase separation in Sections 3 and 5 are basically those of Minlos, Sinai (M.S.) [10]. However, their treatment has been simplified at several points where we make use of the cylindrical boundary conditions, which they do not consider, and of the cluster theory of Section 3, the use of which in many respects shortens their proofs considerably. The results concerning the surface tension are due to Gallavotti, Martin-Löf [5], and the proofs given here are with small modifications the same as those in [5].

Section 2 has been written by S. Miracle-Solé and the other sections by G. Gallavotti and A. Martin-Löf. We are very much indebted to Professor A. Lenard for the invitation to attend the Battelle Summer Rencontres.

1. NOTATIONS AND DEFINITIONS

Let Θ be a finite subset of the infinite square lattice $\mathbb{Z}^2$. We recall that the Ising model in the "box" Θ is defined by associating to each point $x \in \Theta$ a spin variable σ_x taking values ± 1. We shall always suppose that the spins on the boundary of Θ are fixed, and we denote by $\underline{\tau}$ the array specifying their values. To each configuration of spins in Θ, having the specified boundary values $\underline{\tau}$, we associate the weight (Boltzmann-factor):

$$\widetilde{w}_{\underline{\tau}}(\underline{\sigma}) = \exp \frac{\beta}{2} \sum_{\langle x,y\rangle} \sigma_x \sigma_y \quad , \tag{1.1}$$

where the sum runs over all pairs of nearest neighbors in Θ (including the boundary spins), and β is proportional to the inverse temperature. The probability of an allowed spin configuration is then determined by the normalized weight.

For our purpose of investigating the system at low temperature, (i.e. for β large) it will be very convenient to represent a configuration not as an array $\underline{\sigma}$ but by specifying the boundaries separating + and − spins. This can be done as follows. Given a configuration $\underline{\sigma}$, draw for each bond on the lattice having opposite spins at its endpoints a segment of length one perpendicular to the bond and centered at its midpoint. In this way we obtain a family of lines on the lattice shifted $(\frac{1}{2},\frac{1}{2})$ from the original one. The segments separating boundary spins are fixed by $\underline{\tau}$; they and the others form a graph such that each point on the shifted lattice "inside" Θ belongs to 0, 2, or 4 segments. The last case occurs if the spins around the point are arranged as

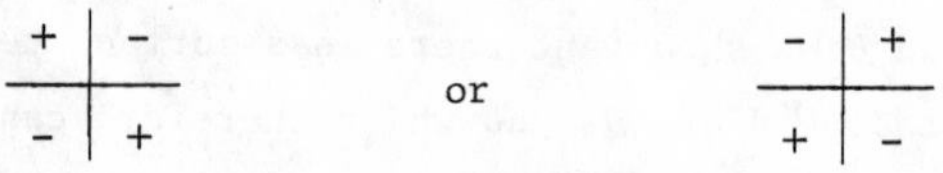

Figure 1

If in this case we modify the lines as follows:

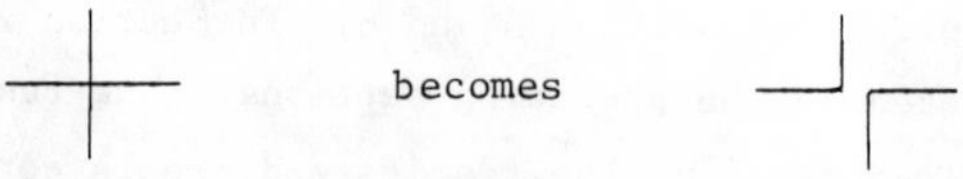

Figure 2

we realize that the set of lines splits into a family of edge-selfavoiding contours, some of which start and end at the fixed segments of the boundary, the others being closed and lying "inside" Θ. (The contours are allowed to touch as in Fig. 2 but not in any other way.) We denote contours by the letter γ and their lengths by $|\gamma|$ and also the number of points in Θ by $|\Theta|$. Often we will denote open contours that start and end at the boundary by the letter η. Any closed contour has at least the length 4 and any open one at least the length 2. We will often denote an equivalence class of congruent contours by (γ). Given the boundary condition $\underline{\tau}$ a spin configuration is uniquely characterized by its associated family of contours, so we can talk about contour configurations instead of spin configurations without ambiguity. The weight of a configuration $(\gamma_1,\ldots,\gamma_n)$ is easily expressed in terms of the contours:

$$\sum_{\langle x,y\rangle}\sigma_x\sigma_y = \text{no. of bonds in } \Theta - 2\sum_1^n|\gamma_i| \quad , \tag{1.2}$$

so we will describe the probability distribution defined by (1.1) by giving to each member of the ensemble $M^{\underline{\tau}}(\Theta)$ of allowed contour configurations $(\gamma_1,\ldots,\gamma_n)$ the weight

$$w_{\underline{\tau}}(\gamma_1,\ldots,\gamma_n) = e^{-\beta\sum_1^n|\gamma_i|} \quad . \tag{1.3}$$

and the probability $w_{\underline{\tau}}(\gamma_1,\ldots,\gamma_n)/Z(M^{\underline{\tau}}(\Theta),\beta)$, where $Z(M^{\underline{\tau}}(\Theta),\beta)$ is the partition function

$$Z(M^{\underline{\tau}}(\Theta),\beta) = \sum_{(\gamma_1,\ldots,\gamma_n)\in M^{\underline{\tau}}(\Theta)} e^{-\beta\sum_1^n|\gamma_i|} \tag{1.4}$$

When the boundary condition is $+1$ or -1 we denote the ensemble by $M^+(\Theta)$ or $M^-(\Theta)$.

More generally, we will also consider other ensembles of configurations and other weights. For any ensemble, M, of configurations in a finite set and any translationally invariant function $\mu(\gamma)$ we define the partition function by:

$$Z(M,\mu) = \sum_{(\gamma_1,\ldots,\gamma_n)\in M} e^{\sum_1^n\mu(\gamma_i)} \tag{1.5}$$

For example, we will consider the canonical ensembles $M^{\underline{\tau}}(\Theta,m)$ of allowed configurations with a given magnetization $\sum_{x\in\Theta}\sigma_x = m|\Theta|$.

In Sections 3, 4, 5 and 6 we will consider the situation where we have periodic boundary conditions in the horizontal direction, i.e. instead of the infinite planar lattice $\mathbb{Z}^2$ we consider an infinitely long cylinder $\Omega_{\infty,N}$ with circumference N. The above considerations are valid for regions $\Theta\subset\Omega_{\infty,N}$ as well; one only has to remember that contours can now also go around the cylinder. Especially we are going to consider a cylindrical region $\Omega\subset\Omega_{\infty,N}$ with flat top and bottom and with height N^δ for some $\delta>1$. For it, the four boundary conditions + on the top, - at the bottom etc., will be denoted by +, - etc. For any of these boundary conditions the number of "big" countours, i.e. contours going around the cylinder, must have a given parity, even for +,+ or −,− and odd for +,− or −,+. We will denote by $M_0^{++}(\Omega)$ etc. the ensembles of contours having a minimal number of such contours, i.e. 0 for +,+ or −,− and 1 for +,− or −,+. More generally, for any region $\Theta \subset \Omega_{\infty,N}$ we denote by $M_0^{+}(\Theta)$ the ensemble of contours defined by the extra restriction that no contour is allowed to go around $\Omega_{\infty,N}$ or any connected component of the complement of Θ. (Θ is allowed to have "holes" in it.) Finally we will also have occasion to consider the ensemble of "c-small" contours in Ω defined by the extra restriction that all closed contours have a length bounded by $c \log|\Omega| = c \log N^{1+\delta}$ for a given c. (We will only consider some fixed value of c, e.g. c = 1/300.) An added subscript c will be used to denote this ensemble. A contour is called "c-large" if it is neither "c-small" nor "big".

Concerning the style used in several of the proofs we remark that the abundance of estimates involving numerical constants is not designed to impress the reader with their high accuracy; they are in fact often quite crude. We think however that it is more informative to give values of constants involved instead of writing the customary const., because then it is easier to keep in mind that they do not after all depend on some other parameters, which are later changed. We have not explicitly worked out an estimate of the critical value of β above which our proofs go through, because we have not strived to make the most careful estimates possible at several points anyhow. The phrase "for β large and N large" often used means: for all β above a value independent of N, and all N above a value possibly dependent on β. Finally we remark that we often freely ignore unimportant rounding off effects due to the fact that the magnetization in a region is an integer $\equiv |\Theta| \bmod 2$, and we write: magnetization = $m|\Theta|$ for any m, height = N^{δ} etc. instead of the more cumbersome correct expressions.

2. THE TRANSLATIONALLY INVARIANT EQUILIBRIUM STATES

In this section we investigate the question of how many phases can coexist in equilibrium. We shall give the following answer to this question: Assume that β is large enough. Then any translationally invariant equilibrium state is a convex linear combination of two pure states describing the up-magnetized and down-magnetized pure phases.

In order to make this statement precise we next introduce some definitions. Let Ω be a finite region on the lattice and let $P_\Omega(\underline{\tau})$ be a given probability distribution over the set of boundary conditions. The corresponding equilibrium state in Ω is then described by the set of correlation functions

$$\langle\sigma_{x_1}\ldots\sigma_{x_n}\rangle_{P_\Omega} = \sum_{\underline{\tau}} \langle\sigma_{x_1}\ldots\sigma_{x_n}\rangle_{\underline{\tau}} P_\Omega(\underline{\tau}) \quad . \tag{2.1}$$

An equilibrium state of the infinite system is defined as a set of correlation functions $\langle\sigma_{x_1}\ldots\sigma_{x_n}\rangle$, which can be written, for a suitable choice of the sequence P_Ω, as

$$\langle \sigma_{x_1} \ldots \sigma_{x_n} \rangle = \lim_{\Omega\to\infty} \langle \sigma_{x_1} \ldots \sigma_{x_n} \rangle_{P_\Omega} \tag{2.2}$$

for all $\{x_1,\ldots,x_n\}$. We say that it is a translationally invariant equilibrium state if furthermore

$$\langle \sigma_{x_1+a} \ldots \sigma_{x_n+a} \rangle = \langle \sigma_{x_1} \ldots \sigma_{x_n} \rangle \tag{2.3}$$

for all a. For β large enough it is known that there are at least two different equilibrium states, which will be denoted by $\langle \sigma_{x_1} \ldots \sigma_{x_n} \rangle_+$ and $\langle \sigma_{x_1} \ldots \sigma_{x_n} \rangle_-$. These states are obtained as

$$\langle \sigma_{x_1} \ldots \sigma_{x_n} \rangle_\pm = \lim_{\Omega\to\infty} \langle \sigma_{x_1} \ldots \sigma_{x_n} \rangle_{\pm,\Omega} \quad . \tag{2.4}$$

They describe the up-magnetized and down-magnetized pure phases. We shall next expose some of the special physical properties of these two states, which justify why one uses this terminology. In the following we will need Lemma 2.1 below. From it the existence of the limits (2.4) follows, with some uniformity in Ω.

Lemma 2.1

If D *is the distance of* $\{x_1,\ldots,x_n\} \subset \Omega$ *from the boundary of* Ω *we have*

$$|\langle \sigma_{x_1} \ldots \sigma_{x_n} \rangle_\pm - \langle \sigma_{x_1} \ldots \sigma_{x_n} \rangle_{\pm,\Omega}| \le f(x_1,\ldots,x_n,D) \tag{2.5}$$

where $f(x_1,\ldots,x_n,D)$ *is a translationally invariant function tending to zero as* $D \to \infty$.

The proof of this lemma is given in Appendix 2A. We also show that the two states $\langle \sigma_{x_1} \ldots \sigma_{x_n} \rangle_\pm$ are translationally invariant, and furthermore that

$$\langle \sigma_{x_1} \ldots \sigma_{x_n} \rangle_+ = (-1)^n \langle \sigma_{x_1} \ldots \sigma_{x_n} \rangle_- \tag{2.6}$$

by the obvious symmetry argument. They are also pure phases in the sense of Section 1, i.e. they are extremal points of the set of all translationally invariant equilibrium states. We can deduce this from the following cluster property:

$$\lim_{d(x_1\ldots x_n;y_1\ldots y_m)\to\infty} \langle \sigma_{x_1} \ldots \sigma_{x_n} \sigma_{y_1} \ldots \sigma_{y_m} \rangle_\pm = \langle \sigma_{x_1} \ldots \sigma_{x_n} \rangle_\pm \langle \sigma_{y_1} \ldots \sigma_{y_m} \rangle_\pm \quad . \tag{2.7}$$

A proof of these facts is given in Appendix 2A. Moreover, let us consider the free energy $f(\beta,h)$ and the equilibrium state $\langle \sigma_{x_1} \ldots \sigma_{x_n} \rangle_h$ in an external field $h \neq 0$. It has recently been proved by Ruelle [12] that this equilibrium state is unique, extremal, translationally invariant and analytic in β,h (as is the free energy) when $h \neq 0$. It has also been proved [9] that

$$\langle \sigma_{x_1} \ldots \sigma_{x_n} \rangle_\pm = \lim_{h\to\pm0} \langle \sigma_{x_1} \ldots \sigma_{x_n} \rangle_h$$
$$\langle \sigma_x \rangle_\pm = \pm m^* \quad , \quad m^* = \frac{\partial f}{\partial h}(\beta,0+) \quad . \tag{2.8}$$

Let us now state the main result of this section.

Theorem 2.1

If $\beta > \log 3$, *any translationally invariant equilibrium state* $\langle\sigma_{x_1}\cdots\sigma_{x_n}\rangle$ *is given for some* α, $0 \le \alpha \le 1$, *by* $\langle\sigma_{x_1}\cdots\sigma_{x_n}\rangle = \alpha\langle\sigma_{x_1}\cdots\sigma_{x_n}\rangle_+ + (1-\alpha)\langle\sigma_{x_1}\cdots\sigma_{x_n}\rangle_-$. (2.9)

Before proving the theorem we make two observations. First, about the value of β above which the theorem holds. It can in fact be proved for $\beta > \log \mu_0$, where μ_0 is the "connective constant" for the self-avoiding walks on the 2-dimensional lattice we consider. It is well known that $\log \mu_1$ for completely self-avoiding walks is an estimate from above of the critical β_c to within 9%. We notice that in order to get the proof of Theorem 2.1 we do not need the techniques which are developed in Section 4. This is the reason why the theorem can be proved so close to the critical temperature.

The second observation concerns the value of the spontaneous magnetization. It is an old question whether its value m^* given in (2.8) coincides with the value

$$m_0 = (1 - (\sinh\beta)^{-4})^{1/8} \tag{2.10}$$

computed from the definition

$$m_0^2 = \lim_{d(x,y)\to\infty} \langle\sigma_x\sigma_y\rangle \quad , \tag{2.11}$$

where $\langle\sigma_x\sigma_y\rangle$ is the limit of the two-spin correlation function with periodic boundary conditions. Since Theorem 2.1 says that $\langle\sigma_x\sigma_y\rangle$ is independent of the boundary conditions, the identity $m_0 = m^*$ for β large enough follows from it. One needs only to apply the last formulae in (2.8) and use the strong cluster property of $\langle\sigma_x\sigma_y\rangle_+$.

For the proof of Theorem 2.1 it is convenient to introduce the averaged correlation functions. They are defined as:

$$\begin{aligned}\langle\overline{\sigma_{x_1}\cdots\sigma_{x_n}}\rangle_{\underline{\tau},\Omega} &= |\Omega|^{-1}\sum_a \langle\sigma_{x_1+a}\cdots\sigma_{x_n+a}\rangle_{\underline{\tau},\Omega} \\ \langle\overline{\sigma_{x_1}\cdots\sigma_{x_n}}\rangle_{P_\Omega} &= |\Omega|^{-1}\sum_a \langle\sigma_{x_1+a}\cdots\sigma_{x_n+a}\rangle_{P_\Omega} \quad ,\end{aligned} \tag{2.12}$$

where the sum runs over all the a's such that $\{x_1+a,\ldots,x_n+a\} \subset \Omega$. The reason for defining the above averages lies in the fact that if $\langle\sigma_{x_1}\cdots\sigma_{x_n}\rangle$ is a set of translationally invariant correlation functions verifying (2.2), (2.3), then one can find a suitable sequence of distributions P_Ω such that for all $\{x_1,\ldots,x_n\}$:

$$\langle\sigma_{x_1}\cdots\sigma_{x_n}\rangle = \lim_{\Omega\to\infty}\langle\overline{\sigma_{x_1}\cdots\sigma_{x_n}}\rangle_{P_\Omega} \quad . \qquad \text{(See [8].)} \tag{2.13}$$

This fact together with the remark that $\langle\overline{\sigma_{x_1}\cdots\sigma_{x_n}}\rangle_{P_\Omega}$ is a convex combination of the $\langle\sigma_{x_1}\cdots\sigma_{x_n}\rangle_{\underline{\tau}}$ will imply Theorem 2.1 if the following lemma holds.

Lemma 2.2

If $\beta > \log 3$ *one can find a family of numbers* $\alpha_{\Omega,\underline{\tau}}$ *such that* $0 \le \alpha_{\Omega,\underline{\tau}} \le 1$ *and such that*

$$|\overline{\langle\sigma_{x_1}\cdots\sigma_{x_n}\rangle_{\underline{\tau},\Omega}} - \alpha_{\underline{\tau},\Omega}\langle\sigma_{x_1}\cdots\sigma_{x_n}\rangle_+ - (1 - \alpha_{\underline{\tau},\Omega})\langle\sigma_{x_1}\cdots\sigma_{x_n}\rangle_-| \le g(x_1\ldots x_n,\Omega) \quad , \tag{2.14}$$

where $g(x_1,\ldots,x_n,\Omega)$ *is a translationally invariant and* $\underline{\tau}$*- independent function tending to zero as* $\Omega \to \infty$.

Let us first describe the physical idea from which the proof of the lemma is obtained. Let Ω be a given square box containing L^2 points, and let $\underline{\tau}$ be a fixed boundary condition. For each spin configuration $X \in M^{\underline{\tau}}(\Omega)$ draw the contours $\gamma_1,\ldots,\gamma_n,\eta_1,\ldots,\eta_s$ associated to X in the way we have described in Section 1. We observe that the open contours $\eta_1,\ldots,\eta_s$ divide the box into $s + 1$ disjoint regions $\Theta_1,\ldots,\Theta_{s+1}$, which are such that either all spins adjacent to the boundary from the inside are all $+1$ or all -1. We call the regions "positive" or "negative" according to whether the first or second case happens. See Fig. 3. Let now $\{x_1,\ldots,x_n\}$

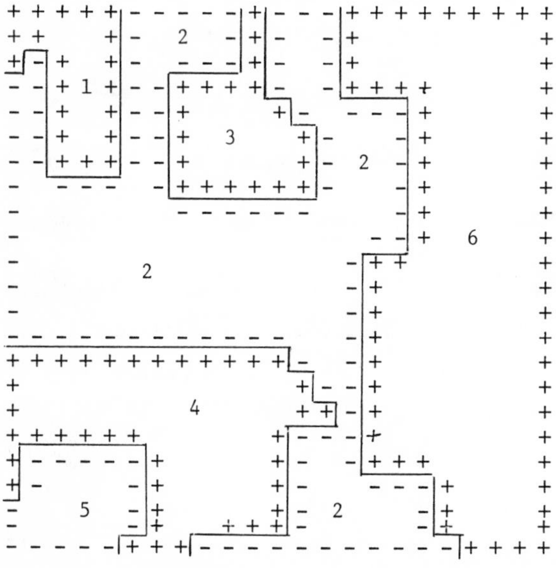

Fig.3. A boundary condition and the η-lines of a spin configuration, (the spins associated with the γ-lines are not drawn). Regions 1, 3, 4, and 6, are positive; 2, and 5, negative; 2 is not connected

be a given set of points inside the "big" box Ω, and suppose that $\beta > \log 3$. Then:

(a) The open contours $\eta_1,\ldots,\eta_s$ have possibly a length of the order of L since they must join points on the boundary, but they tend to be not too long, in order to keep the energy small. They are therefore very far from all but a negligible fraction of the translates of $\{x_1,\ldots,x_n\}$.

(b) A translate $\{x_1 + a,\ldots,x_n + a\}$ is thus almost always in the "middle" of some Θ_i and therefore $\langle\sigma_{x_1+a}\cdots\sigma_{x_n+a}\rangle_{\underline{\tau},\Omega} \approx \langle\sigma_{x_1}\cdots\sigma_{x_n}\rangle_+$ if Θ_i is a positive region or $\approx \langle\sigma_{x_1}\cdots\sigma_{x_n}\rangle_-$ if Θ_i is a negative region.

Lemma 3.1 gives us the precise statement corresponding to the physical remark 2, whereas we can formulate the physical remark 1 by means of the following lemma, which is also proved in Appendix 2A.

Lemma 2.3

If $\beta > \log 3$ *then*

$$P(\underline{\tau},L^{4/3}) = \text{prob}\,\{\sum_1^s |\eta_i| \geq L^{4/3}\} \leq \varepsilon(L) \quad , \tag{2.15}$$

where $\varepsilon(L)$ *is a function independent of* $\underline{\tau}$ *and tending to zero as* $L \to \infty$.

Using the notation (1.3), (1.4) we can compute $\langle\sigma_{x_1}\cdots\sigma_{x_n}\rangle_{\underline{\tau},\Omega}$ as

$$\langle\sigma_{x_1}\cdots\sigma_{x_n}\rangle_{\underline{\tau},\Omega} = \sum_{\substack{\eta_1,\ldots,\eta_s \\ \sum_i|\eta_i|\leq L^{4/3}}} \sum_{\underline{\sigma}}{}^{*} (\sigma_{x_1},\ldots,\sigma_{x_n}) w_{\underline{\tau}}(\underline{\sigma})/Z(M^{\underline{\tau}}(\Omega),\beta) + \varepsilon(x_1,\ldots,x_n,\tau) \quad , \tag{2.16}$$

where the first sum runs over all the sets of possible open contours, and the second sum runs over the spin configurations $\underline{\sigma}$ which have $\eta_1,\ldots,\eta_s$ as associated open contours. The function $\varepsilon(x_1,\ldots,x_n,\underline{\tau})$ is according to the above lemma bounded by:

$$|\varepsilon(x_1,\ldots,x_n,\underline{\tau})| \leq \varepsilon(L) \quad . \tag{2.17}$$

Suppose that $\{x_1,\ldots,x_n\} \subset \Theta_i$, then

$$\sum_{\underline{\sigma}}{}^{*} (\sigma_{x_1},\ldots,\sigma_{x_n}) w_{\underline{\tau}}(\underline{\sigma})/Z(M^{\underline{\tau}}(\Omega),\beta) = P_{\underline{\tau}}(\eta_1,\ldots,\eta_s)\langle\sigma_{x_1}\cdots\sigma_{x_n}\rangle_{\pm,\Theta_i} \quad , \tag{2.18}$$

where the sign has to be chosen to be the same as the one of the region Θ_i, and $P_{\underline{\tau}}(\eta_1,\ldots,\eta_s)$ is the probability of the spin configurations having $\eta_1,\ldots,\eta_s$ as open contours.

This formula follows from the fact that if $\eta_1,\ldots,\eta_s$ are fixed then the probabilities of the spin configurations inside the regions Θ_i are independent and generated by the weights $w_{\pm}(\underline{\sigma})$. Let $N_+(\eta_1,\ldots,\eta_s)$ be the number of points in the positive Θ_i's and put

$$\alpha_{\underline{\tau},\Omega} = \sum_{\underline{\sigma}} P_{\underline{\tau}}(\eta_1,\ldots,\eta_s) N_+(\eta_1,\ldots,\eta_s)/L^2 \quad . \tag{2.19}$$

Denote also by $A(\eta_1,\ldots,\eta_s)$ the set of points at a distance less than $L^{1/3}/2$ from the open contours $\eta_1,\ldots,\eta_s$. Then we find, using (2.16), (2.17), (2.18) and definition (2.12):

$$\left|\overline{\langle\sigma_{x_1}\cdots\sigma_{x_n}\rangle_{\underline{\tau},\Omega}} - \alpha_{\underline{\tau},\Omega}\langle\sigma_{x_1}\cdots\sigma_{x_n}\rangle_+ - (1-\alpha_{\underline{\tau},\Omega})\langle\sigma_{x_1}\cdots\sigma_{x_n}\rangle_-\right|$$

$$\leq \varepsilon(L) + 2f(x_1,\ldots x_n,L^{1/3}) + CL^{5/3}/L^2 \quad , \tag{2.20}$$

where the first term comes from the error term in (2.16), the second comes from the replacement of $\langle\sigma_{x_1+a}\cdots\sigma_{x_n+a}\rangle_{\pm,\Theta_i}$ by $\langle\sigma_{x_1}\cdots\sigma_{x_n}\rangle_{\pm}$ for all the a's such that

$\{x_1 + a,\ldots,x_n + a\}$ does not intersect $A(\eta_1,\ldots,\eta_s)$ and from the use of Lemma 3.1 to estimate the error involved. Finally the third term comes from the contribution of the a's such that $\{x_1 + a,\ldots,x_n + a\}$ intersects $A(\eta_1,\ldots,\eta_s)$. $L^{5/3} = 2\cdot\frac{1}{2}\cdot L^{1/3}\cdot L^{4/3}$ bounds the number of points in $A(\eta_1,\ldots,\eta_s)$, and C is $\underline{\tau}$-independent and depends only on $\max_{i,j} d(x_i,x_j)$. Formula (2.20) proves Lemma 2.2.

We finally remark that the validity of Theorem 2.1 is not restricted to the 2-dimensional Ising model. The same techniques and proofs, (with the appropriate notion of contours) easily extend to any number of dimensions. Furthermore, in the 2-dimensional case, as we already mentioned, the condition $\beta > \log 3$ can be replaced by the weaker condition $\beta > \log \mu_0$. To be convinced of these facts one only needs to examine the proof of Lemma 2.3.

2A. APPENDIX

Proof of Lemma 2.1. The second Griffiths inequality implies that $\langle\sigma_{x_1}\cdots\sigma_{x_n}\rangle_{+,\Omega}$ decreases when Ω increases [9]. Call Q_D a square centered at the barycenter of $x_1,\ldots,x_n$ having side $D^{\frac{1}{2}}$. Then

$$0 \le \langle\sigma_{x_1}\cdots\sigma_{x_n}\rangle_{+,\Omega} - \langle\sigma_{x_1}\cdots\sigma_{x_n}\rangle_{+} \le \langle\sigma_{x_1}\cdots\sigma_{x_n}\rangle_{+,Q_D} - \langle\sigma_{x_1}\cdots\sigma_{x_n}\rangle_{+}$$
$$= f(x_1,\ldots,x_n,D) \quad .$$

The function $f(x_1,\ldots x_n,D)$ is translationally invariant (see Lemma 2.4) and decreases to zero as $D \to \infty$, again by the Griffiths inequality. A similar argument holds for $\langle\sigma_{x_1}\cdots\sigma_{x_n}\rangle_-$.

Lemma 2.4

The states $\langle\sigma_{x_1}\cdots\sigma_{x_n}\rangle_\pm$ *are translationally invariant and verify the relations:*

$$\langle\sigma_{x_1}\cdots\sigma_{x_n}\rangle_+ = (-1)^n\langle\sigma_{x_1}\cdots\sigma_{x_n}\rangle_- \tag{2.6}$$

$$\lim_{d(x_1,\ldots,x_n,y_1,\ldots,y_m)\to\infty} \langle\sigma_{x_1}\cdots\sigma_{x_n}\sigma_{y_1}\cdots\sigma_{y_m}\rangle_\pm = \langle\sigma_{x_1}\cdots\sigma_{x_n}\rangle_\pm\langle\sigma_{y_1}\cdots\sigma_{y_m}\rangle_\pm \tag{2.7}$$

Hence they are extremal translationally invariant and describe "pure phases". (See [8].)

Proof. Let Q_D be defined as above and $Q_D + a$ its translate by a. Obviously $\langle\sigma_{x_1}\cdots\sigma_{x_n}\rangle_{\pm,Q_D} = \langle\sigma_{x_1+a}\cdots\sigma_{x_n+a}\rangle_{\pm,Q_D+a}$. Introduce also the square Q'_D having side D and a region Ω containing Q_D and $Q_D + a$ and lying inside of both Q'_D and $Q'_D + a$. (This is always possible if D is large enough.) Then as in the proof of Lemma 2.1 we have

$$\langle\sigma_{x_1}\cdots\sigma_{x_n}\rangle_{+,Q'_D} \le \langle\sigma_{x_1}\cdots\sigma_{x_n}\rangle_{+,\Omega} \le \langle\sigma_{x_1}\cdots\sigma_{x_n}\rangle_{+,Q_D}$$

and

$$\langle\sigma_{x_1}\cdots\sigma_{x_n}\rangle_{+,Q'_D} = \langle\sigma_{x_1+a}\cdots\sigma_{x_n+a}\rangle_{+,Q'_D+a} \le \langle\sigma_{x_1+a}\cdots\sigma_{x_n+a}\rangle_{+,\Omega} \le \langle\sigma_{x_1+a}\cdots\sigma_{x_n+a}\rangle_{+,Q_D+a}$$

$$= \langle\sigma_{x_1}\cdots\sigma_{x_n}\rangle_{+,Q_D} \quad .$$

Hence

$$|\langle\sigma_{x_1+a}\cdots\sigma_{x_n+a}\rangle_{+,\Omega} - \langle\sigma_{x_1}\cdots\sigma_{x_n}\rangle_{+,\Omega}|$$

$$\le \langle\sigma_{x_1}\cdots\sigma_{x_n}\rangle_{+,Q_D} - \langle\sigma_{x_1}\cdots\sigma_{x_n}\rangle_{+,Q'_D} \le 2f(x_1,\dots,x_n,D) \quad ,$$

and the transitional invariance of $\langle\sigma_{x_1}\cdots\sigma_{x_n}\rangle_+$ follows by passing to the limit $D \to \infty$ in the last inequality. The proof of (2.6) is very easy. It is sufficient to remark that $\langle\sigma_{x_1}\cdots\sigma_{x_n}\rangle_{+,\Omega} = (-1)^n\langle\sigma_{x_1}\cdots\sigma_{x_n}\rangle_{-,\Omega}$ because of the symmetry of the model under the transformation $\sigma_x \to -\sigma_x$. In order to prove (2.7) let us first recall that Griffith's inequality tells us that

$$\langle\sigma_{x_1}\cdots\sigma_{x_n}\sigma_{y_1}\cdots\sigma_{y_m}\rangle_+ \ge \langle\sigma_{x_1}\cdots\sigma_{x_n}\rangle_+\langle\sigma_{y_1}\cdots\sigma_{y_m}\rangle_+ \quad . \tag{2A.1}$$

Suppose that Ω is a square box containing $\{x_1,\dots x_n,y_1,\dots y_m\}$ and (assuming that the horizontal component of a tends to infinity) draw a vertical line v separating the points $x_1,\dots,x_n$ from $y_1,\dots,y_m$. Let us fix the spins at v to be $+1$. This corresponds to putting an infinite positive external field at the points of v. Then by Griffith's inequality the new correlation functions $\langle\sigma_{x_1}\cdots\sigma_{y_m}\rangle^*_\Omega$ verify:

$$\langle\sigma_{x_1}\cdots\sigma_{y_m}\rangle_{+,\Omega} \le \langle\sigma_{x_1}\cdots\sigma_{y_m}\rangle^*_\Omega \quad . \tag{2A.2}$$

But because the two regions Ω_1 and Ω_2 into which the box Ω is divided by v are independent we have:

$$\langle\sigma_{x_1}\cdots\sigma_{y_m}\rangle^*_\Omega = \langle\sigma_{x_1}\cdots\sigma_{x_n}\rangle_{+,\Omega_1}\langle\sigma_{y_1}\cdots\sigma_{y_m}\rangle_{+,\Omega_2} \quad . \tag{2A.3}$$

From (2A.2) and (2A.3) we obtain in the limit $\Omega \to \infty$:

$$\langle\sigma_{x_1}\cdots\sigma_{y_m}\rangle_+ \le \langle\sigma_{x_1}\cdots\sigma_{x_n}\rangle^1_+\langle\sigma_{y_1}\cdots\sigma_{y_m}\rangle^2_+ \tag{2A.4}$$

where the indices 1 and 2 indicate the presence of the boundary v. But when $d \to \infty$ we have:

$$\begin{aligned} \langle\sigma_{x_1}\cdots\sigma_{x_n}\rangle^1_+ &\to \langle\sigma_{x_1}\cdots\sigma_{x_n}\rangle_+ \\ \langle\sigma_{y_1}\cdots\sigma_{y_m}\rangle^2_+ &\to \langle\sigma_{y_1}\cdots\sigma_{y_m}\rangle_+ \end{aligned} \tag{2A.5}$$

and as a consequence of (2A.1), (2A.4) and (2A.5) we obtain (2.7).

Proof of Lemma 2.3. We have

$$P(\underline{\tau},L^{4/3}) = \sum_{\substack{\eta_1,\dots,\eta_s\\ \Sigma|\eta_i|\ge L^{4/3}}} \sum_{\underline{\sigma}}{}^{*}\, w_{\underline{\tau}}(\underline{\sigma})/Z(M^{\underline{\tau}}(\Omega),\beta)$$

$$= \sum_{\substack{\eta_1,\dots,\eta_s\\ \Sigma|\eta_i|\ge L^{4/3}}} e^{-\beta\sum_i|\eta_i|}(Z(M^+(\Theta_1),\beta)\dots Z(M^+(\Theta_{s+1}),\beta))/Z(M^{\underline{\tau}}(\Theta),\beta) \quad , \qquad (2A.6)$$

where we use the notations of Section 1 and observe that $Z(M^+(\Theta),\beta) = Z(M^-(\Theta),\beta)$. We shall establish the following inequalities

$$\sum_{\substack{\eta_1,\dots,\eta_s\\ \Sigma|\eta_i|\ge L^{4/3}}} e^{-\beta\sum_i|\eta_i|} \le \left(\frac{4}{1-3e^{-\beta}}\right)^{2L} \max_{s\le 2L} \binom{L^{4/3}}{s}(3e^{-\beta})^{L^{4/3}} \qquad (2A.7)$$

$$Z(M^+(\Theta_1),\beta)\dots Z(M^+(\Theta_{s+1}),\beta) \le Z(M^+(\Omega),\beta) \qquad (2A.8)$$

$$Z(M^+(\Omega),\beta) \le \left(\exp(8L)\frac{(3e^{-\beta})^4}{1-3e^{-\beta}}\right)Z(M^+(\Omega_1),\beta) \qquad (2A.9)$$

$$Z(M^{\underline{\tau}}(\Omega),\beta) \ge e^{-6\beta L}Z(M^+(\Omega_1),\beta) \quad , \qquad (2A.10)$$

where Ω_1 is a square box concentric with Ω having side $L-2$. From the above inequalities we obtain (2.15) directly from (2A.6) if $\beta > \log 3$. To prove (2A.7) observe that the number of contours starting at a given point and having length ℓ is not larger than 3^{ℓ}, and furthermore we observe that if $\underline{\tau}$ is fixed also the number s of contours $\eta_1,\dots,\eta_s$ is assigned, and there are at most $\binom{2s}{s}$ ways of choosing s starting points among the $2s$ possible ones. Hence, since $2s \le 4L$:

$$\sum_{\substack{\eta_1,\dots,\eta_s\\ \sum_i|\eta_i|\ge L^{4/3}}} e^{-\beta\sum_i|\eta_i|} \le \binom{2s}{s} \sum_{\substack{\ell_1,\dots,\ell_s\\ \Sigma\ell_i\ge L^{4/3}}} (3e^{-\beta})^{\ell_1}\dots(3e^{-\beta})^{\ell_s}$$

$$\le \binom{2s}{s}\binom{L^{4/3}}{s}(3e^{-\beta})^{L^{4/3}} \sum_{\ell'_1,\dots,\ell'_s\ge 0} (3e^{-\beta})^{\ell'_1}\dots(3e^{-\beta})^{\ell'_s}$$

$$\le \left(\frac{4}{1-3e^{-\beta}}\right)^{2L} \max_{s\le 2L}\binom{L^{4/3}}{s}(3e^{-\beta})^{L^{4/3}} \quad .$$

The inequality (2A.8) is obvious from the definition of $Z(M^+(\Omega),\beta)$. The inequality (2A.9) can be proved in the following way:

$$Z(M^+(\Omega),\beta) = \sum_{\gamma_1,\dots,\gamma_n\subset\Omega} e^{-\beta\sum_i|\gamma_i|} \le \left(\sum_{\gamma_1\dots\gamma_n\subset\Omega_1} e^{-\beta\sum_i|\gamma_i|}\right)\left(\sum_{\gamma'_1\dots\gamma'_n\not\subset\Omega_1} e^{-\beta\Sigma|\gamma'_i|}\right) =$$

$$= Z(M^+(\Omega_1),\beta)\left(\sum_{\gamma'_1\dots\gamma'_n\not\subset\Omega_1} e^{-\beta\sum_i|\gamma'_i|}\right) \quad ,$$

where $\gamma'_1,\dots,\gamma'_n \not\subset \Omega_1$ means that more of the contours lies in Ω_1. This last sum is bounded by:

$$\sum_{r=1}^{\infty} \frac{1}{r!}\Big(\sum_{\gamma'\not\subset\Omega_1} e^{-\beta|\gamma'|}\Big)^r \le \sum_{0}^{\infty} \frac{1}{r!}\Big(8L \sum_{\gamma'\ni 0} e^{-\beta|\gamma'|}\Big)^r \le \exp 8L \sum_{\ell=y}^{\infty} (3e^{-\beta})^{\ell} = \exp 8L \frac{(3e^{-\beta})^4}{1-3e^{-\beta}} ,$$

where $\sum_{\gamma'\ni 0}$ denotes the sum over all contours containing a fixed point, and (2A.9) follows. Finally, to prove (2A.10) restrict the sum defining $Z(M^{\tau}(\Omega),\beta)$ to a few terms, namely to those in which the contours $\eta_1,\dots,\eta_s$ are fixed, while the contours $\gamma_1,\dots,\gamma_n$ are put in Ω_1. We take the set of open contours to be such that they isolate the + spins of $\underline{\tau}$ and run parallel to the boundary of Ω and next to it. Then $\sum_i |\eta_i| \le 6L$, and therefore (2A.10) follows.

3. DESCRIPTION OF THE PHASE SEPARATION AND DEFINITION OF THE SURFACE TENSION

In this section we describe in more detail the properties of the phase separation taking place in the ensemble $M^{+-}(\Omega,m)$ described in Section 1 for $m = \alpha m^* + (1-\alpha)(-m^*)$, $0 < \alpha < 1$. We also state the definition of surface tension we use. The proofs of the occurrence of the phase separation and the existence of surface tension are given in Sections 5 and 6.

Consider the cylinder Ω with circumference N and height N^{δ}, $\delta > 1$, introduced in Section 1, and let the subensemble $\widetilde{M}_0^{+-}(\Omega,m)$ of $M^{+-}(\Omega,m)$ be defined by the restrictions:

(a) There is only one "big" contour, λ, going around the cylinder, and its length is bounded by $(2 \log 3)\, N/\beta$.

(b) The area of the region Ω_λ above λ is restricted by $\big||\Omega_\lambda| - \alpha|\Omega|\big| \le a|\Omega|^p$, and hence a similar bound holds for the area below λ.

(c) The total magnetization of the region Ω_λ is restricted by $\big|M_\lambda - m^*|\Omega_\lambda|\big| \le a|\Omega|^p$, and hence a similar bound holds for the magnetization below λ.

(d) The total length of the c-large contours is bounded by N/β.

Theorem 3.1

If β *is large enough and* $0 < \alpha < 1$, $\delta > 1$, *then the probability of* $\widetilde{M}_0^{+-}(\Omega,m)$ *in* $M^{+-}(\Omega,m)$ *converges to* 1 *as* $N \to \infty$, i.e.

$$\lim_{N\to\infty} \frac{Z(\widetilde{M}_0^{+-}(\Omega,m),\beta)}{Z(M^{+-}(\Omega,m),\beta)} = 1 \qquad (3.1)$$

for any $a > 0$ *and a suitable* p, $0 < p < 1$. *(Any* p *satisfying* $1 > p$, $p > (1 + c \log 3)/2$, $p > 2/(1+\delta)$, $p > (1 + 1/(1+\delta))/2$ *is "suitable").*

Theorem 3.1 thus says that a phase separation as described in Section 1 takes place with very high probability in the ensemble $M^{+-}(\Omega,m)$.

This picture of the phase separation is the basis for the following definition of the surface tension between the co-existing phases. A basic property of the partition function of any thermo-dynamic system is the extensivity of its logarithm. E.g. $|\Omega|^{-1} \log Z(M^{+-}(\Omega,m),\beta) \to -\beta f(\beta,m)$ as $|\Omega| \to \infty$, where $f(\beta,m)$ is the limiting free energy per unit volume. Surface effects are manifested in terms proportional to the area of the surface between interacting phases in the deviation $\log Z + \beta f(\beta,m)|\Omega|$. For the ensemble $M^{+-}(\Omega,m)$, where there is typically one surface between the two phases and one at each end of the cylinder, one would therefore expect to have an asymptotic relation:

$$\log Z(M^{+-}(\Omega,m),\beta) = -\beta f(\beta,m)|\Omega| + \tau N + 2\tau' N + o(N) \quad , \qquad (3.2)$$

τ being the surface tension between the two phases and τ' that between each phase and the fixed spins. τ is the quantity we want to study and τ' a "spurious" contribution associated directly with the boundary condition. To extract τ we compare (3.2) to the corresponding expression expected for an ensemble consisting only of one phase, e.g. $M^{++}(\Omega,m^*)$:

$$\log Z(M^{++}(\Omega,m^*),\beta) = -\beta f(\beta,m^*)|\Omega| + 2\tau' N + o(N) \quad . \qquad (3.3)$$

In fact, in Section 5 we are going to see that typically there is not going to be any big contour in this case, and for symmetry reasons the contributions from the bases in (3.2) and (3.3) should be the same. $f(\beta,m^*) = f(\beta,-m^*)$ also, for symmetry reasons, so, because $f(\beta,m)$ should be the sum of the contribution from each phase, $f(\beta,m) = \alpha f(\beta,m^*) + (1-\alpha)f(\beta,-m^*) = f(\beta,m^*)$. This should allow us to extract τ and define it by the relation

$$\tau = \lim_{N\to\infty} N^{-1} \log \frac{Z(M^{+-}(\Omega,m),\beta)}{Z(M^{++}(\Omega,m^*),\beta)} \quad . \qquad (3.4)$$

Indeed, in Section 6 we prove

Theorem 3.2

If β *is large enough and* $0 < \alpha < 1$, $\delta > 1$, *then the limit* (3.4) *exists and can also be expressed as the limit of a partition function over the possible shapes of the line of separation* λ:

$$\tau = \lim_{N\to\infty} N^{-1} \log \sum_{\substack{(\lambda) \\ |\lambda| \le N(1+(2\log 3)/\beta)}} e^{-\beta|\lambda| + \mu(\lambda,\beta)} \quad , \qquad (3.5)$$

where $\mu(\lambda,\beta)$ *is a certain weight function (defined in* (6.6)*).* τ *is thus independent of* α *and directly associated with the line of separation between the phases.*

The proof of Section 6 could easily be extended to show also the existence of τ' defined in (3.3) thereby fully justifying (3.3) and (3.4), see [5]. The reason why we consider a very long cylinder $(\delta > 1)$ is that we can then easily exclude spurious boundary effects, which could occur for α small if λ comes near the ends of the cylinder as will be seen in the proof. (See also comment in Section 7.)

4. CLUSTER THEORY FOR A PURE PHASE

In this section we are going to study in some detail the ensemble $M_0^+(\Theta)$ of configurations defined in Section 1 for a region Θ on the infinitely long cylinder $\Omega_{\infty,N}$ or on the planar lattice $\mathbb{Z}^2$. We derive a convenient "virial expansion" of the partition function, which allows us to study its dependence on the region Θ, and we also derive estimates for the probability distributions of groups of contours.

We have seen in Section 1 that a configuration $X \in M_0^+(\Theta)$ is characterized as a family $(\gamma_1,\ldots,\gamma_n)$ of closed self-avoiding contours in Θ, and that its probability is given by:

$$P_{M_0^+(\Theta)}(X) = \begin{cases} e^{\sum_1^n \mu(\gamma_i)} / Z(M_0^+(\Theta),\mu) & \text{if } (\gamma_1,\ldots,\gamma_n) \text{ are compatible} \\ 0 & \text{otherwise} \quad . \end{cases} \tag{4.1}$$

We are of course mainly interested in the special weight function $\mu(\gamma) = -\beta|\gamma|$, but we also need to consider other weights, so we carry through the discussion for a general translationally invariant weight restricted by $\mu(\gamma) \le -b|\gamma|$ for some constant $b > 0$. For definiteness we consider configurations on an infinite cylinder $\Omega_{\infty,N}$ below, but keep in mind that the arguments are also valid for an infinite planar lattice.

The starting point of our discussion is the observation that the Boltzmann factor $\phi(\gamma_1,\ldots,\gamma_n)$, can be expressed in terms of a "pair interaction" $f(\gamma,\gamma')$ between the contours as follows:

$$\phi(\gamma_1,\ldots,\gamma_n) = e^{\sum_1^n \mu(\gamma_i)} \prod_{i<j} f(\gamma_i,\gamma_j) \quad , \tag{4.2}$$

where $f(\gamma,\gamma')$ is defined by

$$f(\gamma,\gamma') = \begin{cases} 1 & \text{if } \gamma,\gamma' \text{ compatible} \\ 0 & \text{otherwise} \quad . \end{cases} \tag{4.3}$$

Observe that this is only true if we do not allow big contours or contours going around "holes" of Θ. We can thus consider the system as a "gas" of contours with "interaction" determined by $f(\gamma,\gamma')$ and "chemical potentials" $\mu(\gamma)$. We can then apply to this "gas" the theory of "low activity" cluster expansions and obtain convenient expressions for the quantities of interest valid for "low activities" $e^{\mu(\gamma)} = e^{-\beta|\gamma|}$, i.e. for low temperatures. We use the "algebraic method" to treat the expansion as described in [11, p86] for an ordinary gas and in [6] for a lattice gas.

In order to introduce the "algebraic method" we consider the set of finite configurations $X = (\gamma_1,\ldots,\gamma_n)$ of not big contours on $\Omega_{\infty,N}$. $\gamma_1,\ldots,\gamma_n$ are allowed to be incompatible and even to coincide. More formally, a configuration is a function $X(\gamma)$ with non-negative integer values such that $N(X) = \sum_\gamma X(\gamma) < \infty$. $X(\gamma)$ is the "multiplicity" of γ in X. Configurations can be added in the obvious sense: $(X_1 + X_2)(\gamma) = X_1(\gamma) + X_2(\gamma)$. We also consider the space F of real-valued functions of configurations $\phi(X)$ such that $\sup_{N(X)=n} |\phi(X)| < \infty$ for all n, and its subspaces F_0 and F_1 of functions such that $\phi(\emptyset) = 0$ and $\phi(\emptyset) = 1$ respectively. If ϕ_1 and $\phi_2 \in F$ we can define their convolution product by:

$$(\phi_1 * \phi_2)(X) = \sum_{X_1+X_2=X} \phi_1(X_1)\phi_2(X_2) \quad . \tag{4.4}$$

The sum is finite since X is finite and $\phi_1 * \phi_2 \in F$ also. Next we define the exponential corresponding to the convolution for $\phi \in F_0$:

$$(\mathrm{Exp}\ \phi)(X) = \sum_{n\geq 0} \frac{\phi^{n*}(X)}{n!} = 1(X) + \sum_{n\geq 1} \frac{1}{n!} \sum_{X_1+\ldots+X_n=X} \phi(X_1)\ldots\phi(X_n) \quad . \tag{4.5}$$

$\phi^{0*}(X) = 1(X)$ is defined to be 1 if $X = \emptyset$ and 0 otherwise. For each X the sum in (4.5) is finite because $\phi \in F_0$, and $\mathrm{Exp}\ \phi \in F_1$. We also define the corresponding logarithmic function for $\phi \in F_1$ as follows. If $\phi = 1 + \phi_0$ with $\phi_0 \in F_0$ then:

$$(\mathrm{Log}\ \phi)(X) = \sum_{n\geq 1} \frac{(-1)^{n+1}}{n} \phi_0^{n*}(X) = \sum_{n\geq 1} \frac{(-1)^{n+1}}{n} \sum_{X_1+\ldots+X_n=X} \phi_0(X_1)\ldots\phi_0(X_n). \tag{4.6}$$

Again, each sum is finite and $\mathrm{Log}\ \phi \in F_0$. We also see that $\mathrm{Exp}\ \mathrm{Log}\ \phi_1 = \phi_1$ and $\mathrm{Log}\ \mathrm{Exp}\ \phi_0 = \phi_0$ for $\phi_0 \in F_0$, $\phi_1 \in F_1$. The main reason for introducing this convolution product is, as we are going to see, the following product property. If $\chi(X)$ is a character function in the sense that $\chi(X_1 + X_2) = \chi(X_1)\chi(X_2)$, i.e. if $\chi(X) = \prod_\gamma z(\gamma)^{X(\gamma)}$ for some function $z(\gamma)$, (which relation we write as $\chi(X) = z^X$) and if $\sum_X |\phi_i(X) z^X| < \infty$, $i = 1, 2$, then

$$\sum_X (\phi_1 * \phi_2)(X) z^X = (\sum_X \phi_1(X) z^X)(\sum_X \phi_2(X) z^X) \tag{4.7}$$

and

$$\sum_X (\mathrm{Exp}\ \phi)(X) z^X = \exp(\sum_X \phi(X) z^X) \tag{4.8}$$

if $\phi \in F_0$.

Especially we are going to use (4.8) when

$$z(\gamma) = \begin{cases} 1 & \text{if } \gamma \text{ lies in some region } \Theta \text{ and does not go} \\ & \text{around any of its "holes".} \\ 0 & \text{otherwise} \end{cases}$$

so that we get

$$\sum_{X\subset\Theta} (\mathrm{Exp}\ \phi)(X) = \exp(\sum_{X\subset\Theta} \phi(X)) \quad . \tag{4.9}$$

($X \subset \Theta$ means that all the contours of X lie in Θ).

We also introduce a "derivation" operator D_X on F defined by:

$$(D_X\phi)(Y) = \phi(X + Y)\,\frac{(X + Y)!}{Y!} \tag{4.10}$$

with $X! = \prod_\gamma X(\gamma)!$. In terms of it the following "Taylor's formula" is valid:

$$\sum_X \phi(X)(u + v)^X = \sum_X \frac{u^X}{X!} \sum_Y (D_X\phi)(Y)\, v^Y \quad ,$$

from which the following rules can easily be proved:

$$D_\gamma(\phi_1 * \phi_2) = (D_\gamma\phi_1) * \phi_2 + \phi_1 * (D_\gamma\phi_2)$$

$$\frac{D_X(\phi_1 * \phi_2)}{X!} = \sum_{X_1+X_2=X} \left(\frac{D_{X_1}\phi_1}{X_1!}\right) * \left(\frac{D_{X_2}\phi_2}{X_2!}\right) \tag{4.11}$$

$$D_\gamma(\mathrm{Exp}\,\phi) = (D_\gamma\phi) * (\mathrm{Exp}\,\phi)$$

$$\frac{D_X(\mathrm{Exp}\,\phi)}{X!} = \left(\sum_{n\geq 1} \frac{1}{n!} \sum_{\substack{\sum_1^n x_i = x \\ x_i = 0}} \left(\frac{D_{X_1}\phi}{X_1!}\right) * \dots * \left(\frac{D_{X_n}\phi}{X_n!}\right)\right) * (\mathrm{Exp}\,\phi) \quad .$$

We now use the notions above to analyze the Boltzmann factor $\phi \in F_1$ defined in (4.2). Let $\phi^T \in F_0$ be defined by $\phi^T = \mathrm{Log}\,\phi$, so that $\phi = \mathrm{Exp}\,\phi^T$. Then

$$Z(M_0^+(\Theta),\mu) = \sum_{X\subset\Theta} \phi(X) = \exp\Bigl(\sum_{X\subset\Theta} \phi^T(X)\Bigr) \quad , \tag{4.12}$$

so we see that the logarithm of the partition function can be expanded in terms of ϕ^T. The important feature of this expansion is that $\phi^T(X)$ does not depend on the region Θ and can be estimated in a suitable way, and this will allow us to study how the partition function depends on Θ.

We can obtain a "graphological" formula for $\phi^T(X)$ like the one for the Ursell functions used in the theory of the Mayer expansion as follows:

If we define $\tilde{\phi}^T(\gamma_1,\dots,\gamma_n) = \phi^T(X)X!$ for all the $n!/X!$ ordered sequences $(\gamma_1,\dots,\gamma_n) = X$ we can write

$$\sum_X \phi^T(X) z^X = \sum_{n\geq 0} \sum_{n(X)=n} z^X \phi^T(X) \sum_{\substack{\gamma_1,\dots,\gamma_n \\ (\gamma_1,\dots,\gamma_n)=X}} \frac{X!}{n!}$$

$$= \sum_{n\geq 0} \frac{1}{n!} \sum_{\gamma_1,\dots,\gamma_n} z(\gamma_1)\dots z(\gamma_n)\tilde{\phi}^T(\gamma_1,\dots,\gamma_n) \quad , \tag{4.13}$$

which expression is used e.g. in [11]. In terms of $\tilde{\phi}^T$ the expression for the convolution can be found by considering

$$\Bigl(\sum_{X_1} \phi_1^T(X_1) z^{X_1}\Bigr)\Bigl(\sum_{X_2} \phi_2^T(X_2) z^{X_2}\Bigr) =$$

$$= \sum_{n_1,n_2\geq 0} \frac{1}{n_1!n_2!} \sum_{\substack{\gamma'_1,\dots,\gamma'_{n_1} \\ \gamma^2_1,\dots,\gamma^2_{n_2}}} z(\gamma'_1)\dots z(\gamma^2_{n_2})\tilde{\phi}^T_1(\gamma'_1,\dots,\gamma'_{n_1})\tilde{\phi}^T_2(\gamma^2_1,\dots,\gamma^2_{n_2})$$

$$= \sum_{n\geq 0} \frac{1}{n!} \sum_{n_1+n_2=n} \frac{n!}{n_1!n_2!} \sum_{\gamma_1\dots\gamma_n} z(\gamma_1)\dots z(\gamma_n)\tilde{\phi}^T_1(\gamma_1,\dots,\gamma_{n_1})\tilde{\phi}^T(\gamma_{n_1+1},\dots,\gamma_{n_2}) \quad .$$

But for any partition of $N = \{1,\dots,n\}$ into $N_1 \cup N_2$ with $|N_1| = n_1$, $|N_2| = n_2$ the last sum can be written

$$\sum_{\gamma_1,\dots,\gamma_n} z(\gamma_1)\dots z(\gamma_n)\tilde{\phi}^T_1(\gamma_i;\ i \in N_1)\tilde{\phi}^T_2(\gamma_i;\ i \in N_2)$$

by a suitable change of dummy variables. Because there are $n!/n_1!n_2!$ such partitions the sum with n_1,n_2 given can be written:

$$\sum_{\gamma_1\dots\gamma_n} z(\gamma_1)\dots z(\gamma_n) \sum_{\substack{N_1\cup N_2=N \\ |N_1|=n_1;\,|N_2|=n_2}} \tilde{\phi}^T_1(\gamma_i;\ i \in N_1)\tilde{\phi}^T_2(\gamma_i;\ i \in N_2) \quad ,$$

and we get:

$$(\sum_{X_1} \phi^T_1(X_1)z^{X_1})(\sum_{X_2} \phi^T_2(X_2)z^{X_2}) = \sum_X z^X(\phi_1 * \phi_2)(X)$$

$$= \sum_{n\geq 0} \frac{1}{n!} \sum_{\gamma_1,\dots,\gamma_n} z(\gamma_1)\dots z(\gamma_n)(\tilde{\phi}^T_1 \cdot \tilde{\phi}^T_2)(\gamma_1,\dots,\gamma_n) \tag{4.14}$$

with

$$(\tilde{\phi}^T_1 \cdot \tilde{\phi}^T_2)(\gamma_1,\dots,\gamma_n) = \sum_{N_1\cup N_2=N} \tilde{\phi}^T_1(\gamma_i;\ i \in N_1)\tilde{\phi}^T_2(\gamma_i;\ i \in N_2) \quad . \tag{4.15}$$

This function being symmetric in $\gamma_1,\dots,\gamma_n$ we can conclude that $\widetilde{(\phi^T_1 * \phi^T_2)} = \tilde{\phi}^T_1 \cdot \tilde{\phi}^T_2$ and by induction that $\widetilde{((\phi^T)^{n*})} = (\tilde{\phi}^T)^{no}$ for all n. This means that $(\mathrm{Exp}\ \phi^T)(X)$ is also given by the expression:

$$\phi(X) = (\mathrm{Exp}\ \phi^T)(X) = (X!)^{-1} \sum_{m=0}^{\infty} \tilde{\phi}^{Tmo}\frac{(\gamma_1,\dots,\gamma_n)}{m!} =$$

$$= (X!)^{-1} \sum_{n=0}^{\infty} \frac{1}{m!} \sum_{N_1\cup\dots\cup N_m=N} \tilde{\phi}^T(\gamma_i;\ i \in N_1)\dots\tilde{\phi}^T(\gamma_i;\ i \in N_m)$$

$$= (X!)^{-1} \sum_{N_1\cup\dots\cup N_m=N}{}' \tilde{\phi}^T(\gamma_i;\ i \in N_1)\dots\tilde{\phi}^T(\gamma_i;\ i \in N_m) \quad , \tag{4.16}$$

where the last sum is over all different partitions of N into any number of parts. This formula is useful for finding ϕ^T when ϕ is defined in terms of a pair interaction as in (4.2) as we will now see. In (4.2) write $f(\gamma,\gamma') = 1 + g(\gamma,\gamma')$ with

$$g(\gamma,\gamma') = \begin{cases} 0 & \text{if } \gamma,\gamma' \text{ compatible} \\ -1 & \text{if } \gamma,\gamma' \text{ incompatible} \end{cases} \tag{4.17}$$

and expand the product: ($\phi = \tilde{\phi}$ because $\phi(X) = 0$ when $X! \neq 1$)

$$\tilde{\phi}(\gamma_1,\ldots,\gamma_n) = e^{\sum_1^n \mu(\gamma_i)} \prod_{1\le i<j\le n} (1 + g(\gamma_i,\gamma_j))$$

$$= e^{\sum_1^n \mu(\gamma_i)} \sum_G \prod_{\{i,j\}\in G} g(\gamma_i,\gamma_j) \quad . \tag{4.18}$$

The last summation is over all subgraphs G on N. Each such G induces a partition of N into connected components and isolated points, so if we define $g(M)$ for any $M \subseteq N$ by:

$$g(M) = \begin{cases} e^{\sum_M \mu(\gamma_i)} \sum_{C_M} \prod_{C_M} g(\gamma_i,\gamma_j) & \text{if } M \ge 2 \\ e^{\mu(\gamma_i)} & \text{if } M = \{i\} \\ 0 & \text{if } M = \emptyset \end{cases} \tag{4.19}$$

where the sum is over all connected graphs on M, we realize that

$$\tilde{\phi}(\gamma_1,\ldots,\gamma_n) = \sum'_{N_1\cup\ldots\cup N_m = N} g(N_1)\ldots g(N_m) \quad , \tag{4.20}$$

and we see from (4.16) that $g(M) = \tilde{\phi}^T(\gamma_i;\ i \in M)$ for $M \subseteq N$. We thus finally get the following formula for $\phi^T(X) = \phi^T(\gamma_1,\ldots,\gamma_n)$: Construct the graph G with vertices $\{1,\ldots,n\}$ and edges $\{i,j\}$ corresponding to incompatible pairs $\{\gamma_i,\gamma_j\}$. Then

$$\phi^T(\gamma_1,\ldots,\gamma_n) = (X!)^{-1} e^{\sum_1^n \mu(\gamma_i)} \sum_{C\subseteq G} (-1)^{\#\text{ of edges in } C} \quad , \tag{4.21}$$

where the sum is over-all connected subgraphs of G visiting all the points $\{1,\ldots,n\}$. From this expression we see that $\phi^T(\gamma_1,\ldots,\gamma_n) = 0$ if G is not connected, i.e. if $(\gamma_1,\ldots,\gamma_n)$ can be split into two groups such that every γ in one is compatible with every γ in the other, which fact will be used repeatedly in the following. We also see that ϕ^T is transitionally invariant, and that if $(\gamma_1,\ldots,\gamma_n)$ is a configuration on the cylinder $\Omega_{\infty,N}$, which can be drawn on the infinite planar lattice as well without changing the "compatibilities", then ϕ^T is the same for the two configurations. This happens e.g. if $(\gamma_1,\ldots,\gamma_n)$ does not "encircle" $\Omega_{\infty,N}$.

We next derive a "Kirkwood-Salsburg" equation like that used in [6], which will allow us to get convenient estimates for ϕ^T and the correlation functions $\rho_\Theta(X)$ defined by: $\rho_\Theta(X) = P(\text{the contours in } X \text{ are present})$

$$= \frac{\sum_{Y\subset\Theta} \phi(X + Y)}{\sum_{Y\subset\Theta} \phi(Y)} = \sum_{Y\subset\Theta} (\phi^{-1} * D_X\phi)(Y) \tag{4.22}$$

(ϕ^{-1} defined by $\phi^{-1} * \phi = 1$ is well defined if $\phi(\emptyset) \neq 0$, and $\phi(X) = 0$ if $X! \neq 1$). Define $\Delta_X(Y)$ by

$$\Delta_X(Y) = (\phi^{-1} * D_X\phi)(Y) = \sum_{Y_1+Y_2=Y} \phi^{-1}(Y_1)\phi(X + Y_2) \quad . \tag{4.23}$$

Then a recursive equation for $\Delta_X(Y)$ can be derived as follows. From (4.23) we see that $\Delta_X(Y) = 0$ if X is not a compatible set. Consider $\Delta_{\gamma+X}(Y)$ for $\gamma + X$ compatible. From (4.2) follows that

$$\phi(\gamma + X + Y_2) = e^{\mu(\gamma)}\phi(X + Y_2) \prod_{\gamma'\in Y_2} (1 + g(\gamma,\gamma'))$$

$$= e^{\mu(\gamma)}\phi(X + Y_2) \sum_{S\subseteq Y_2}{}^{*} (-1)^{N(S)}$$

if Y_2 is without "multiplicities", and the sum is over sets S, all of whose elements are incompatible with γ. We then get:

$$\Delta_{\gamma+X}(Y) = \sum_{Y_1+Y_2=Y} \phi^{-1}(Y_1)\phi(\gamma + X + Y_2)$$

$$= e^{\mu(\gamma)} \sum_{Y_1+Y_2=Y} \phi^{-1}(Y_1)\phi(X + Y_2) \sum_{S\subseteq Y_2}{}^{*} (-1)^{N(S)} \quad , \tag{4.24}$$

because only Y_2's without multiplicities contribute. Put $Y_2 = S + Y_3$. Then:

$$\Delta_{\gamma+X}(Y) = e^{\mu(\gamma)} \sum_{S\subseteq Y}{}^{*} (-1)^{N(S)} \sum_{Y_1+Y_3=Y-S} \phi^{-1}(Y_1)\phi(X + S + Y_3)$$

$$= e^{\mu(\gamma)} \sum_{S\subseteq Y}{}^{*} (-1)^{N(S)}\Delta_{S+X}(Y - S) \quad . \tag{4.25}$$

In the sum $S = \emptyset$ is to be included, and $\Delta_\emptyset(Y) = 1(Y)$. Observe that this equation determines $\Delta_X(Y)$ with $N(X) + N(Y) = m + 1$ in terms of $\Delta_X(Y)$ with $N(X) + N(Y) = m$ for $m = 0, 1, \ldots,$ successively. This makes it possible to derive the following useful estimate:

Let I_m be defined by:

$$I_m = \sup_{\substack{\gamma_1,\ldots,\gamma_n \\ m\geq n\geq 1}} \sum_{\substack{Y \\ N(Y)=m-n}} |\Delta_{\gamma_1,\ldots,\gamma_n}(Y)| e^{\frac{b}{2}\sum_i|\gamma_i|} \quad . \tag{4.26}$$

We can then deduce from (4.25):

$$\sum_{\substack{Y \\ N(Y)+N(X)=m}} |\Delta_{\gamma+X}(Y)| e^{\frac{b}{2}(|\gamma|+|X|)} \leq$$

$$\le \sum_{\substack{Y\\ N(Y)+N(X)=m}} \; \sum_{S\subseteq Y}{}^{*} |\Delta_{X+S}(Y - S)| e^{\frac{b}{2}(|\gamma|+|X|)-b|\gamma|}$$

$$\le \sum_{S}{}^{*} I_m e^{-\frac{b}{2}(|\gamma|+|S|)} \le I_m e^{-\frac{b|\gamma|}{2}} \sum_{n\ge 0} \frac{1}{n!}\Big(\sum_{\sigma \text{ inters. } \gamma} e^{-\frac{b|\sigma|}{2}}\Big)^n$$

$$\le I_m e^{-\frac{b|\gamma|}{2}} \exp\Big(\sum_{p\in\gamma}\sum_{\sigma\ni p} e^{-\frac{b|\sigma|}{2}}\Big) \le I_m e^{-\frac{b|\gamma|}{2}} \exp\Big(|\gamma| \sum_{\ell=4}^{\infty} (3e^{-\frac{b}{2}})^{\ell}\Big)$$

$$\le I_m \exp|\gamma|\Big(-\frac{b}{2} + 3^4 e^{-2b}(1 - 3e^{-\frac{b}{2}})^{-1}\Big) \tag{4.27}$$

if $3e^{-b/2} < 1$. (Here and at several other instances we use the fact that the number of different contours of length ℓ that go through a given point is less than 3^{ℓ}.) If $3e^{-b/2} \le 1/2$ it is easy to see that the last expression in (4.27) is $\le I_m e^{-(1.8)b}$ because $|\gamma| \ge 4$. We can conclude that

$$I_{m+1} \le I_m e^{-(1.8)m} \quad \text{for} \quad m \ge 1 \tag{4.28}$$

if $3e^{-b/2} \le 1/2$. Because:

$$I_1 = \sup_{\gamma} |\Delta_{\gamma}(\emptyset)| e^{\frac{b|\gamma|}{2}} = \sup_{\gamma} |(\phi^{-1} * D_{\gamma}\phi)(\emptyset)| e^{\frac{b|\gamma|}{2}}$$

$$= \sup_{\gamma} |\phi(\gamma)| e^{\frac{b|\gamma|}{2}} \le \sup_{|\gamma|\ge 4} e^{-\frac{b|\gamma|}{2}} = e^{-2b} \tag{4.29}$$

we see from (4.28) that:

$$I_m \le e^{-(1.8)mb} \tag{4.30}$$

if $3e^{-b/2} \le 1/2$.

This bound allows us to estimate ϕ^T as follows. From (4.11) we see that

$$\Delta_{\gamma}(X) = (\phi^{-1} * D_{\gamma}\phi)(X) = D_{\gamma}\phi^T(X) = \phi^T(\gamma + X)\frac{(\gamma + X)!}{X!} \tag{4.31}$$

and we can derive a bound for the quantity $\sum_X |\phi^T(\gamma + X)|$, which will be very useful:

$$\sum_X |\phi^T(\gamma + X)| \le \sum_{m=1}^{\infty} \sum_{N(X)=m-1} |\Delta_{\gamma}(X)| \le \sum_{m=1}^{\infty} e^{-\frac{b|\gamma|}{2}-(1.8)mb}$$

$$\le e^{-\frac{b|\gamma|}{2}-(1.8)b} (1 - e^{-(1.8)b})^{-1} \le (1.1)e^{-\frac{b|\gamma|}{2}-(1.8)b} \tag{4.32}$$

if $3e^{-b/2} \le 1/2$. We can also bound $\sum_{X\ni p} |\phi^T(X)|$ for any given point p:

$$\sum_{X\ni p} |\phi^T(X)| \le \sum_{\gamma\ni p}\sum_{X} |\phi^T(\gamma + X)| \le (1.1) \sum_{\gamma\ni p} e^{-\frac{b|\gamma|}{2}-(1.8)b}$$

$$\le (1.1)e^{-(1.8)b} \sum_{\ell=4}^{\infty} (3e^{-\frac{b}{2}})^{\ell} \le (2.2)e^{-(3.8)b} \qquad (4.33)$$

if $3e^{-b/2}$.

These bounds also allow us to estimate $\sum_{\substack{X\ni p\\ XiQ}} |\phi^T(X)|$, where p is a point and Q any set of points, and XiQ means that X intersects Q. Let d be the distance between p and Q, and divide the above sum into two parts according to whether $N(X) \ge d^{\frac{1}{2}}$ or $N(X) < d^{\frac{1}{2}}$. The first part can be estimated as in (4.32) using (4.30):

$$\text{first part} \le \sum_{\gamma\ni p}\sum_{N(X)\ge d^{\frac{1}{2}}} |\phi^T(\gamma + X)| \le \sum_{\gamma\ni p} e^{-\frac{b|\gamma|}{2}} \sum_{m\ge d^{\frac{1}{2}}} e^{-(1.8)mb}$$

$$\le (1.1)e^{-(1.8)d^{\frac{1}{2}}b} \sum_{\ell=4}^{\infty} (3e^{-\frac{b}{2}})^{\ell} \le (2.2)e^{-2b-(1.8)d^{\frac{1}{2}}b} \quad . \qquad (4.34)$$

If $N(X) < d^{\frac{1}{2}}$ we can conclude that the longest contour in X, $\bar{\gamma}$, has a length $\ell \ge d^{\frac{1}{2}}$ if $\phi^T(X) \ne 0$. This is true because if $\phi^T(X) \ne 0$ we know that the contours in X form one overlapping group, so that

$$d \le \text{length of } X \le \ell \cdot N(X) \le \ell d^{\frac{1}{2}}, \text{ and } \ell \ge d^{\frac{1}{2}} \quad .$$

We also have $d(p,\bar{\gamma}) \le \ell \cdot N(X) \le \ell \cdot d^{\frac{1}{2}}$ for the same reason, so $\bar{\gamma}$ must intersect the square with side $2\ell d^{\frac{1}{2}}$ centered at p. The second part can therefore be estimated as follows using (4.32):

$$\text{second part} \le \sum_{\ell\ge d^{\frac{1}{2}}} \sum_{\substack{|\bar{\gamma}|=\ell\\ d(p,\bar{\gamma})\le \ell d^{\frac{1}{2}}}} \sum_{X} |\phi^T(\bar{\gamma} + X)| \le \sum_{\ell\le d^{\frac{1}{2}}} \sum_{\substack{|\bar{\gamma}|=\ell\\ d(p,\bar{\gamma})\le \ell d^{\frac{1}{2}}}} (1.1)e^{-(1.8)b-\frac{b|\gamma|}{2}}$$

$$\le (1.1)e^{(-1.8)b} \sum_{\ell\ge d^{\frac{1}{2}}} (4\ell^2 d)(3e^{-\frac{b}{2}})^{\ell} \le 60e^{-(1.8)b}d^2(3e^{-\frac{b}{2}})^{d^{\frac{1}{2}}} \qquad (4.35)$$

if $3e^{-b/2} \le 1/2$, and we finally find that

$$\sum_{\substack{X\ni p\\ XiQ}} |\phi^T(X)| \le (60)e^{-(1.8)b}(d^2 + 1)(3e^{-\frac{b}{2}})^{d^{\frac{1}{2}}(p,Q)} \qquad (4.36)$$

if $3e^{-b/2} \le 1/2$.

(Remark in proof: (4.36) can actually be improved so that $d(p,Q)$ occurs in the exponent instead of $d^{\frac{1}{2}}(p,Q)$. This is easily seen because from (4.21) follows that if

$b > b_0$ and $3e^{-\frac{-b_0}{2}} \le 1/2$ then $\phi^T(\gamma_1,\dots,\gamma_n) = e^{\sum_1^n u(\gamma_i)+b_0|\gamma_i|}\phi_0^T(\gamma_1,\dots,\gamma_n)$ with ϕ_0^T defined by the weight $\mu_0(\gamma) = -b_0|\gamma|$. ϕ_0^T can be estimated by (4.36), so the corresponding estimate for ϕ^T can be improved by a factor $e^{-(b-b_0)d(p,Q)}$. We do not need to use this sharper bound however.)

For the proof of Lemma 5.4 in Appendix 5A we need an estimate of the quadratic form

$$Q = \sum_{\gamma_1,\gamma_2\subset\Theta} \Delta\mu(\gamma_1) \frac{\partial^2 \log Z(M^+_{0,c}(\Theta),\beta)}{\partial\mu(\gamma_1)\partial\mu(\gamma_2)} \Delta\mu(\gamma_2)$$

which we now derive. (Remember that $M^+_{0,c}(\Theta)$ is the sub-ensemble of $M^+_0(\Theta)$ having only c-small contours.) Observe that the partition function for this ensemble can also be expressed in terms of ϕ^T as in (4.9). We only redefine $z(\gamma)$ as:

$$z_c(\gamma) = \begin{cases} z(\gamma) & \text{if } \gamma \text{ is c-small} \\ 0 & \text{otherwise} \end{cases} \tag{4.37}$$

and get

$$Z(M^+_{0,c}(\Theta,\beta) = \sum_X \phi(X) z_c^X = \exp \sum_X \phi^T(X) z_c^X \quad . \tag{4.38}$$

We can thus use the same expansions as before only adding the restriction on the length in the summations (which we denote by $\sum^c_{X\subset\Theta}$). From the definition of $\phi(X)$ in (4.2) we see that

$$\frac{\partial\phi(X)}{\partial\mu(\gamma)} = \frac{\partial^2\phi(X)}{\partial\mu^2(\gamma)} = \begin{cases} \phi(X) & \text{if } X = \gamma + Y \text{ for some } Y \\ 0 & \text{otherwise} \end{cases}$$

$$\frac{\partial^2\phi(X)}{\partial\mu(\gamma_1)\partial\mu(\gamma_2)} = \begin{cases} \phi(X) & \text{if } X = \gamma_1 + \gamma_2 + Y \text{ for some } Y \\ 0 & \text{otherwise} \end{cases} \tag{4.39}$$

if $\mu_1 \neq \mu_2$, so that we get (remembering the definition of the correlation functions in (4.22):

$$Q = \sum^c_{\gamma\subset\Theta} (\Delta\mu(\gamma))^2[\rho_{\Theta,c}(\gamma) - \rho^2_{\Theta,c}(\gamma)]$$
$$+ \sum^c_{\substack{\gamma_1,\gamma_2\subset\Theta \\ \gamma_1\neq\gamma_2}} \Delta\mu(\gamma_1)[\rho_{\Theta,c}(\gamma_1,\gamma_2) - \rho_{\Theta,c}(\gamma_1)\rho_{\Theta,c}(\gamma_2)]\Delta\mu(\gamma_2) \quad . \tag{4.40}$$

From (4.22) we get the important bound

$$\rho_{\Theta,c}(\gamma) \le e^{-b|\gamma|} \quad , \tag{4.41}$$

because all terms in the numerator contain this factor, and the remaining sum is contained in the denominator. (The same bound and argument is also true for $\pi_{\Theta,c}(\gamma)$, the probability that γ is an outer contour.) Hence the first term, Q′, in Q can directly be bounded by

$$Q' \leq \sum_{\gamma\subset\Theta}^{c} (\Delta\mu(\gamma))^2 e^{-b|\gamma|} \leq |\Theta| \sum_{(\gamma)}^{c} (\Delta\mu(\gamma))^2 e^{-b|\gamma|} \tag{4.42}$$

if $\Delta\mu(\gamma)$ is translationally invariant. To bound the second term, Q″, we need to estimate the clustering property of $\rho_{\Theta,c}(\gamma_1,\gamma_2)$ as $d(\gamma_1,\gamma_2) \to \infty$. From (4.22) follows that

$$\rho_{\Theta,c}(\gamma_1,\gamma_2) - \rho_{\Theta,c}(\gamma_1)\rho_{\Theta,c}(\gamma_2) = \sum_Y z_c^Y \Delta_{\gamma_1,\gamma_2}(Y) - (\sum_Y z_c^Y \Delta_{\gamma_1}(Y))(\sum_Y z_c^Y \Delta_{\gamma_2}(Y))$$

$$= \sum_Y z_c^Y [\Delta_{\gamma_1,\gamma_2}(Y) - (\Delta_{\gamma_1} * \Delta_{\gamma_2})(Y)] \quad . \tag{4.43}$$

The last term can be expressed in terms of ϕ^T, because as in (4.31) we have:

$$D_{\gamma_1}\phi^T = \phi^{-1} * D_{\gamma_1}\phi = \Delta_{\gamma_1} \tag{4.44}$$

so that (using 4.11):

$$D_{\gamma_1\gamma_2}\phi^T = \phi^{-1} * D_{\gamma_1\gamma_2}\phi - \phi^{-2} * D_{\gamma_1}\phi * D_{\gamma_2}\phi = \Delta_{\gamma_1\gamma_2} - \Delta_{\gamma_1} * \Delta_{\gamma_2} \quad . \tag{4.45}$$

To estimate $\sum_{Y\subset\Theta}^{c} |D_{\gamma_1\gamma_2}\phi^T(Y)|$ in terms of $d(\gamma_1,\gamma_2)$ we proceed as in the derivation of (4.36), and then we need to estimate also $D_{\gamma_1\gamma_2\gamma_3}\phi^T(Y)$ It can be expressed in terms of $\Delta_X(Y)$ if we differentiate (4.45) once more: ($\gamma_1,\gamma_2,\gamma_3$ are all different)

$$D_{\gamma_1\gamma_2\gamma_3}\phi^T = \phi^{-1} * D_{\gamma_1\gamma_2\gamma_3}\phi - \phi^{-2} * (D_{\gamma_1}\phi * D_{\gamma_2\gamma_3}\phi$$

$$+ D_{\gamma_2}\phi * D_{\gamma_1\gamma_3}\phi + D_{\gamma_3}\phi * D_{\gamma_1\gamma_2}\phi) + 2\phi^{-3} * D_{\gamma_1}\phi * D_{\gamma_2}\phi * D_{\gamma_3}\phi$$

$$= \Delta_{\gamma_1\gamma_2\gamma_3} - \Delta_{\gamma_1} * \Delta_{\gamma_2\gamma_3} - \Delta_{\gamma_2} * \Delta_{\gamma_1\gamma_3} - \Delta_{\gamma_3} * \Delta_{\gamma_1\gamma_2} + 2\Delta_{\gamma_1} * \Delta_{\gamma_2} * \Delta_{\gamma_3} \quad . \tag{4.46}$$

Any term appearing in (4.45) and (4.46) can be estimated using (4.30), which says:

$$\sum_{\substack{Y \\ N(Y)=m-N(X)}} |\Delta_X(Y)| \leq e^{-\frac{b|X|}{2}-(1.8)mb} \quad . \tag{4.47}$$

We thus get for a typical term:

$$\sum_{\substack{Y \\ N(Y)=m-\sum_1^n N(X_i)}} |(\Delta_{X_1} * \ldots * \Delta_{X_n})(Y)|$$

$$\leq \sum_{\substack{Y \\ N(Y)=m-\sum_1^n N(X_i)}} \sum_{Y_1+\ldots+Y_n=Y} |\Delta_{X_1}(Y_1)|\ldots|\Delta_{X_n}(Y_n)| \leq$$

$$\leq \sum_{m_1+\ldots+m_n=m} \sum_{\substack{N(Y_i)=m_i-N(X_i) \\ i=1,\ldots,n}} |\Delta_{X_1}(Y_1)|\ldots|\Delta_{X_n}(Y_n)|$$

$$\leq \sum_{m_1+\ldots+m_n=m} e^{-\frac{b}{2}\sum_1^n |X_i|-(1.8)mb} \leq m^n e^{-\frac{b}{2}\sum_1^n |X_i|-(1.8)mb} , \tag{4.48}$$

and hence:

$$\sum_{N(Y)=m-2} |D_{\gamma_1\gamma_2}\phi^T(Y)| \leq 2m^2 e^{-\frac{b}{2}(|\gamma_1|+|\gamma_2|)-(1.8)mb} \tag{4.49}$$

$$\sum_{N(Y)=m-3} |D_{\gamma_1\gamma_2\gamma_3}\phi^T(Y)| \leq 6m^3 e^{-\frac{b}{2}(|\gamma_1|+|\gamma_2|+|\gamma_3|)-(1.8)mb} . \tag{4.50}$$

Now we can turn to the quantities in (4.43) $(d = d(\gamma_1,\gamma_2))$:

$$|\rho_{\Theta,c}(\gamma_1,\gamma_2) - \rho_{\Theta,c}(\gamma_1)\rho_{\Theta,c}(\gamma_2)| \leq \sum_Y |D_{\gamma_1\gamma_2}\phi^T(Y)|$$

$$\leq \sum_{N(Y)\geq d^{\frac{1}{2}}} + \sum_{N(Y)<d^{\frac{1}{2}}} \tag{4.51}$$

The first sum is bounded using (4.49):

$$\sum_{N(Y)\geq d^{\frac{1}{2}}} \leq 2e^{-\frac{b}{2}(|\gamma_1|+|\gamma_2|)} \sum_{d^{\frac{1}{2}}+2}^{\infty} m^2 e^{-(1.8)mb} \tag{4.52}$$

For the second we use the argument leading to (4.35): The longest member of Y, $\bar{\gamma}$, must have $|\bar{\gamma}| \geq d^{\frac{1}{2}}$ and $d(\gamma_1,\bar{\gamma}) \leq |\bar{\gamma}|d^{\frac{1}{2}}$ $(|\gamma_1| \leq |\gamma_2|$ say).
Then:

$$\sum_{N(X)<d^{\frac{1}{2}}} \leq \sum_{\ell\geq d^{\frac{1}{2}}} \sum_{\substack{|\bar{\gamma}|=\ell \\ d(\gamma_1,\bar{\gamma})\leq \ell d^{\frac{1}{2}}}} \sum_Y |D_{\gamma_1\gamma_2\bar{\gamma}}\phi^T(Y)|$$

$$\leq \sum_{\ell\geq d^{\frac{1}{2}}} 4\ell^2 d|\gamma_1| \sum_{\substack{|\bar{\gamma}|=\ell \\ \bar{\gamma}\ni 0}} 6e^{-\frac{b}{2}(|\gamma_1|+|\gamma_2|+|\bar{\gamma}|)} \sum_1^{\infty} m^3 e^{-(1.8)mb}$$

$$\leq 24d|\gamma_1|(\sum_1^{\infty} m^3 e^{-(1.8)mb})(\sum_{\ell\geq d^{\frac{1}{2}}} \ell^2(3e^{-\frac{b}{2}})^{\ell})e^{-\frac{b}{2}(|\gamma_1|+|\gamma_2|)} . \tag{4.53}$$

To continue it is convenient to have a simple estimate of sums of the type $S_{N,p} = \sum_N^{\infty} n^p a^n$ with $0 < a < 1$.

Lemma 4.1

$$S_{N,p} \le \frac{a^N p!(1 + pN^p)}{(1 - a)^{p+1}} \quad \textit{for any integers} \quad N,p \ge 0.$$

Proof. The mean value theorem tells us that $(n + 1)^p - n^p \le p(n + 1)^{p-1}$ for $n,p \ge 0$, hence $\sum_{N-1}^{\infty} ((n + 1)^p - n^p)a^{n+1} \le \sum_{N-1}^{\infty} p(n + 1)^{p-1}a^{n+1}$ for $N \ge 1$ and $S_{N,p} - aS_{N,p} - (N - 1)^p a^N \le pS_{N,p-1}$, so we have the recursion:

$$S_{N,p} \le \frac{pS_{N,p-1}}{(1 - a)} + \frac{(N - 1)^p a^N}{(1 - a)}$$

$$\text{for} \quad p = 0,1,\ldots \quad , \quad N = 1,2,\ldots \quad , \tag{4.54}$$

$$S_{N,0} = \frac{a^N}{1 - a} \quad \text{for} \quad N \ge 0 \quad .$$

It gives

$$S_{N,p} \le \frac{a^N p!}{(1 - a)^{p+1}} \sum_0^p \frac{(N - 1)^q (1 - a)^q}{q!} \le \frac{a^N p!(1 + p(N - 1)^p)}{(1 - a)^{p+1}} \le \frac{a^N p!(1 + pN^p)}{(1 - a)^{p+1}} \tag{4.55}$$

for $p \ge 0$, $N \ge 1$.

For $p \ge 1$, $N = 0$ we have

$$S_{0,p} = S_{1,p} \le \frac{ap!}{(1 - a)^{p+1}} \le \frac{a^0 p!(1 + 0^p)}{(1 - a)^{p+1}} \quad ,$$

and for $p = 0$: $S_{N,0} = \frac{a^N}{1 - a} = \frac{a^N 0!(1 + 0N^0)}{(1 - a)}$,

so the inequality is true for all $N,p \ge 0$.

It is now easy to bound the series in (4.52) and (4.53) if $3e^{-b/2} \le 1/2$, which implies that $e^{-(1.8)b} \le 1/500$.

$$\sum_{N(X) \ge d^{\frac{1}{2}}} \le (0.2)e^{-b}(1 + d)(500)^{-d^{\frac{1}{2}}} e^{-\frac{b}{2}(|\gamma_1|+|\gamma_2|)} \tag{4.56}$$

$$\sum_{N(X) < d^{\frac{1}{2}}} \le 1600e^{-b}d^2 2^{-d^{\frac{1}{2}}}|\gamma_1| e^{-\frac{b}{2}(|\gamma_1|+|\gamma_2|)} \quad , \tag{4.57}$$

and we get:

$$|\rho_{\Theta,c}(\gamma_1,\gamma_2) - \rho_{\Theta,c}(\gamma_1)\rho_{\Theta,c}(\gamma_2)|$$

$$\le 1600e^{-b}(|\gamma_1||\gamma_2|)^{\frac{1}{2}} e^{-\frac{b}{2}(|\gamma_1|+|\gamma_2|)}(1 + d^2(\gamma_1,\gamma_2))2^{-d^{\frac{1}{2}}(\gamma_1,\gamma_2)} \quad . \tag{4.58}$$

With the help of (4.58) we can estimate Q″, and we get:

$$Q'' \leq \sum_{\gamma_1,\gamma_2\subset\Theta}^{c} (1600)e^{-b}|\Delta\mu(\gamma_1)||\Delta\mu(\gamma_2)|(|\gamma_1||\gamma_2|)^{\frac{1}{2}}e^{-\frac{b}{2}(|\gamma_1|+|\gamma_2|)}$$

$$\times (1 + d^2(\gamma_1,\gamma_2))2^{-d^{\frac{1}{2}}(\gamma_1,\gamma_2)}$$

$$\leq (1600)e^{-b} \sum_{\gamma_1\subset\Theta}^{c} |\Delta\mu(\gamma_1)||\gamma_1|^{\frac{1}{2}}e^{-\frac{b}{2}|\gamma_1|} \sum_{(\gamma_2)}^{c} |\Delta\mu(\gamma_2)||\gamma_2|^{\frac{1}{2}}e^{-\frac{b}{2}|\gamma_2|}$$

$$\times \sum_{\gamma_2\in(\gamma_2)} (1 + d^2(\gamma_1,\gamma_2))2^{-d^{\frac{1}{2}}(\gamma_1,\gamma_2)} \tag{4.59}$$

The innermost sum is bounded by:

$$|\gamma_1||\gamma_2| \sum_1^{\infty} (4d)(d^2 + 1)2^{-d^{\frac{1}{2}}} \leq |\gamma_1||\gamma_2| \sum_1^{\infty} 4n^2(n^4 + 1)2n2^{-(n-1)}$$

$$\leq |\gamma_1||\gamma_2|(2.1)10^7 \quad , \tag{4.60}$$

so we get

$$Q'' \leq (3.5)10^{10}e^{-b}|\Theta|(\sum_{(\gamma)}^{c} |\Delta\mu(\gamma)||\gamma|^{3/2}e^{-\frac{b|\gamma|}{2}})^2$$

$$\leq (3.5)10^{10}e^{-b}|\Theta|(\sum_{(\gamma)}^{c} (\Delta\mu(\gamma))^2e^{-\frac{b|\gamma|}{2}})(\sum_{(\gamma)} |\gamma|^3e^{-\frac{b|\gamma|}{2}}) \quad .$$

The last sum is bounded by $\sum_0^{\infty} \ell^3 2^{-\ell} \leq 100$, so, remembering the estimate of Q', we finally get:

$$Q \leq 4 \cdot 10^{12} \cdot e^{-b}|\Theta|(\sum_{(\gamma)}^{c} (\Delta\mu(\gamma))^2e^{-\frac{b|\gamma|}{2}}) \quad , \tag{4.61}$$

which is the estimate needed. (Actually the restriction on the length of the contours was not used in the proof, so a similar estimate is valid in the unrestricted ensemble.)

5. THE PHASE SEPARATION

In this section we give a proof of the phase separation as described in Theorem 3.1 using the method of proof used by M.S. adapted to our situation. The proof depends on several estimates of various probabilities, which we formulate as a series of lemmas. Their proofs are given in detail in [5] and will not be repeated here except Lemma 5.4, which is proved in Appendix 5A.

The first lemma says that in all the ensembles of interest we need only consider the "minimal" ensembles having a minimal number of big contours, because their probabilities converge to 1:

Lemma 5.1

$$\lim_{N\to\infty} \frac{Z(M_0^{++}(\Omega),\beta)}{Z(M^{++}(\Omega),\beta)} = 1 \tag{5.1}$$

$$\lim_{N\to\infty} \frac{Z(M_0^{+-}(\Omega),\beta)}{Z(M^{+-}(\Omega),\beta)} = 1 \tag{5.2}$$

$$\lim_{N\to\infty} \frac{Z(M_0^{+-}(\Omega,m),\beta)}{Z(M^{+-}(\Omega,m),\beta)} = 1 \tag{5.3}$$

if β *is large and* $m = (2\alpha - 1)m^*$ *with* $0 < \alpha < 1$.

To prove that some set $E \subset M_0^{+-}(\Omega,m)$ has a small probability in the "difficult" ensemble $M_0^{+-}(\Omega,m)$ the following argument will allow us to consider the "simpler" ensemble $M_0^{+-}(\Omega)$ instead:

$$\frac{Z(E,\beta)}{Z(M_0^{+-}(\Omega,m),\beta)} = \frac{Z(E,\beta)}{Z(M_0^{+-}(\Omega),\beta)}\,\frac{Z(M_0^{+-}(\Omega),\beta)}{Z(M_0^{+-}(\Omega,m),\beta)} \quad , \tag{5.4}$$

so if we have an upper bound on the last ratio we can conclude that the left hand side goes to zero if the probability of E in $M_0^{+-}(\Omega)$ goes to zero fast enough.

Lemma 5.2

$$\frac{Z(M_0^{+-}(\Omega),\beta)}{Z(M_0^{+-}(\Omega,m),\beta)} \le D(\alpha,\beta)N^{2\delta+3}e^{N\delta'(\beta)} \tag{5.5}$$

for some constants $D(\alpha,\beta)$, $\delta'(\beta)$ *if* β *is large and* $m = (2\alpha - 1)m^*$. $\delta'(\beta)$ *goes to zero exponentially as* $\beta \to \infty$.

"Fast enough" in the above argument is thus e.g. $P_{M_0^{+-}(\Omega)}(E) \le e^{-N^{1+\varepsilon}}$ for some $\varepsilon > 0$. We also need some bound on the length of the big contour λ always present in $M_0^{+-}(\Omega,m)$ and of the c-large contours:

Lemma 5.3

In $M_0^{+-}(\Omega,m)$ *the probability that* $|\lambda| - N \le (2 \log 3)N/\beta$ *and the total length of the* c-*large contours is less than* N/β *tends to* 1 *as* $N \to \infty$ *for* β *large*, $m = (2\alpha - 1)m^*$.

We finally need the following important estimate of the fluctuations of the total magnetization $M(X)$ in the ensemble $M_{0,c}^{+}(\Theta)$ of c-small contours in any large subregion Θ of Ω:

Lemma 5.4

Let $\Theta \subset \Omega$ *be a region such that* $|\Theta| \geq k|\Omega|$ *and* $|\partial\Theta| \leq k|\Theta|^{\frac{1}{2}}$ *for some* $k > 0$. *Then*

$$P_{M^{+}_{0,c}(\Theta)}\left(M(X) - m^*|\Theta|| \geq t|\Theta|^{p}\right) \leq 3e^{-\frac{t^2|\Theta|^{2p-1}}{400\delta(\beta)}} \tag{5.6}$$

with $\delta(\beta) = 10^{14}e^{-\beta}$ *if* t *and* p *are restricted by:*

$$(1 + c \log 3)/2 < p < 1 \ , \ 2\delta^{\frac{1}{2}}(\beta) \leq t \leq 2\delta(\beta)|\Theta|^{(1-p)/2} \quad ,$$

and if $|\Theta|$ *and* β *are large. (e.g. if* $3e^{-\beta-1/2} \leq 1/2$, $3e^{-\beta} \leq e^{-3/2c}$ *and* $|\Theta|$ *large.)* The lemma thus says that the probability of "large" fluctuations (larger than const. $|\Theta|^{\frac{1}{2}}$) have a bound $\exp -\frac{(t|\Theta|^{p})^2}{(\text{const.})|\Theta|}$ as one would expect for "normal" fluctuations. (Θ is allowed to have "holes" in it, but no contour of a configuration in $M^{+}_{0,c}(\Theta)$ encircles any of them.) With the help of these lemmas we can give a proof of the phase separation along the following lines. Consider first the fluctuations in the area of the region Ω_λ above λ, the big contour of $X \in M^{+-}_{0}(\Omega,m)$. We can assume that the bounds of Lemma 5.3 are satisfied by X. If $|\Omega_\lambda|$ deviates much from $\alpha|\Omega|$, e.g. if $|\Omega_\lambda| \geq \alpha|\Omega| + a|\Omega|^{p}$, then either above or below λ the total magnetization M_λ or $m|\Omega| - M_\lambda$ will deviate much from the "expected" values $m^*|\Omega_\lambda|$ or $-m^*(|\Omega| - |\Omega_\lambda|)$. Because of the bound on the length of the c-large contours this deviation cannot come from the region enclosed by them or by c-small contours which enclose large contours. Hence it comes from the regions formed only by c-small contours and its probability can be effectively estimated using Lemma 5.4 and shown to be "small enough" on the scale of Lemma 5.2.

The above argument can be made precise as follows. Let E be the subset of $M^{+-}_{0}(\Omega,m)$ defined by the restrictions $|\Omega_\lambda| \geq \alpha|\Omega| + a|\Omega|^{p}$, $|\lambda| - N \leq (2 \log 3)N/\beta$ and $|\text{c-large contours}| \leq N/\beta$. For any configuration $X \in E$ let $\gamma_1,\ldots,\gamma_n$ be the c-large outer contours (if any are present in X) and $\gamma'_1,\ldots,\gamma'_{n'}$ those c-small contours that enclose a c-large contour (if any are present in X) and let A be the area enclosed by all of them. Because a contour has at least the length 4 we see that $4(n + n') \leq |\text{c-large contours}| \leq N/\beta$ and $A \leq (|\gamma_1| + \ldots + |\gamma_n|)^2/16 + n'(c \log |\Omega|)^2/16 \leq ((N/\beta)^2 + N(c \log |\Omega|)^2/4\beta)/16 \leq N^2$ if N is large and β not too small. The magnetization inside $\Gamma = (\gamma_1,\ldots,\gamma'_{n'})$, M_0, is thus also bounded by $N^2 = |\Omega|^{2/(1+\delta)}$, so if $p > 2/(1 + \delta)$ it is much smaller than $|\Omega|^{p}$. Let Θ_1 and Θ_2 be the regions outside Γ above and below λ and let M_1 and M_2 be their magnetizations respectively. Let also A be split into $A_1 + A_2$ by λ. Because $M_0 + M_1 + M_2 = m|\Omega| = (2\alpha - 1)m^*|\Omega|$ and $A + |\Theta_1| + |\Theta_2| = |\Omega|$ we have

$$(M_1 - m^*|\Theta_1|) + (M_2 - m^*|\Theta_2|) = (2\alpha - 1)m^*|\Omega| - M_0 - m^*|\Theta_1| +$$

$$+ m^*(|\Omega| - |\Theta_1| - A) = 2m^*(\alpha|\Omega| - |\Omega_\lambda|) + m^*A_1 - m^*A_2 - M_0 \quad . \tag{5.7}$$

Consider now the two subsets of E defined by

$$E_1: |M_1 - m^*|\Theta_1\| \geq m^*a|\Omega|^p \quad \text{and} \quad E_2: |M_1 - m^*|\Theta_1\| < m^*a|\Omega|^p \quad .$$

From (5.7) we see that E_2 implies that

$$M_2 + m^*|\Theta_2| \leq m^*(\alpha|\Omega| - |\Omega_\lambda|) + 2N^2 \quad , \tag{5.8}$$

which implies that

$$M_2 + m^*|\Theta_2| \leq -m^*a|\Omega|^p + 2N^2 \quad , \tag{5.9}$$

and also because $M_2 \geq -|\Theta_2|$:

$$-|\Theta_2| + m^*|\Theta_2| \leq m^*(|\Omega| - |\Omega_\lambda| - |\Omega|(1 - \alpha)) + 2N^2$$

$$= m^*(|\Theta_2| + A^2 - |\Omega|(1 - \alpha)) + 2N^2 \quad , \tag{5.10}$$

so $|\Theta_2| \geq |\Omega|m^*(1 - \alpha) - 3N^2 \geq k|\Omega|$ for some $k > 0$ in this case. These consideration allow us to estimate the probability of E in $M_0^{+-}(\Omega)$ as follows:

$$P(E) = P(E_1) + P(E_2) = \sum_{\lambda,\Gamma} [P(E_1|\lambda,\Gamma) + P(E_2|\lambda,\Gamma)]P(\lambda,\Gamma)$$

$$\leq \sum_{\lambda,\Gamma} [P(|M_1 - m^*|\Theta_1\| \geq m^*a|\Omega|^p|\lambda,\Gamma)$$

$$+ P(|M_2 + m^*|\Theta_2\| \geq m^*a|\Omega|^p|\lambda,\Gamma)]P(\lambda,\Gamma) \quad . \tag{5.11}$$

But when λ and Γ are fixed these last probabilities are computed in the ensembles $M_{0,c}^{+}(\Theta_1)$ and $M_{0,c}^{+}(\Theta_2)$ respectively, because Θ_1 and Θ_2 only contain c-small contours. Moreover, the conditions on Θ_1 and Θ_2 for the use of Lemma 5.4 are fulfilled (uniformly in λ,Γ), so we get:

$$P_{M_0^{+-}(\Omega)}(E) \leq 6 \exp(-\text{const.}\ |\Omega|^{2p-1} \tag{5.12}$$

if $(1 + c \log 3)/2 < p < 1$ and $p > 2/(1 + \delta)$. We thus see using the bound of Lemma 5.2 that $P_{M_0^{+-}(\Omega,m)}(E) \to 0$ also, if $|\Omega|^{2p-1} = N^{(1+\delta)(2p-1)} > N$, i.e. if $p > (1 + 1/(1 + \delta))/2$. Using Lemma 5.3 we then finally see that $P_{M_0^{+-}(\Omega,m)}(|\Omega_\lambda| \geq \alpha|\Omega| + a|\Omega|^p) \to 0$ also. The case $|\Omega_\lambda| \leq \alpha|\Omega| - a|\Omega|^p$ is treated in the same way.

The fluctuations of the magnetization above and below λ are estimated quite analogously: Consider M_λ the magnetization of Ω_λ e.g. and define E this time by the restrictions: $X \in M_0^{+-}(\Omega,m)$, $|M_\lambda - m^*|\Omega_\lambda\| \geq a|\Omega|^p$, $|\Omega_\lambda| > k|\Omega|$, $|\Omega| - |\Omega_\lambda| > k|\Omega|$ for a suitable $k > 0$, $|\lambda| - N \leq (2 \log 3)N/\beta$ and $|$c-large contours$| \leq N/\beta$. Because as before $|M_\lambda - M_1| \leq N^2$, $|\Omega_\lambda| - |\Theta_1| \leq N^2$ we have $|M_1 - m^*|\Theta_1\| \geq a|\Omega|^p - 2N^2$ in E, and $P(E|\lambda,\Gamma) \leq P(M_1 - m^*|\Theta_1\| \geq a|\Omega|^p - 2N^2|\lambda,\Gamma) \leq 3 \exp(-\text{const.}\ |\Omega|^{2p-1})$ by Lemma 5.4, so that $P_{M_0^{+-}(\Omega,m)}(E) \to 0$ for the same p-values as above. From what was just proved we know that $P_{M_0^{+-}(\Omega,m)}(|\Omega_\lambda| > k|\Omega|,\ |\Omega| - |\Omega_\lambda| > k|\Omega|) \to 1$ for a suitable $k > 0$, however, so we know that

$$P_{M_0^{+-}(\Omega,m)}(|M_\lambda - m^*|\Omega_\lambda\| \geq a|\Omega|^p) \to 0 \quad . \tag{5.13}$$

The same argument applies to the magnetization below λ, and Theorem 3.1 is proved.

5A. APPENDIX

Proof of Lemma 5.4. The proof uses the fact that the total number of - spins, $N(X)$, of a configuration X is the sum of contributions from the regions inside the outer contours (i.e. those not surrounded by any other contour). For a given configuration of outer contours, Γ, these contributions are independent random variables and those coming from congruent regions have equal distributions. Thus $N(X)$ fluctuates for two reasons, one because the number of outer contours in each congruence class (γ), $K_{(\gamma)}(\Gamma(X))$, is random, and the other because the contributions from the regions belonging to the various congruence classes are also random. The fluctuations for given outer contours can easily be estimated using the independence of the contributions. The fluctuations in the $K_{(\gamma)}$ are more complicated to estimate, and we consider them first. We want to estimate the deviations $S_{(\gamma)}(\Gamma(X))$ defined by:

$$|\Theta|^p S_{(\gamma)}(\Gamma(X)) = K_{(\gamma)}(\Gamma(X)) - \langle K_{(\gamma)}(\Gamma(X))\rangle_{M_{0,c}^{+}(\Theta)} \quad . \tag{5A.1}$$

The contribution to $N(X)$ due to the fact that $S_{(\gamma)} \neq 0$ is clearly bounded by:

$$|\Theta|^p \sum_{(\gamma)} (\text{area of } \gamma)|S_{(\gamma)}| \leq |\Theta|^p \sum_{(\gamma)} |\gamma|^2 |S_{(\gamma)}| \equiv |\Theta|^p S \quad , \tag{5A.2}$$

and therefore we want to estimate the fluctuations of $S(\Gamma(X))$ defined in (5A.2). This can be done using the following "saddle point" technique, which is often useful for the probability distributions of "exponential type" common in statistical mechanics. (We suggest, however, that the reader first studies how the theorem follows from the estimate after (5A.23).)

The probability in $M_{0,c}^{+}(\Theta)$ of a possible configuration of outer contours Γ is given by:

$$\begin{aligned} P_\mu(\Gamma) &= \prod_{\gamma\in\Gamma} e^{\mu(\gamma)} Z_c^-(\gamma,\mu)/Z(M_{0,c}^{+}(\Theta),\mu) \\ &= \prod_{(\gamma)} [e^{\mu(\gamma)} Z_c^-(\gamma,\mu)]^{K_{(\gamma)}(\Gamma)} / Z(M_{0,c}^{+}(\Theta,\mu) \quad , \end{aligned} \tag{5A.5}$$

where $Z_c^-(\gamma,\mu)$ denotes the partition function of the ensemble of configurations in the region enclosed by γ such that all spins along the inside of γ are -1. We consider for the moment an arbitrary translationally invariant weight $\mu(\gamma)$ instead of $-\beta|\gamma|$. $P_\mu(\Gamma)$ is thus of "exponential type"

$$P_\mu(\Gamma) = \exp(\sum_{(\gamma)} a_{(\gamma)}(\mu) K_{(\gamma)}(\Gamma) - f(a(\mu))) \tag{5A.6}$$

with

$$a_{(\gamma)}(\mu) = \mu(\gamma) + \log Z_c^-(\gamma,\mu) \tag{5A.7}$$

and

$$\exp f(a(\mu)) = \sum_{\Gamma} \exp \sum_{(\gamma)} a_{(\gamma)}(\mu) K_{(\gamma)}(\Gamma) = Z(M_{0,c}^+(\Theta),\mu) \quad . \tag{5A.8}$$

The generating function $\langle \exp \sum_{(\gamma)} \Delta a_{(\gamma)} K_{(\gamma)} \rangle_\mu$ can thus be written for any numbers $\Delta a_{(\gamma)}$:

$$\langle \exp \sum_{(\gamma)} \Delta a_{(\gamma)} K_{(\gamma)} \rangle_\mu = \exp(f(a + \Delta a) - f(a)) \quad . \tag{5A.9}$$

For any possible values $K_{(\gamma)}$ we thus have the following inequality:

$$P_\mu(K_{(\gamma)}) \exp \sum_{(\gamma)} \Delta a_{(\gamma)} K_{(\gamma)} \le \exp(f(a + \Delta a) - f(a)) \tag{5A.10}$$

for all Δa. By a judicious choice of Δa we can thus estimate $P_\mu(K_{(\gamma)})$ by:

$$P_\mu(K_{(\gamma)}) \le \exp(f(a + \Delta a) - f(a) - \sum_{(\gamma)} \Delta a_{(\gamma)} K_{(\gamma)}) \quad . \tag{5A.11}$$

For $K_{(\gamma)}$ near the average $\langle K_{(\gamma)} \rangle_\mu = \partial f(a)/\partial a_{(\gamma)}$ we can find a good choice of $\Delta a_{(\gamma)}$ by considering the first two terms in the Taylor expansion of the exponent in (5A.11): We have $K_{(\gamma)} = \partial f(a(\mu))/\partial a_{(\gamma)} + |\Theta|^p S_{(\gamma)}$ for $\mu(\gamma) = -\beta|\gamma|$, and put $\Delta a_{(\gamma)} = a_{(\gamma)}(\mu + \Delta\mu) - a_{(\gamma)}(\mu)$ and expand to second order in $\Delta\mu(\gamma)$. The 0th order term vanishes, the first order term is $-\sum_{(\gamma)} |\Theta|^p S_{(\gamma)} da_{(\gamma)}(\mu)$ and the second order term is $1/2(d^2 f(a(\tilde\mu)) - \sum_{(\gamma)} K_{(\gamma)} d^2 a_{(\gamma)}(\tilde\mu))$ for some $\tilde\mu$ between μ and $\mu + \Delta\mu$. Consider the second order term first. Because

$$f(a(\mu)) = \log Z(M_{0,c}^+(\Theta),\mu) = \log \sum_{\gamma_1,\ldots,\gamma_n \subset \Theta}^{c} e^{\sum_1^n \mu(\gamma_i)} \tag{5A.12}$$

we see as explained in (4.40) that

$$\begin{aligned} d^2 f(a(\tilde\mu)) = & \sum_{\gamma \subset \Theta}^{c} (\Delta\mu(\gamma))^2 [\tilde\rho_{\Theta,c}(\gamma) - \tilde\rho_{\Theta,c}^2(\gamma)] \\ & + \sum_{\substack{\gamma_1,\gamma_2 \subset \Theta \\ \gamma_1 \ne \gamma_2}}^{c} \Delta\mu(\gamma_1) [\tilde\rho_{\Theta,c}(\gamma_1,\gamma_2) - \tilde\rho_{\Theta,c}(\gamma_1)\tilde\rho_{\Theta,c}(\gamma_2)] \quad . \end{aligned} \tag{5A.13}$$

We now make the "judicious choice":

$$\Delta\mu(\gamma) = \begin{cases} (\operatorname{Sgn} S_{(\gamma)}) |\gamma|^2 t & \text{if } |\gamma| \le c \log |\Omega| \\ 0 & \text{otherwise} \end{cases} \tag{5A.14}$$

for some t with $0 \le t \le (c \log |\Omega|)^{-1}$.

This restriction ensures that $|\Delta\mu(\gamma)| \le |\gamma|$, so that $\mu(\gamma) + \Delta\mu(\gamma) \le -(\beta - 1)|\gamma|$ and $\tilde\mu(\gamma) \le -(\beta - 1)|\gamma|$. Then we can estimate (5A.13) using (4.61) with $b = \beta - 1$ and get:

$$\frac{1}{2} d^2 f \le (2 \cdot 10^{12}) e^{-(\beta-1)} t^2 |\Theta| \sum_{(\gamma)} |\gamma|^2 e^{-(\frac{\beta-1}{2})|\gamma|} \tag{5A.15}$$

if $3e^{-(\beta-1/2)} \le 1/2$. The last sum can be estimated using Lemma 4.1: $\sum_{(\gamma)} \le 16$ if $3e^{-(\beta-1/2)} \le 1/2$, so we get:

$$\frac{1}{2} d^2 f \le 10^{14} e^{-\beta} |\Theta| t^2 \tag{5A.16}$$

if $3e^{-(\beta-1/2)} \le 1/2$. Similarly $d^2 a_{(\gamma)}(\tilde{\mu})$ can be expressed as a quadratic form similar to (5A.13) with the correlation functions in the ensemble inside γ mentioned above. $d^2 a_{(\gamma)}(\tilde{\mu})$ is thus non-negative, because the quadratic form is non-negative definite, and $\sum_{(\gamma)} K_{(\gamma)} d^2 a_{(\gamma)}(\tilde{\mu}) \ge 0$.

Consider now the first order term. In it we have

$$da_{(\gamma)}(\mu) = \Delta\mu(\gamma) + d \log Z_c^-(\gamma,\mu) = \Delta\mu(\gamma) + \sum_{\gamma_1 \text{ inside } \gamma}^{c} \Delta\mu(\gamma_1)\rho^-_{\gamma,c}(\gamma_1) \quad , \tag{5A.17}$$

where $\rho^-_{\gamma,c}(\gamma_1)$ is the correlation function in the ensemble inside γ mentioned above. As shown in (4.41) we have $\rho^-_{\gamma,c}(\gamma_1) \le e^{-\beta|\gamma_1|}$, so the sum can be estimated by:

$$\sum_{\gamma_1 \text{ ins. } \gamma} t|\gamma_1|^2 e^{-\beta|\gamma_1|} \le \frac{t|\gamma|^2}{16} \sum_{4}^{\infty} \ell^2 (3e^{-\beta})^{\ell} \le t|\gamma|^2 e^{-2\beta} \tag{5A.18}$$

if $3e^{-\beta/2} \le 1/2$. Its contribution is thus estimated by:

$$\sum_{(\gamma)} |\Theta|^p |s_{(\gamma)}| t|\gamma|^2 e^{-2\beta} = t|\Theta|^p e^{-2\beta} S \le \frac{t|\Theta|^p S}{2} \quad \text{if} \quad 3e^{-\frac{\beta}{2}} \le 1/2 \quad ,$$

and finally:

$$-\sum_{(\gamma)} |\Theta|^p s_{(\gamma)} da_{(\gamma)}(\mu) \le -\sum_{(\gamma)} |\Theta|^p |s_{(\gamma)}| \, |\gamma|^2 t + \frac{t|\Theta|^p}{2} \le -\frac{2|\Theta|^p S}{2} \quad . \tag{5A.19}$$

Forgetting the negative second order term and choosing t in an optimal way we thus get the following bound for $P(K_{(\gamma)})$:

$$P(K_{(\gamma)}) \le \exp \min_{0 \le t \le (c \log |\Omega|)^{-1}} \left[\frac{-t|\Theta|^p S}{2} + \delta(\beta)|\Theta| t^2\right] \quad , \tag{5A.20}$$

where we have put $\delta(\beta) = 10^{14} e^{-\beta}$.

The min. occurs for $t = S|\Theta|^p / 4|\Theta|\delta(\beta)$ if this quantity $\le (c \log |\Omega|)^{-1}$, so we finally get:

$$P(K_{(\gamma)}) \le \exp - \frac{S^2 |\Theta|^{2p-1}}{16\delta(\beta)} \tag{5A.21}$$

for $S \le 4\delta(\beta)|\Theta|^{1-p}(c \log |\Omega|)^{-1}$ and $3e^{-(\beta-1/2)} \le 1/2$. To estimate the fluctuations in S we first note that $K_{(\gamma)} \le |\Theta|$ and $|\gamma| \le c \log |\Omega|$ if $P(K_{(\gamma)}) \ne 0$, so the number of sequences $\{K_{(\gamma)}\}$ with $P(K_{(\gamma)}) \ne 0$ is bounded by:

$$|\Theta|^{\left(\sum_{4}^{c \log |\Omega|} 3\ell\right)} \le \exp(2 \log |\Theta|) |\Omega|^{c \log 3} \quad .$$

We can thus conclude that:

$$P(S(\Gamma(X)) \geq T) \leq \sum_{\substack{\{K_{(\gamma)}\} \\ S(\Gamma)\geq T}} \exp \text{Min}[\ldots]$$

$$\leq \exp((2 \log |\Omega|)|\Omega|^{c \log 3} - \frac{T^2|\Theta|^{2p-1}}{16\delta(\beta)}) \tag{5A.22}$$

if $T \leq 4\delta(\beta)|\Theta|^{1-p}(c \log |\Omega|)^{-1}$, because $\text{Min}[\ldots]$ is a decreasing function of S as is easily checked. If $|\Theta| > k|\Omega|$ and $2p - 1 > c \log 3$, which is true if $p > (1 + c \log 3)/2$, and if $T^2/\delta(\beta)$ is bounded below e.g. by 1 then the negative term dominates when $|\Theta|$ is large, so we can finally conclude that the following estimate is valid:

$$P_{M^+_{0,c}(\Theta)}(S(\Gamma(X)) \geq T) \leq \exp - \frac{T^2|\Theta|^{2p-1}}{20\delta(\beta)} \tag{5A.23}$$

if $\delta^{\frac{1}{2}}(\beta) \leq T \leq 4\delta(\beta)|\Theta|^{1-p}(c \log |\Omega|)^{-1}$, $1 > p > (1 + c \log 3)/2$, $|\Theta| > k|\Omega|$ for some $k > 0$, $3e^{-(\beta-1/2)} \leq 1/2$ and $|\Theta|$ is large.

We now come to the study of the fluctuations in $N(X)$ for a given configuration $\Gamma = (\gamma_1,\ldots,\gamma_k)$ of outer contours with "occupation numbers" $\{K_{(\gamma)}\}$. We make use of the following straightforward estimate:

<u>Lemma 5A.1</u>

Let $n_1,\ldots,n_k$ *be independent random variables which are all bounded,* $|n_i| \leq B_i$, $i = 1, \ldots, k$, *and let* $N = \sum_1^k n_i$, $B^2 = \sum_1^k B_i^2$. *Then*

$$P(|N - \langle N\rangle| \geq a) \leq 2e^{-\frac{a^2}{2B^2}} \tag{5A.24}$$

<u>Proof</u>. Let $f_i(t) = \log \langle e^{tn_i}\rangle$ and $f(t) = \log \langle e^{tN}\rangle = \sum_1^k f_i(t)$, and denote "canonical averages" by $\langle g(n_i)\rangle_t \equiv \langle g(n_i)e^{tn_i}\rangle/\langle e^{tn_i}\rangle$. Because $\langle N\rangle = f'(0)$ we have $\langle e^{t(N-\langle N\rangle)}\rangle = e^{f(t)-tf'(0)}$, and as in the previous argument we get the inequality $e^{ta}P((N - \langle N\rangle) \geq a) \leq e^{f(t)-tf'(0)}$ for any $t \geq 0$, so that for a judicious choice of t we get:

$$P(N - \langle N\rangle \geq a) \leq e^{f(t)-tf'(0)-ta} \quad . \tag{5A.25}$$

If we expand the exponent to second order in t we get $-ta + t^2/2\, f''(\tilde{t})$ for some $\tilde{t}$ between 0 and t. $f''(\tilde{t})$ can be written as $\sum_1^k \langle n_i^2\rangle_{\tilde{t}} - \langle n_i\rangle_{\tilde{t}}^2$ and is thus bounded by $\sum_1^k B_i^2 = B^2$ for any $\tilde{t}$. We thus see that

$$P(N - \langle N\rangle \geq a) \leq \exp \operatorname*{Min}_{t\geq 0}\left(\frac{B^2t^2}{2} - at\right) = e^{-\frac{a^2}{2B^2}} \quad , \tag{5A.26}$$

and the "judicious choice" is $t = a/B^2$.

In the same way we see that

$$P(N - \langle N\rangle \leq -a) \leq e^{-\frac{a^2}{2\beta^2}} \tag{5A.27}$$

and the lemma is proved.

In our context n_i is the contribution to $N(X)$ from the region inside γ_i. It is clearly bounded by the area of γ_i, which is bounded by $B_i = |\gamma_i|^2/16$, and we get $B^2 = \sum_i |\gamma_i|^4/16 = \sum_{(\gamma)} K_{(\gamma)}|\gamma|^4/16^2$. We thus see (because $|\gamma| \leq c \log |\Omega|$ if $K_{(\gamma)} \neq 0$) that

$$\begin{aligned} B^2 &\leq \sum_{(\gamma)} \langle K_{(\gamma)}\rangle |\gamma|^4/16^2 + \frac{(c \log |\Omega|)^2}{16^2} |\Theta|^p \sum_{(\gamma)} |\gamma|^2 |S_{(\gamma)}| \\ &= \sum_{\gamma\subset\Theta} \pi_{\Theta,c}(\gamma)|\gamma|^4/16^2 + \frac{(c \log |\Omega|)^2 |\Theta|^p S}{16^2} \\ &\leq \frac{|\Theta|}{16^2} \left[\sum_4^\infty \ell^4 (3e^{-\beta})^\ell + (c \log |\Omega|)^2 |\Theta|^{p-1} S\right] \end{aligned} \tag{5A.28}$$

using the estimate $\pi_{\Theta,c}(\gamma) \leq e^{-\beta|\gamma|}$ (4.41) of the probability that γ is an outer contour. This means that if we consider configurations Γ satisfying e.g. $S(\Gamma) \leq \delta(\beta)|\Theta|^{1-p'}$ for some p' with $p < p' < 1$ then the last term becomes uniformly small for $|\Theta| > k|\Omega|$ large, so we can say that $B^2(\Gamma) \leq \delta(\beta)|\Theta|$ if $S(\Gamma) \leq \delta(\beta)|\Theta|^{1-p'}$, $3e^{-\beta/2} \leq 1/2$, $|\Theta| > k|\Omega|$ and $|\Theta|$ is large. For any such configuration we thus conclude from the lemma that

$$P\{|N(X) - \langle N(X)\rangle_p| > \frac{t|\Theta|^p}{2} |\Gamma\} \leq 2e^{-\frac{t^2|\Theta|^{2p}}{8\delta(\beta)|\Theta|}} \quad . \tag{5A.29}$$

If now we consider values of t such that $\delta^{\frac{1}{2}}(\beta) \leq t/2 \leq \delta(\beta)|\Theta|^{1-p'}$ and Γ such that $S(\Gamma) \leq t/2$ then because

$$\begin{aligned} |\langle N(X)\rangle_\Gamma - \langle N(X)\rangle_{M^+_{0,c}(\Theta)}| &\leq \sum_{(\gamma)} (\text{area of } \gamma)\, K_{(\gamma)}(\Gamma) - \langle K_{(\gamma)}(\Gamma(X))\rangle_{M^+_{0,c}(\Theta)}| \\ &\leq |\Theta|^p S(\Gamma) \leq t|\Theta|^p/2 \quad , \end{aligned} \tag{5A.30}$$

so that $|N(X) - \langle N(X)\rangle_{M^+_{0,c}(\Theta)}| \geq t|\Theta|^p$ implies that $|N(X) - \langle N(X)\rangle_\Gamma| \geq t|\Theta|^p/2$, we see from (5A.29) that

$$P\{|N(X) - \langle N(X)\rangle_{M^+_{0,c}(\Theta)}| \geq t|\Theta|^p|\Gamma\} \leq 2e^{-\frac{t^2|\Theta|^{2p-1}}{8\delta(\beta)}} \tag{5A.31}$$

uniformly in Γ and t. Summing over all possible Γ we then get the following estimate using (5A.23):

$$P_{M^+_{0,c}(\Theta)}\{|N(X) - \langle N(X)\rangle_{M^+_{0,c}(\Theta)}| \geq t|\Theta|^p\} \leq P_{M^+_{0,c}(\Theta)}(S(\Gamma(X)) \geq t/2)$$

$$+ \sum_{\substack{\Gamma_i \\ S(\Gamma)\leq t/2}} P_{M^+_{0,c}(\Theta)}(\Gamma)\,2e^{-\frac{t^2|\Theta|^{2p-1}}{8\delta(\beta)}} \leq 3e^{-\frac{t^2|\Theta|^{2p-1}}{80\delta(\beta)}} \tag{5A.32}$$

if $p' = (1 + p)/2$ e.g., $1 > p > (1 + c \log 3)/2$, $\delta^{\frac{1}{2}}(\beta) \leq t/2 \leq \delta(\beta)|\Theta|^{1-p'}$, $3e^{-(\beta-1/2)} \leq 1/2$, $|\Theta| > k|\Omega|$ and $|\Theta|$ large. (Note that the last restriction on t implies that needed to apply (5A.23) if $|\Theta|$ is large.)

To prove Lemma 5.4 we finally have to estimate the difference $\langle N(X)\rangle_{M^+_{0,c}(\Theta)} - |\Theta|(1 - m^*)/2$. Using the estimate $\rho(\gamma) \leq e^{-\beta|\gamma|}$ (4.41) valid in the ensemble $M^+_0(\Theta)$ also we get:

$$1 - P_{M^+_0(\Theta)}(M^+_{0,c}(\Theta)) \leq \sum_{\substack{\gamma\subset\Theta \\ |\gamma|\geq c \log |\Omega|}} e^{-\beta|\gamma|} \leq |\Theta| \sum_{c \log |\Omega|}^{\infty} (3e^{-\beta})^{\ell}$$

$$\leq 2|\Theta|(3e^{-\beta})^{c \log |\Omega|} \leq 2|\Omega|^{1+c \log(3e^{-\beta})} \tag{5A.33}$$

if $3e^{-\beta/2} \leq 1/2$, so it goes to zero if β is large, and because

$$\langle N(X)\rangle_{M^+_0(\Theta)} = P_{M^+_0(\Theta)}(M^+_{0,c}(\Theta))\langle N(X)\rangle_{M^+_{0,c}(\Theta)}$$

$$+ \left(1 - P_{M^+_0(\Theta)}(M^+_{0,c}(\Theta))\right)\langle N(X)\rangle_{\overline{M^+_{0,c}}(\Theta)} \tag{5A.34}$$

we get

$$|\langle N(X)\rangle_{M^+_0(\Theta)} - \langle N(X)_{M^+_{0,c}(\Theta)}| \leq 4|\Theta||\Omega|^{1+c \log(3e^{-\beta})}$$

$$\leq 4|\Omega|^{2+c \log(3e^{-\beta})} \leq 4|\Omega|^{\frac{1}{2}} \tag{5A.35}$$

e.g. if $3e^{-\beta} \leq e^{-3/2c}$, i.e. if β is large. Moreover, it is shown in [5] that

$$|\langle N(X)\rangle_{M^+_0(\Theta)} - |\Theta|(1 - m^*)/2| \leq \text{const } |\partial\Theta| \tag{5A.36}$$

if $3e^{-\beta/2} \leq 1/2$. Because $|\Omega|^{\frac{1}{2}}/t|\Theta|^p \to 0$ as $|\Theta| \to \infty$ for the values we consider these estimates show that for $|\Theta|$ large

$$P_{M^+_{0,c}(\Theta)}\{N(X) - |\Theta|(1 - m^*)/2| \geq t|\Theta|^p\} \leq 3e^{-\frac{t^2|\Theta|^{2p-1}}{100\delta(\beta)}} \tag{5A.37}$$

when the above restrictions are fulfilled, and Lemma 5.4 is proved because the magnetization is $|\Theta| - 2N(X)$.

6. THE SURFACE TENSION

In this section we show that the definition of the surface tension given in Section 3 is allowed in that the limit (3.4) exists and is independent of α. In the course of the proof the partition functions appearing in (3.4) will be approximated by simpler objects, and in the end the surface tension will appear as the thermodynamic limit of a partition function of the ensemble of big contours λ, each one having a weight of the form $e^{-\beta|\lambda|+\mu(\lambda,\beta)}$ with $|\mu(\lambda,\beta)| \le (2.2)|\lambda|e^{-(3.8)\beta}$. We need yet another estimate similar to Lemma 5.2 which is also proved in [5].

Lemma 6.1

Let $\Theta \subset \Omega$ *be a cylinder whose bases are not necessarily flat but are restricted by the condition that their lengths do not exceed* $2N$ *each and their distance is at least* kN^{δ} *for some* $k > 0$. *Then if* $0 \le m^* - m \le AN|\Theta|^{-1}$ *for some* $A > 0$ *we have:*

$$1 \ge \frac{Z(M_0^{+}(\Theta,m),\beta)}{Z(M_0^{+}(\Theta),\beta)} \ge D(\beta)|\Theta|^{-\frac{1}{2}}e^{-d(\beta)N^{\frac{1}{2}}} \tag{6.1}$$

for some constants $D(\beta)$, $d(\beta)$ *if* β *is large.*

The significance of this lemma for us will be that the logarithms of the two partition functions differ by an amount which is small compared to a "surface term" $\tau \cdot N$.

Consider now the formula (3.4). Lemma 5.1 and 6.1 show that we can replace $Z(M^{++}(\Omega,m^*),\beta)$ and $Z(M^{+-}(\Omega,m),\beta)$ by $Z(M_0^{++}(\Omega),\beta)$ and $Z(M_0^{+-}(\Omega,m),\beta)$ without changing τ. Furthermore, we consider the ensemble $\widetilde{M}_0^{+-}(\Omega,m)$ of Theorem 3.1 and get:

$$Z(\widetilde{M}_0^{+-}(\Omega,m),\beta) \le \widetilde{\sum_{\lambda}} e^{-\beta|\lambda|} \sum_{m^+,m^-} Z(M_0^{++}(\Omega_\lambda,m^+),\beta)Z(M_0^{--}(\Omega'_\lambda,m^-),\beta)$$

$$\le Z(M_0^{+-}(\Omega,m),\beta) \tag{6.2}$$

where $\widetilde{\sum_{\lambda}}$ denotes the sum over the allowed λ's, and m^+,m^- and the magnetizations of the regions Ω_λ and Ω'_λ above and below λ are restricted by $m^+|\Omega_\lambda| + m^-|\Omega'_\lambda| = m|\Omega|$. By Theorem 3.1 the ratio of the left and right term in (6.2) tends to 1 as $N \to \infty$, so we can replace $Z(M_0^{+-}(\Omega,m),\beta)$ by the middle term $\widetilde{Z}$ in the definition of τ.

We now obtain a lower bound on $\widetilde{Z}$ by restricting the sum further. For each λ in $\widetilde{\sum_{\lambda}}$ we can find another one λ' congruent to λ by shifting it vertically until $0 \le |\Omega_{\lambda'}| - \alpha|\Omega| < 2N$, because when λ is shifted one step $|\Omega_\lambda|$ changes by at most $|\lambda| \le N(1 + (2 \log 3)/\beta) < 2N$ when β is not too small. We denote by $\widetilde{\sum_{(\lambda)}}$ the sum obtained by picking one such translate, λ', for each shape (λ) restricted by $|\lambda| \le N(1 + (2 \log 3)/\beta)$. We then only pick one term in the sum $\sum_{m^+,m^-}$, namely that having $m^- = -m^*$. It has $m^+|\Omega_\lambda| = m|\Omega| + m^*(|\Omega| - |\Omega_\lambda|) = (2\alpha - 1)m^*|\Omega| +$

$+ m^*|\Omega| - m^*|\Omega_\lambda|$, so $0 \le (m^* - m^+)|\Omega_\lambda| = 2m^*(|\Omega_\lambda| - \alpha|\Omega|) \le 4N$, and both $Z(M_0^{++}(\Omega_\lambda, m^+),\beta)$ and $Z(M_0^{--}(\Omega'_\lambda, m^-),\beta)$ can be estimated by Lemma 6.1:

$$\tilde{Z} \ge \tilde{\sum_{(\lambda)}} e^{-\beta|\gamma|} Z(M_0^{++}(\Omega_\lambda, m^+),\beta) Z(M_0^{--}(\Omega'_\lambda, m^-),\beta)$$

$$\ge \frac{D^2(\beta) e^{-d(\beta)N^{\frac{1}{2}}}}{2\sqrt{\alpha(1-\alpha)}|\Omega|} \tilde{\sum_{(\lambda)}} e^{-\beta|\lambda|} Z(M_0^{++}(\Omega_\lambda),\beta) Z(M_0^{--}(\Omega'_\lambda),\beta) \tag{6.3}$$

$\tilde{Z}$ can on the other hand be estimated by:

$$\tilde{Z} \le \tilde{\sum_{\lambda}} e^{-\beta|\lambda|} Z(M_0^{++}(\Omega_\lambda),\beta) Z(M_0^{--}(\Omega'_\lambda),\beta)$$

$$= \tilde{\sum_{(\lambda)}} e^{-\beta|\lambda|} \tilde{\sum_{\lambda' \in (\lambda)}} Z(M_0^{++}(\Omega_{\lambda'}),\beta) Z(M_0^{--}(\Omega'_{\lambda'}),\beta) \quad . \tag{6.4}$$

The last sum is over all allowed λ' vertically congruent to (λ). Because of the restrictions defining $\tilde{M}_0^{+-}(\Omega,m)$ the distance from any such λ' to the bases of Ω is larger than kN^δ for some $k > 0$. This fact will allow us to find an expression for the product of the two partition functions independent of the position of λ'. Using the expression (4.12) for the partition function in terms of $\phi^T(X)$ we get:

$$Z(M_0^{++}(\Omega_{\lambda'}),\beta) Z(M_0^{--}(\Omega'_{\lambda'}),\beta) = \exp\Big\{ \sum_{X \subset \Omega_{\lambda'}} \phi^T(X) + \sum_{X \subset \Omega'_{\lambda'}} \phi^T(X) \Big\}$$

$$= Z(M^{++}(\Omega),\beta) \exp\Big\{- \sum_{\substack{X \subset \Omega_{\infty,N} \\ Xi\lambda'}} \phi^T(X) + \sum_{\substack{Xi\lambda' \\ Xi\partial\Omega}} \phi^T(X)\Big\} \quad , \tag{6.5}$$

where $Xi\lambda'$ means that X intersects λ'. The last sum in (6.5) can be estimated using (4.36):

$$\sum_{\substack{Xi\lambda' \\ Xi\partial\Omega}} |\phi^T(X)| \le (\tfrac{3}{4})^{k^{\frac{1}{2}}N^{\delta/2}} \le (\tfrac{3}{4})^{N^{\frac{1}{2}}} \quad \text{if} \quad 3e^{\frac{\beta}{2}} \le 1/2$$

and N is large uniformly in λ'.

Putting

$$\mu(\lambda,\beta) = \sum_{\substack{Xi\lambda \\ X \subset \Omega_{\infty,N}}} \phi^T(X) \tag{6.6}$$

we have thus obtained the two bounds:

$$\frac{D^2(\beta) e^{-(\frac{3}{4})^{N^{\frac{1}{2}}} - d(\beta)N^{\frac{1}{2}}}}{2\sqrt{\alpha(1-\alpha)}|\Omega|} Z(M_0^{++}(\Omega),\beta) \tilde{\sum_{(\lambda)}} e^{-\beta|\lambda| - \mu(\lambda,\beta)}$$

$$\le \tilde{Z} \le N^\delta e^{(\frac{3}{4})^{N^{\frac{1}{2}}}} Z(M_0^{++}(\Omega),\beta) \tilde{\sum_{(\lambda)}} e^{-\beta|\lambda| - \mu(\lambda,\beta)} \tag{6.6a}$$

which show that the expression for τ can also be taken to be:

$$\tau = \lim_{N\to\infty} N^{-1} \log \sum_{(\lambda)}^{\sim} e^{-\beta|\lambda|-\mu(\lambda,\beta)} \quad . \tag{6.7}$$

From (4.33) follows that

$$|\mu(\lambda,\beta)| \le |\lambda|(2.2)e^{-(3.8)\beta} \equiv \varepsilon(\beta)|\lambda| \tag{6.8}$$

if $3e^{-\beta/2} \le 1/2$, so the crude bounds

$$\sum_{\ell=N}^{\infty} e^{-(\beta+\varepsilon(\beta))\ell} \le \sum_{\lambda}^{\sim} \le \sum_{\ell=N}^{\infty} 3e^{-(\beta-\varepsilon(\beta))\ell} \tag{6.9}$$

show that

$$-\beta - \varepsilon(\beta) \le \tau \le -\beta + \varepsilon(\beta) + \log 3 \quad . \tag{6.10}$$

Put $\sum\limits_{(\lambda)}^{\sim} e^{-\beta|\lambda|-\mu(\lambda,\beta)} = S_N(\beta)$. To show that the limit (6.7) exists we will use the well known subadditivity argument for $\log S_N(\beta)$ and show that $S_N(\beta)S_M(\beta) \le S_{N+M}(\beta)$, at least approximatively. To this end we note the following properties of the paths appearing in any of the sums, $S_N(\beta)$ e.g. Let β be not too small, so that $(2 \log 3)/\beta < 1/2$ and hence $|\lambda| - N < N/2$ for all terms in $S_N(\beta)$. For any λ we can then find a column C on the cylinder with the following properties:

(a) The strip of width one immediately to the right of C only contains one horizontal step of λ.

(b) The strip of width $2N^{1/3}$ centered at C contains a portion of $|\lambda|$ at most $N^{1/2}$ long. (This can be seen as follows. The number of horizontal steps of λ which are simple in the sense that there are no other horizontal steps above or below it is at least $N/2$, because for each group of "multiple" steps at least one unit of the excess length $|\lambda| - N$ is "consumed". Consider for each simple step the strip of width $2N^{1/3}$ centered at its left end, and let L be the shortest length of λ contained in any of these strips. Let M be the maximal size of a family of *disjoint* such strips. Then any simple step has a horizontal distance at most $N^{1/3}$ from this maximal family, so if the maximal strips are widened to $4N^{1/3}$ their union will contain all simple steps. Hence $M \cdot 4N^{1/3} \ge$ the width of the union $\ge N/2$. Moreover, $LM \le$ the length in the maximal family $\le 3N/2$, so $L \le 3N/2M \le 12N^{1/3} < N^{1/2}$ if N is large, and we can take C as the column containing the left end of any simple step whose strip contains the length L.)

Using property (a) we can now construct a mapping F which associates to any pair (λ_N,λ_M) coming from $S_N(\beta)$ and $S_M(\beta)$ a λ_{N+M} included in $S_{N+M}(\beta)$ as follows. "Open up" λ_N and λ_M at some C_N and C_M as described above and join them together on a cylinder with circumference $N + M$ to a closed path λ_{N+M} (first λ_N and then λ_M e.g. starting from a fixed origin). λ_{N+M} will be allowed in $S_{N+M}(\beta)$ because of (a). Moreover, at most NM pairs can be mapped on the same λ_{N+M}. Because of restriction (b) $\mu(\lambda,\beta)$ will be nearly additive in this process:

$$\mu(\lambda_N,\beta) = \sum_{\substack{X i \lambda_N \\ X\subset\Omega_{\infty,N}}} \phi^T(X) = \sum_{\substack{X i\lambda_N \\ X \not\!i C_N \\ X\subset\Omega_{\infty,N}}} \phi^T(X) + \sum_{\substack{X i\lambda_N \\ X i C_N \\ X\subset\Omega_{\infty,N}}} \phi^T(X) \quad . \tag{6.11}$$

The last sum can be bounded by considering those X that intersect λ_N in the strip around C_N and the others separately using (4.33) and (4.36).

$$\sum_{\substack{X i\lambda_N \\ X i C_N \\ X\subset\Omega_{\infty,N}}} |\phi^T(X)| \le (2.2)(N^{\frac{1}{2}} + (\tfrac{3}{4})^{N^{1/6}}) \tag{6.12}$$

e.g. if $3e^{-\beta/2} \le 1/2$ and N is not too small. A similar estimate is valid for $\mu(\lambda_M,\beta)$, and for $\mu(\lambda_{N+M},\beta)$ we get:

$$\mu(\lambda_{N+M},\beta) = \sum_{\substack{X \text{ betw. } C_N \text{ and } C_M \\ X i\lambda_{M+M}, X\subset\Omega_{\infty,N+M}}} \phi^T(X) + \sum_{\substack{X \text{ betw. } C_M \text{ and } C_N \\ X i\lambda_{N+M}, X\subset\Omega_{\infty,N+M}}} \varphi^T(X)$$

$$+ \sum_{\substack{X i C_M \text{ or } C_N \\ X i\lambda_{N+M} \\ X\subset\Omega_{\infty,N+M}}} \phi^T(X) \tag{6.13}$$

with a similar estimate for the last term. The first two terms in (6.13) are equal to the corresponding terms in $\mu(\lambda_N,\beta)$ and $\mu(\lambda_M,\beta)$, because $\phi^T(X)$ does not depend on the cylinder unless X encircles it as explained after (4.21). We thus see that

$$|\mu(\lambda_{N+M},\beta) - \mu(\lambda_N,\beta) - \mu(\lambda_M,\beta)| \le 5(N^{\frac{1}{2}} + M^{\frac{1}{2}}) \tag{6.14}$$

e.g. uniformly in (λ_N,λ_M).

These considerations give us the desired approximative subadditivity of $\log S_N(\beta)$ as follows:

$$S_{N+M}(\beta) = \tilde{\sum_{(\lambda_{N+M})}} e^{-\beta|\lambda_{N+M}|-\mu(\lambda_{N+M},\beta)} \ge \sum_{\substack{(\lambda_{N+M}) \\ \varepsilon \text{ range } F}} e^{-\beta|\lambda_{N+M}|-\mu(\lambda_{N+M},\beta)}$$

$$\ge (NM)^{-1} \tilde{\sum_{(\lambda_N),(\lambda_M)}} e^{-\beta(|\lambda_N|+|\lambda_M|)-\mu(\lambda_N,\beta)-\mu(\lambda_M,\beta)-5(N^{\frac{1}{2}}+M^{\frac{1}{2}})}$$

$$\ge (NM)^{-1} e^{-5(N^{\frac{1}{2}}+M^{\frac{1}{2}})} S_N(\beta) S_M(\beta) \tag{6.15}$$

from which the existence of the limit (6.7) follows [11].

We finally add the following remarks concerning the length of $|\lambda|$. If we define the "microcanonical" partition function corresponding to $S_N(\beta)$ by

$$Q_N(\varepsilon,\beta) = \sum_{\substack{(\lambda) \\ |\lambda|=(1+\varepsilon)N}}^{\sim} e^{-\mu(\lambda,\beta)} \quad , \tag{6.16}$$

then by an argument similar to that above one can show that

$$\sigma(\varepsilon,\beta) = \lim_{N\to\infty} N^{-1} \log Q_N(\varepsilon) \tag{6.17}$$

exists and is convex in ε and is related to $\tau(\beta)$ by the usual Legendre relation:

$$\tau(\beta) = -\beta + \operatorname*{Sup}_{0\le\varepsilon}(-\beta\varepsilon + \sigma(\varepsilon,\beta)) \quad . \tag{6.18}$$

If the sup is attained at a unique point $\varepsilon_0(\beta)$ then it is easy to see that $|\lambda|/N \to 1 + \varepsilon_0(\beta)$ in probability as $N \to \infty$. It can be shown [5] that for small ε $\sigma(\varepsilon,\beta) = -\varepsilon \log \varepsilon + 0(\varepsilon)$ and hence that $\varepsilon_0(\beta) = 0(e^{-\beta})$, which gives a measure of how "straight" the line of separation is.

7. CONCLUDING REMARKS

The technique of Section 2 can easily be applied to an antiferromagnetic Ising model with nearest neighbor interaction and zero external field to show that there is only one translationally invariant equilibrium state. It is known that there are at least two states which are not translationally invariant. In view of Dobrushin's result [4] that a small external field does not change the states drastically one would expect that the uniqueness persists in the presence of such a field. As already mentioned the method of proof extends directly to 3 or more dimensions. It probably also extends to the case of an interaction between more than nearest neighbors if the neighbor interaction dominates the sum of the others. For a general ferromagnetic interaction the result is probably true, but it is not clear how to prove it by the technique used here.

Concerning the surface tension problem we remark that the generalization to 3 dimensions is not obviously straightforward. In this case it would be natural to consider e.g. the following boundary condition: Ω is a rectangular box of size $N \times N \times H$, and the upper half of it is surrounded by $+$ spins and the lower half by $-$ spins. Then a "big" surface of separation is present in every configuration, and its boundary is fixed. Its area will only exceed N^2 by a small amount $\delta(\beta)N^2$, but it is not easy to rule out that it sticks to the boundary", so that a portion of order N^2 is located near it. That can cause trouble when one tries to carry through a subadditivity proof as that leading to (6.15) by joining together 4 boxes $N\times N\times H$ to one box $2N \times 2N \times H$.

The above problem is probably related to the problem of determining the magnitude of the fluctuations of the surface of separation with the following probability assumptions: Consider as above the class of surfaces having a fixed square boundary of size $N \times N$ and give to each surface λ the relative probability $e^{-\beta(\text{area of } \lambda)}$. One can then ask for the probability $P_q(n)$ that a vertical line through the point q in the square intersects the surface at height n. If one considers only the class of surfaces that intersect each vertical line only once one can prove that if β is large

$$P_q(n) \leq \left(\sum_{K=4}^{\infty} (3e^{-\beta})^k k^2\right)^n \tag{7.1}$$

for all q, so the surface is very "rigid" in this case. The analogous problem in 2 dimensions, where λ becomes a line of separation, was considered by Temperley [13]. In this case, it is easy to see that the successive vertical steps of λ are independent and equally distributed at random variables. The fluctuations far from the ends are therefore easily determined by the central theorem, and one sees that $P_q(n) \sim N^{\frac{1}{2}}$ if $n = O(N^{\frac{1}{2}})$ and the distance of q from the ends is $O(N)$. The "surface" has thus large fluctuations and is not "rigid" in 2 dimensions. As Temperley pointed out, if one computes the "surface tension" defined by:

$$\tau'' = \lim_{N\to\infty} N^{-1} \log \sum_{\lambda} e^{-\beta|\lambda|} = -\beta - \log \tan h\, \beta/2 \tag{7.2}$$

one gets the value computed by Onsager for the Ising model, which is equal to τ, (see below). There is thus a cancellation of the errors involved in replacing the true class of paths by the restricted class and in neglecting the weight $\mu(\lambda,\beta)$ defined in (6.6). This cancellation is not really understood, except that one can check explicitly that the first few terms in an expansion in $e^{-\beta}$ are identical.

As mentioned above one can prove that our τ has the same value as that computed by Onsager from a grand canonical definition [1], [2]. Moreover, it is also equal to $\lim N^{-1} \log Z(M^{+-}(\Omega),\beta)/Z(M^{++}(\Omega),\beta)$ if $\delta > 1$, as can be seen by an explicit calculation [2]. *A priori* this is not obvious because the big contour present in $M^{+-}(\Omega)$ is not prevented from being near the ends of the cylinder as it is in $M^{+-}(\Omega,m)$, and this could cause extra boundary effects. The fact that the two surface tensions are equal seems to indicate that the ratio of the probability of finding λ near the ends to that of finding it near the middle is not larger than $e^{o(N)}$.

The question of how much the phase boundary fluctuates is closely related to the question of the existence of non-translationally invariant equilibrium states. Consider the boundary condition just described. If λ fluctuates much a finite group of spins, $\{\sigma_{x_1},\ldots,\sigma_{x_n}\}$ situated far from the boundary of Ω will be far from λ too with high probability and thus $\langle\sigma_{x_1}\cdots\sigma_{x_n}\rangle_\lambda$ given λ will be equal to either $\langle\sigma_{x_1}\cdots\sigma_{x_n}\rangle_+$ or $\langle\sigma_{x_1}\cdots\sigma_{x_n}\rangle_-$, so the state will be a weighted average of these two states. If λ is very rigid, however, it can have a positive probability of going between some spins in the group, and $\langle\sigma_{x_1}\cdots\sigma_{x_n}\rangle_\lambda$ can take several other values with positive probability. Then $\langle\sigma_{x_1}\cdots\sigma_{x_n}\rangle$ is a weighted average not only of the above two states. The previous discussion of the fluctuations supports the belief that in 2 dimensions there are only two extremal Gibbs states, whereas in 3 dimensions at low temperatures this is not the case.

REFERENCES

[1] Abraham, D., Gallavotti, G., Martin-Löf, A., *Lettere Nuovo Cimento* 2, 143 (1971).

[2] Abraham, D., Gallavotti, G., Martin-Löf, A., "Surface Tension in the Two Dimensional Ising Model." Preprint (1971).

[3] Dobrushin, R. L., *Funct. Anal. Appl.* 2, 292 (1968).

[4] Dobrushin, R. L., *Funct. Anal. Appl.* 2, 302 (1968).

[5] Gallavotti, G., Martin-Löf, A., *Comm. Math. Phys.* 25, 87 (1972).

[6] Gallavotti, G., Miracle-Solé, S., *Comm. Math. Phys.* 7, 274 (1968). The paper however contains a combinatorial error in the last section; the error has been copied in [5] but it is corrected in section 4 of the present paper. See also Shen, C. Y., *J. Math. Phys.* 13 754 (1972) or, for an alternative correction, Gallavotti, G., *Comm. Math. Phys.* 27, 103 (1972).

[7] Gallavotti, G., Miracle-Solé, S., *Phys. Rev.* 5B, 2555 (1972).

[8] Lanford, O., Ruelle, D., *Comm. Math. Phys.* 13, 194 (1969).

[9] Lebowitz, J. L., Martin-Löf, S., *Comm. Math. Phys.* 25, 276 (1972).

[10] Minlos, R. A., Sinai, Ja., I. *Math. U.S.S.R. - Sbornik* 2, 355 (1967), II. *Trans. Moscow Math. Soc.* 19, 121 (1968).

[11] Ruelle, D., *Statistical Mechanics*. Benjamin, N.Y., (1969).

[12] Ruelle, D., *Ann. Phys.* 69, 364 (1972).

[13] Temperley, H. N. V., *Phys. Rev.* 103, 1 (1956).

STATES AND AUTOMORPHISMS OF OPERATOR ALGEBRAS
STANDARD REPRESENTATIONS AND THE KUBO-MARTIN-SCHWINGER BOUNDARY CONDITION

Masamichi Takesaki*

INTRODUCTION

Suppose $\mathcal{B}$ denotes the von Neumann algebra $\mathcal{B}(K)$ of all bounded operators on a Hilbert space K. Then it is well-known that $\mathcal{B}$ is the conjugate space of the Banach space $\mathbf{J}$ of all nuclear operators on K, where the duality of $\mathcal{B}$ and $\mathbf{J}$ is given by the bilinear form: $(x,y) \in \mathcal{B} \times \mathbf{J} \mapsto \mathrm{Tr}(xy) \in \mathbb{C}$. A linear functional on $\mathcal{B}$ is said to be *normal* if it is given by an element of $\mathbf{J}$. Let φ be a normal state of $\mathcal{B}$. Then, by definition, there exists a self-adjoint, positive nuclear operator h_φ with $\varphi(x) = \mathrm{Tr}(xh_\varphi)$, $x \in \mathcal{B}$, and $\mathrm{Tr}(h_\varphi) = 1$. Let e_φ be the projection of K onto the closure of the range of h_φ. Then we have $\varphi(x) = \varphi(e_\varphi x e_\varphi)$, $x \in \mathcal{B}$. Therefore, the state φ is considered as the composed map: $x \in \mathcal{B} \mapsto e_\varphi x e_\varphi \in e_\varphi \mathcal{B} e_\varphi \mapsto \varphi(e_\varphi x e_\varphi)$. Since the map: $x \in \mathcal{B} \mapsto e_\varphi x e_\varphi \in e_\varphi \mathcal{B} e_\varphi$ is relatively simple, we may assume that $e_\varphi = 1$. Therefore, in this situation h_φ is non-singular, that is, we can consider the inverse h_φ^{-1} of h_φ, which is unbounded though if $\dim K = \infty$.

Suppose for a while, K is of dimension n, $n < +\infty$, and $h_\varphi = 1/n$. In this case, we have

$$\varphi(xy) = \varphi(yx), \quad x,y \in \mathcal{B} \quad . \tag{1.1}$$

Such a state is called a (finite) *trace*. In this particular case, we denote φ by τ. Considering $\mathcal{B}$ as a vector space, we denote $\mathcal{B}$ by $\mathcal{H}_\tau$ and by $\eta_\tau(x)$ an element $x \in \mathcal{B}$ regarded as a vector in $\mathcal{H}_\tau$. Define the inner product in $\mathcal{H}_\tau$ by

$$\langle \eta_\tau(x) | \eta_\tau(y) \rangle = \tau(y^* x), \ x,y \in \mathcal{B} \quad . \tag{1.2}$$

Then we get a Hilbert space $\mathcal{H}_\tau$ of dimension n^2. Define the two actions π_τ and π'_τ of $\mathcal{B}$ on $\mathcal{H}_\tau$ by

$$\begin{cases} \pi_\tau(a)\eta_\tau(x) = \eta_\tau(ax) \ ; \\ \pi'_\tau(a)\eta_\tau(x) = \eta_\tau(xa) \ , \quad a,x \in \mathcal{B} \quad . \end{cases} \tag{1.3}$$

Then we see easily that π_τ is a faithful representation of $\mathcal{B}$ and π'_τ is a faithful anti-representation of $\mathcal{B}$. Due to the associativity of the multiplication in $\mathcal{B}$, $\pi_\tau(\mathcal{B})$ and $\pi'_\tau(\mathcal{B})$ commute. Suppose an operator a on $\mathcal{H}_\tau$ commutes with $\pi_\tau(\mathcal{B})$. Since the map: $x \in \mathcal{B} \mapsto \eta_\tau(x) \in \mathcal{H}_\tau$ is surjective, we can find an element $a_0 \in \mathcal{B}$ with $a\eta_\tau(1) = \eta_\tau(a_0)$. Then we have, for any $x \in \mathcal{B}$,

$$a\eta_\tau(x) = a\pi_\tau(x)\eta_\tau(1) = \pi_\tau(x)a\eta_\tau(1) = \pi_\tau(x)\eta_\tau(a_0) = \eta_\tau(xa_0) = \pi'_\tau(a_0)\eta_\tau(x) \quad .$$

Hence $a = \pi'_\tau(a_0)$. Therefore, we get

---

* Department of Mathematics, University of California, Los Angeles, California. 90024.

$$\pi_\tau(\mathcal{B})' = \pi'_\tau(\mathcal{B}) \quad . \tag{1.4}$$

Now, let J_τ denote the conjugate linear operator in $\mathcal{H}_\tau$ defined by $J_\tau \eta_\tau(x) = \eta_\tau(x^*)$, $x \in \mathcal{B}$. Then it is easily seen that J_τ is an isometry of $\mathcal{H}_\tau$ onto $\mathcal{H}_\tau$ and that $J_\tau^2 = 1$. Furthermore, we have, for each $a \in \mathcal{B}$,

$$J_\tau \pi_\tau(a) J_\tau = \pi'_\tau(a^*) \quad . \tag{1.5}$$

Thus, combining (5) with (4), we get

$$\begin{cases} J_\tau \pi_\tau(\mathcal{B}) J_\tau = \pi_\tau(\mathcal{B})' \ ; \\ J_\tau \pi_\tau(\mathcal{B})' J_\tau = \pi_\tau(\mathcal{B}) \ . \end{cases} \tag{1.6}$$

Now, consider the cyclic representation π_φ of $\mathcal{B}$ induced by an arbitrary faithful normal state φ of $\mathcal{B}$. In this case, we regard $\mathcal{B}$ again as a vector space; denote it by $\mathcal{H}_\varphi$ and by $\eta_\varphi(x)$ the vector in $\mathcal{H}_\varphi$ corresponding to $x \in \mathcal{B}$. We define the inner product in $\mathcal{H}_\varphi$ by

$$\langle \eta_\varphi(x) | \eta_\varphi(y) \rangle = \varphi(y^* x), \ x, y \in \mathcal{B} \quad . \tag{1.7}$$

The cyclic representation π_φ of $\mathcal{B}$ on $\mathcal{H}_\varphi$ is defined by

$$\pi_\varphi(a)\eta_\varphi(x) = \eta_\varphi(ax), \ a, x \in \mathcal{B} \quad . \tag{1.8}$$

Then we get

$$\varphi(a) = \varphi(1^* a 1) = \langle \pi_\varphi(a)\eta_\varphi(1) | \eta_\varphi(1) \rangle \quad .$$

We shall look at the commutant $\pi_\varphi(\mathcal{B})'$ of $\pi_\varphi(\mathcal{B})$. Noticing the equality:

$$\varphi(a) = \mathrm{Tr}(ah_\varphi) = n\tau(ah_\varphi) = \tau((\sqrt{n}\, h_\varphi^{\frac12}) a (\sqrt{n}\, h_\varphi^{\frac12})) = \langle \pi_\tau(a)\eta_\tau(\sqrt{n}\, h_\varphi^{\frac12}) | \eta_\tau(\sqrt{n}\, h_\varphi^{\frac12}) \rangle \ ,$$

define the map U of $\mathcal{H}_\varphi$ onto $\mathcal{H}_\tau$ by

$$U\eta_\varphi(x) = \sqrt{n}\, \eta_\tau(x h_\varphi^{\frac12}), \quad x \in \mathcal{B}. \tag{1.9}$$

Then it is not hard to show that U is an isometry of $\mathcal{H}_\varphi$ onto $\mathcal{H}_\tau$ such that

$$U\pi_\varphi(a)U^* = \pi_\tau(a), \ a \in \mathcal{B} \ .$$

Therefore, putting $J_\varphi = U^* J_\tau U$, we find a conjugate linear isometry J_φ of $\mathcal{H}_\varphi$ onto $\mathcal{H}_\varphi$ with $J_\varphi^2 = 1$ such that

$$J_\varphi \pi_\varphi(\mathcal{B}) J_\varphi = \pi_\varphi(\mathcal{B})' \quad .$$

Thus, we get the following:

<u>Theorem 1</u>

If K is finite dimensional, then for any faithful normal state φ of $\mathcal{B}$ $(=\mathcal{B}(K))$, there exists a conjugate linear isometry J_φ, with $J_\varphi^2 = 1$, of the representation space $\mathcal{H}_\varphi$ of the cyclic representation π_φ induced by φ onto $\mathcal{H}_\varphi$ itself such that

$$J_\varphi \pi_\varphi(\mathcal{B}) J_\varphi = \pi_\varphi(\mathcal{B})' \quad . \tag{1.10}$$

Definition 2. A conjugate linear isometry J of a Hilbert space onto itself with $J^2 = 1$ is called a *unitary involution*. A representation $\{\pi, \mathcal{H}\}$ of a C^*-algebra A is said to be *standard* if there exists a unitary involution J on $\mathcal{H}$ such that

(a) $J\pi(A)''J = \pi(A)'$;

(b) J commutes with every central projection in $\pi(A)''$.

A von Neumann algebra $\{M, \mathcal{H}\}$ is said to be *standard* if the identity representation of M is standard.

In Theorem 1, the existence of the unitary involution J_τ is, in essence, due to the commutativity, see (1). What is the relation between $\varphi(xy)$ and $\varphi(yx)$ in general? Of course, $\varphi(xy) \neq \varphi(yx)$ unless φ is a trace. But we have

$$\varphi(xy) = \mathrm{Tr}(xyh_\varphi) = \mathrm{Tr}(yh_\varphi x) \quad ;$$
$$\varphi(yx) = \mathrm{Tr}(yxh_\varphi) \quad .$$

Therefore, we should look at $h_\varphi x$ and xh_φ. Then we find a bridge $h_\varphi^{(1-t)}xh_\varphi^t$, $0 \le t \le 1$, between $h_\varphi x$ and xh_φ, which lies in $\mathbf{J}$ even if $\dim K = \infty$. From this, one can conclude that for any pair x,y in $\mathcal{B}$ there exists a bounded function $F_{x,y}(\alpha)$ holomorphic in and continuous on the strip; $0 \le \mathrm{Im}\,\alpha \le 1$, with boundary values:

$$F_{x,y}(t) = \varphi(h_\varphi^{it}xh_\varphi^{-it}y) \quad ;$$

$$F_{x,y}(t+i) = \varphi(yh_\varphi^{it}xh_\varphi^{-it}), \quad t \in \mathbb{R} \quad .$$

Definition 3. Let A be a C^*-algebra equipped with a one parameter group σ_t of automorphisms. A state (or more generally positive linear functional) φ of A is said to satisfy the *Kubo-Martin-Schwinger (KMS) boundary condition* at $\beta > 0$ for σ_t if

(a) φ is invariant under σ_t and

(b) for any pair x,y in A there exists a bounded function $F_{x,y}(\alpha)$ holomorphic in and continuous on the strip, $0 \le \mathrm{Im}\,\alpha \le \beta$, such that

$$F_{x,y}(t) = \varphi(\sigma_t(x)y) \quad ;$$
$$F_{x,y}(t+i\beta) = \varphi(y\sigma_t(x)), \quad t \in \mathbb{R} \quad . \tag{1.11}$$

Replacing σ_t by $\sigma_{\beta t}$, we may assume $\beta = 1$.

If A has a unit, then the σ_t-invariance of φ follows automatically from condition (b). Indeed, the periodicity and the boundedness of $F_{x,1}(\alpha)$ implies that $F_{x,1}(t) = \varphi \cdot \sigma_t(x)$ is constant.

From the arguments preceding Definition 3, we conclude

Theorem 4

A faithful normal state φ of $\mathcal{B}$ satisfies the KMS-condition at $\beta = 1$ for the one parameter inner automorphism group $\sigma_t(x) = h_\varphi^{it}xh_\varphi^{-it}$, $x \in \mathcal{B}$, $t \in \mathbb{R}$.

The main purposes of this series of lectures are to show the following results and to find further applications:

(a) The cyclic representation of a C^*-algebra induced by a KMS-state is standard;

(b) A von Neumann algebra $\{M,\mathcal{H}\}$ with a cyclic and separating vector is standard;

(c) To any faithful normal state φ of a von Neumann algebra M, there corresponds uniquely a one parameter group σ_t^φ of automorphisms of M for which φ satisfies the KMS-condition at $\beta = 1$. The automorphism group σ_t^φ is called the *modular* automorphism group of M *associated with* φ;

(d) For a von Neumann subalgebra N of M and a faithful normal state φ of M, there exists a normal projection ε of norm one (called the *conditional expectation*) of M to N such that

$$\varphi(\varepsilon(x)y) = \varphi(xy), \quad x \in M, \quad y \in N \ ,$$

if and only if N is invariant under σ_τ^φ;

(e) The modular automorphism group σ_τ^φ associated with a faithful normal state φ of M is inner if and only if M is semi-finite;

(f) Any normal KMS-state of M for the modular automorphism group σ_t^φ is, roughly speaking, obtained by the multiplication of φ by a central element of M.

1. A VON NEUMANN ALGEBRA WITH A CYCLIC AND SEPARATING VECTOR

Suppose $\mathcal{H}$ is a complex Hilbert space. by $\mathcal{B}(\mathcal{H})$, we denote the algebra of all bounded operators on $\mathcal{H}$. For each subset S of $\mathcal{B}(\mathcal{H})$, we denote by S' the set of all $x \in \mathcal{B}(\mathcal{H})$ commuting with every operator in S. The set S' is called the *commutant* of S. The commutant S' of any subset S of $\mathcal{B}(\mathcal{H})$ is closed under the linear operation and the multiplication, that is, S' is an algebra over the complex number field $\mathbb{C}$. Since the relation $S \subset \mathbf{J}$ implies $S' \supset \mathbf{J}'$, we have

$$S \subset S'' = S^{(IV)} = \dots \ ;$$

$$S' = S''' = S^{(V)} = \dots \ .$$

<u>Definition 1.1</u>. A self-adjoint subalgebra M of $\mathcal{B}(\mathcal{H})$ is called a *von Neumann algebra* if $M = M''$.

We study a von Neumann algebra $\{M,\mathcal{H}\}$ admitting a vector ξ_0 such that $[M\xi_0] = \mathcal{H}$, where $[M]$ denotes the closed subspace spanned by M for any subset M of $\mathcal{H}$. Such a vector ξ_0 is said to be *cyclic* for M. A vector ξ_0 in $\mathcal{H}$ is said to be *separating* for M if $x\xi_0 \neq 0$ for every non-zero $x \in M$. The following is known, see [4; Prop. 5 in p.6].

Proposition 1.2

A vector in $\mathcal{H}$ *is separating for a von Neumann algebra* M *on* $\mathcal{H}$ *if and only if it is cyclic for the commutant* M' *of* M.

Let $K = [M'\xi_0]$ and e be the projection of $\mathcal{H}$ onto K. Then K is invariant under M', so that e commutes with M'; hence it falls in M. Consider the algebras eMe and $M'e$ as subalgebras of $\mathcal{B}(K)$. Then it is also known that

$$(eMe)' = M'e \quad \text{and} \quad (M'e)' = eMe \quad ,$$

where the commutants are considered in $\mathcal{B}(K)$, see [4; Prop. 1, p. 16]. Since ξ_0 is separating for M', the map: $x \in M' \mapsto xe \in M'e$ is an isomorphism. Therefore in order to study the commutant M' of M it suffices to investigate the commutant $M'e$ of eMe on the Hilbert space K. It is easy to see that ξ_0 is a cyclic vector in K for eMe; hence ξ_0 is cyclic and separating in K for eMe. Thus, we will consider only von Neumann algebras with a cyclic and separating vector.

Suppose A is a C^*-algebra equipped with a one parameter group σ_t of automorphisms. Let φ be a KMS-state of A for σ_t, where we mean by a KMS-state a state satisfying the KMS-condition at $\beta = 1$ unless β is specified otherwise. Let $\{\pi_\varphi, \mathcal{H}_\varphi, \xi_\varphi\}$ be the cyclic representation of A induced by φ. This representation is characterized within unitary equivalence by the fact that

$$[\pi_\varphi(A)\xi_\varphi] = \mathcal{H}_\varphi \quad ; \quad \varphi(x) = \langle \pi_\varphi(x)\xi_\varphi | \xi_\varphi \rangle \quad , \quad x \in A \quad .$$

Let M be the von Neumann algebra $\pi_\varphi(A)''$ generated by $\pi_\varphi(A)$. Since φ is σ_t-invariant, there exists the unitary operator $U(t)$ for each $t \in \mathbb{R}$ such that

$$U(t)\pi_\varphi(x)\xi_\varphi = \pi_\varphi(\sigma_t(x))\xi_\varphi \quad , \quad x \in A \quad .$$

Then we have

$$\pi_\varphi \cdot \sigma_t(x) = U(t)\pi_\varphi(x)U(t)^{-1} \quad , \quad x \in A \quad .$$

It follows that $U(t)MU(t)^{-1} = M$, $t \in \mathbb{R}$. Hence $U(t)$ gives rise to an automorphism $\tilde{\sigma}_t$ of M as $\tilde{\sigma}_t(x) = U(t)xU(t)^{-1}$, $x \in M$. Define the state $\tilde{\varphi}$ of M by $\tilde{\varphi}(x) = \langle x\xi_\varphi | \xi_\varphi \rangle$, $x \in M$. Then we claim the following:

Proposition 1.3

In the above situation, $\tilde{\varphi}$ *satisfies the* KMS*-condition for* $\tilde{\sigma}_t$ *and* ξ_φ *is separating for* M.

Proof. Let x and y be arbitrary elements in M. Then there exist sequences $\{x_n\}$ and $\{y_n\}$ in A such that

$$\lim_{n\to\infty} \|x\xi_\varphi - \pi_\varphi(x_n)\xi_\varphi\| = 0 \ , \ \lim_{n\to\infty} \|x^*\xi_\varphi - \pi_\varphi(x_n^*)\xi_\varphi\| = 0 \quad ;$$

$$\lim_{n\to\infty} \|y\xi_\varphi - \pi_\varphi(y_n)\xi_\varphi\| = 0 \ , \ \lim_{n\to\infty} \|y^*\xi_\varphi - \pi_\varphi(y_n^*)\xi_\varphi\| = 0 \quad .$$

By the KMS-condition, there exists, for each $n = 1, 2, \ldots$, a bounded function $F_n(\alpha)$ holomorphic in and continuous on the strip, $0 \le \operatorname{Im} \alpha \le 1$, such that

$$F_n(t) = \varphi(\sigma_t(x_n)y_n) \quad \text{and} \quad F_n(t + i) = \varphi(y_n\sigma(x_n)) \quad .$$

Then we get

$$F_n(t) = \varphi(\sigma_t(x_n)y_n) = \langle \pi_\varphi \cdot \sigma_t(x_n)\pi_\varphi(y_n)\xi_\varphi|\xi_\varphi\rangle$$

$$= \langle U(t)\pi_\varphi(x_n)U(t)^{-1}\pi_\varphi(y_n)\xi_\varphi|\xi_\varphi\rangle = \langle \pi_\varphi(y_n)\xi_\varphi|U(t)\pi_\varphi(x_n^*)\xi_\varphi\rangle \quad ,$$

so that $F_n(t)$ converges uniformly in t to the functions:

$$f(t) = \langle y\xi_\varphi|U(t)x^*\xi_\varphi\rangle = \langle U(t)xU(t)^{-1}y\xi_\varphi|\xi_\varphi\rangle = \tilde{\varphi}(\tilde{\sigma}_t(x)y) \quad .$$

Similarly $F_n(t + i)$ converges uniformly in t to the function $g(t) = \tilde{\varphi}(y\tilde{\sigma}_t(x))$. By the Phragmen-Lindröf theorem, $F_n(\alpha)$ converges uniformly in the strip to a function $F(\alpha)$ defined on the strip; hence $F(\alpha)$ is bounded, holomorphic in and continuous on it. It is obvious that $F(t) = f(t)$ and $F(t + i) = g(t)$; thus $\tilde{\varphi}$ is a KMS-state for $\tilde{\sigma}_t$.

Suppose $x\xi_\varphi = 0$ for some $x \in M$. For any $y \in M$, there exists a bounded function F holomorphic in and continuous on the strip such that

$$F(t) = \tilde{\varphi}(\tilde{\sigma}_t(y^*)x^*xy) \quad ;$$

$$F(t + i) = \tilde{\varphi}(x^*xy\tilde{\sigma}_t(y^*)) \quad .$$

Then we have

$$F(t + i) = \langle x^*xy\tilde{\sigma}_t(y^*)\xi_\varphi|\xi_\varphi\rangle = \langle xy\tilde{\sigma}_t(y^*)\xi_\varphi|x\xi_\varphi\rangle = 0$$

for every $t \in \mathbb{R}$. Hence F must be identically zero. Thus

$$0 = F(0) = \varphi(y^*xxy) = \|xy\xi_\varphi\|^2 \quad ,$$

which means that $xM\xi_\varphi = 0$; hence $x = 0$. Therefore, ξ_φ is separating for M. Q.E.D.

2. THE POLAR DECOMPOSITION OF THE INVOLUTION

From now on, we consider an arbitrary but fixed von Neumann algebra $\{M, \mathcal{H}\}$ with a cyclic and separating vector ξ_0. Let $A = M\xi_0$ and $A' = M'\xi_0$. Since the maps: $x \in M \mapsto x\xi_0 \in A$ and $y \in M' \mapsto y\xi_0 \in A'$ are both bijective, we can consider the inverse maps π and π' of the above maps respectively, that is, we write $x = \pi(x\xi_0)$, $x \in M$, and $y = \pi'(y\xi_0)$, $y \in M'$. For each pair $(\xi,\eta) \in A \times \mathcal{H}$, we define its product by:

$$\xi\eta = \pi(\xi)\eta \quad . \tag{2.1}$$

Similarly, we define the product of each pair $(\xi,\eta) \in \mathcal{H} \times A'$ by

$$\xi\eta = \pi'(\eta)\xi \quad . \tag{2.1'}$$

Since M and M' commute, the products defined in (2.1) and (2.1') are coherent and follow the associative law. Hence A and A' turn out to be algebras with these products. The map π and π' are an isomorphism of A onto M and an anti-isomorphism of A' onto M' respectively.

We define the densely defined conjugate linear operators S_0 on A and F_0 on A' by

$$\begin{cases} S_0\xi = \pi(\xi)^*\xi_0 \ , \ \xi \in A & ; \\ F_0\eta = \pi'(\eta)^*\xi_0 \ , \ \eta \in A' & . \end{cases} \tag{2.2}$$

Then we have, for each $\xi \in A$ and $\eta \in A'$,

$$\begin{aligned} \langle S_0\xi|\eta\rangle &= \langle \pi(\xi)^*\xi_0|\pi'(\eta)\xi_0\rangle = \langle \xi_0|\pi(\xi)\pi'(\eta)\xi_0\rangle \\ &= \langle \xi_0|\pi'(\eta)\pi(\xi)\xi_0\rangle = \langle \pi'(\eta)^*\xi_0|\xi\rangle = \langle F_0\eta|\xi\rangle \quad . \end{aligned}$$

It follows that $S_0^* \supset F_0$ and $F_0^* \supset S_0$, where we should remark that S_0 and F_0 are both conjugate linear so that the adjoint operators S_0^* and F_0^* appear always the same side of the inner product as S_0 and F_0. Let S denote the closure S_0^{**} of S_0 and F denote the adjoint S_0^* of S_0. The domain of S is denoted by $\mathcal{D}^\#$ and that of F is denoted by $\mathcal{D}^@$. For convenience, we often denote $S\xi$, $\xi \in \mathcal{D}^\#$, by $\xi^\#$ and $F\eta$, $\eta \in \mathcal{D}^@$, by $\eta^@$ respectively. It is now easy to show that A (resp. A') is an involutive algebra with the involution: $\xi \mapsto \xi^\#$ (resp. $\eta \mapsto \eta^@$).

Remark. A vector $\xi \in \mathcal{H}$ falls in $\mathcal{D}^\#$ if and only if there exists a sequence $\{\xi_n\}$ in A such that $\xi = \lim_{n\to\infty} \xi_n$ and $\{\xi_n^\#\}$ is a Cauchy sequence in $\mathcal{H}$. If this is the case, then $\xi^\#$ is obtained by $\xi^\# = \lim_{n\to\infty} \xi_n^\#$.

Lemma 2.1

The conjugate linear closed operator F *is the closure of* F_0*, that is, a vector* $\eta \in \mathcal{H}$ *belongs to* $\mathcal{D}^@$ *if and only if there exists a sequence* $\{\eta_n\}$ *in* A' *such that* $\eta = \lim_{n\to\infty} \eta_n$ *and* $\{\eta_n^@\}$ *is a Cauchy sequence of* $\mathcal{H}$*. If this is the case, then* $\eta^@$ *is given by* $\eta^@ = \lim_{n\to\infty} \eta^@$.

Proof. Let η be an arbitrary but fixed element of $\mathcal{D}^@$. Define operators a_0 and b_0 on A by

$$a_0\xi = \pi(\xi)\eta \ \text{ and } \ b_0\xi = \pi(\xi)\eta^@ \ , \ \xi \in A \quad .$$

Then we have, for each pair x, y in M,

$$\begin{aligned} \langle a_0x\xi_0|y\xi_0\rangle &= \langle x\eta|y\xi_0\rangle = \langle \eta|x^*y\xi_0\rangle \\ &= \langle \eta|(y^*x\xi_0)^\#\rangle = \langle y^*x\xi_0|\eta^@\rangle = \langle x\xi_0|y\eta^@\rangle = \langle x\xi_0|b_0y\xi_0\rangle \quad ; \end{aligned}$$

hence we get $a_0^* \supset b_0$ and $b_0^* \supset a_0$. Therefore, a_0 and b_0 are both preclosed. Let

a denote the closure a_0^{**} of a_0. We claim that a is affiliated with M' in the sense that every unitary operator u in M commutes with a. For any x in M, we have

$$(ua_0u^{-1})x\xi_0 = ua_0(u^{-1}x\xi_0) = u(u^{-1}x\eta) = x\eta = a_0x\xi_0 \quad ,$$

so that $ua_0u^{-1} = a_0$ for every unitary $u \in M$; hence $uau^{-1} = (ua_0u^{-1})^{**} = a_0^{**} = a$.

Now, let $a = uh = ku$ be the left and right polar decompositions of a, where $h = (a^*a)^{\frac{1}{2}}$ and $k = (aa^*)^{\frac{1}{2}}$. Since a is affiliated with M', h and k are both affiliated with M' and u falls in M'. Note that $uhu^* = k$ and $u^*ku = h$. Let

$$h = \int_0^\infty \lambda dp(\lambda) \quad \text{and} \quad k = \int_0^\infty \lambda dq(\lambda)$$

be the spectral decompositions of h and k respectively. Put

$$h_n = \int_0^n \lambda dp(\lambda) \quad \text{and} \quad k_n = \int_0^n \lambda dq(\lambda) \quad .$$

Then h_n and k_n are bounded; so fall in M'. Since ξ_0 falls in the domains $\mathcal{D}(a)$ of a and $\mathcal{D}(a^*)$ of a^*, ξ_0 lies in the domains $\mathcal{D}(h)$ of h and $\mathcal{D}(k)$ of k simultaneously. Hence we have

$$h\xi_0 = \lim_{n\to\infty} h_n\xi_0 \quad \text{and} \quad k\xi_0 = \lim_{n\to\infty} k_n\xi_0 \quad .$$

Put $\eta_n = uh_n\xi_0 \in A'$. Then, since $uh_nu^* = k_n$, we get

$$\eta_n^@ = h_nu^*\xi_0 = u^*uh_nu^*\xi_0 = u^*k_n\xi_0 \quad ;$$

thus

$$\eta = a\xi_0 = uh\xi_0 = \lim_{n\to\infty} uh_n\xi_0 = \lim_{n\to\infty} \eta_n \quad ;$$

$$\eta^@ = a^*\xi_0 = u^*k\xi_0 = \lim_{n\to\infty} u^*k_n\xi_0 = \lim_{n\to\infty} \eta_n^@ \quad .$$

Hence η is approximated by a sequence in A' with respect to the graph norm of the operator F, which means that F is the closure of its restriction F_0 to A'. Q.E.D.

We define the inner products in $\mathcal{D}^\#$ and $\mathcal{D}^@$ respectively as follows:

$$\begin{cases} \langle\xi_1|\xi_2\rangle_\# = \langle\xi_1|\xi_2\rangle + \langle\xi_2^\#|\xi_1^\#\rangle, \quad \xi_1,\xi_2 \in \mathcal{D}^\# \quad ; \\ \langle\eta_1|\eta_2\rangle_@ = \langle\eta_1|\eta_2\rangle + \langle\eta_2^@|\eta_1^@\rangle, \quad \xi_1,\eta_2 \in \mathcal{D}^@ \quad . \end{cases} \tag{2.3}$$

The norms defined above are exactly the norms in the graphs of S and F respectively, so that $\mathcal{D}^\#$ and $\mathcal{D}^@$ are both complete with these new norms since S and F are closed.

Remark. For a subspace M (not necessarily closed) of $\mathcal{D}^\#$ (resp. $\mathcal{D}^@$), S (resp. F) is the closure of its restriction to M if and only if M is dense in $\mathcal{D}^\#$ (resp. $\mathcal{D}^@$) with respect to the new norm.

Since $S_0 = S_0^{-1}$ and $F_0 = F_0^{-1}$, we have

$$S = S^{-1} \quad \text{and} \quad F = F^{-1} \quad . \tag{2.4}$$

Hence, $\mathcal{D}^{\#}$ (resp. $\mathcal{D}^{@}$) is invariant under the map: $\xi \mapsto \xi^{\#}$ (resp. $\eta \mapsto \eta^{@}$).

Remark. If an injective densely defined operator T_0 is preclosed and its inverse operator T_0^{-1} is also preclosed, then the closure T of T_0 is injective and the inverse T^{-1} is the closure of T_0^{-1}. However, the injectivity of a preclosed operator T_0 does not necessarily imply the injectivity of its closure T even if the operator T_0 is bounded.

Put

$$\Delta = FS = S^*S \quad . \tag{2.5}$$

Then Δ is a linear, non-singular, self-adjoint, positive operator on $\mathcal{H}$, but not necessarily bounded. As in the usual theory of closed operators, we get the polar decomposition of S:

$$S = J\Delta^{\frac{1}{2}} \quad .$$

Since S is conjugate linear and $\Delta^{\frac{1}{2}}$ is linear, J must be a conjugate linear isometry. Since S is non-singular, J is actually a conjugate linear unitary operator of $\mathcal{H}$ onto $\mathcal{H}$ itself. Since $S^{-1} = \Delta^{-\frac{1}{2}}J^{-1}$, we have $\Delta^{-\frac{1}{2}}J^{-1} = J\Delta^{\frac{1}{2}}$ by (2.4); hence $J\Delta^{\frac{1}{2}}J = \Delta^{-\frac{1}{2}}$. Since $S = \Delta^{-\frac{1}{2}}J^{-1}$ is the right polar decomposition of S, we get

$$J = J^{-1} \quad , \quad \Delta^{-\frac{1}{2}} = (SS^*)^{\frac{1}{2}} = (SF)^{\frac{1}{2}}.$$

Thus we get the following:

Lemma 2.2

The operator J *is a unitary involution of* $\mathcal{H}$ *and*

$$\begin{cases} S = J\Delta^{\frac{1}{2}} = \Delta^{-\frac{1}{2}}J & ; \\ F = J\Delta^{-\frac{1}{2}} = \Delta^{\frac{1}{2}}J & ; \\ J\Delta J = \Delta^{-1} & . \end{cases} \tag{2.6}$$

For each $x \in \mathcal{B}(\mathcal{H})$, we define the operator x^T by $x^T = Jx^*J$. Then the map: $x \mapsto x^T$ is an anti-automorphism of $\mathcal{B}(\mathcal{H})$ with period two.

The domain $\mathcal{D}(\Delta^{\frac{1}{2}})$ of $\Delta^{\frac{1}{2}}$ is precisely the domain $\mathcal{D}^{\#}$ of S and the domain $\mathcal{D}(\Delta^{-\frac{1}{2}})$ of $\Delta^{-\frac{1}{2}}$ is also the domain $\mathcal{D}^{@}$ of F. Furthermore, the inner products in $\mathcal{D}^{\#}$ and $\mathcal{D}^{@}$ are given by the following:

$$\begin{cases} \langle \xi_1 | \xi_2 \rangle_{\#} = \langle \xi_1 | \xi_2 \rangle + \langle \Delta^{\frac{1}{2}}\xi_1 | \Delta^{\frac{1}{2}}\xi_2 \rangle, \quad \xi_1, \xi_2 \in \mathcal{D}^{\#} & ; \\ \langle \eta_1 | \eta_2 \rangle_{@} = \langle \eta_1 | \eta_2 \rangle + \langle \Delta^{-\frac{1}{2}}\eta_1 | \Delta^{-\frac{1}{2}}\eta_2 \rangle, \quad \xi_1, \eta_2 \in \mathcal{D}^{@} & . \end{cases} \tag{2.7}$$

Definition 2.3. The self-adjoint positive operator Δ is called the *modular operator* of $\{M,\mathcal{H},\xi_0\}$.

3. BOUNDED ELEMENTS

In this section, we discuss a criterion for an element ξ of $\mathcal{D}^{\#}$ to be in $A = M\xi_0$. We fix an arbitrary element ξ in $\mathcal{D}^{\#}$. Define operators a_0 and b_0 on $A' = M'\xi_0$ by $a_0 x\xi_0 = x\xi$ and $b_0 x\xi_0 = x\xi^{\#}$ for every $x \in M'$. Then as in the proof of Lemma 2.1, we have $a_0^* \supset b_0$ and $a_0 \subset b_0^*$, and a_0 commutes with every unitary operator in M'. Hence, the closure $a = a_0^{**}$ of a_0 is affiliated with M. Let $a = uh = ku$ be the left and right polar decompositions of a, where $h = (a^*a)^{\frac{1}{2}}$ and $k = (aa^*)^{\frac{1}{2}}$. Let K denote the algebra of all continuous functions on the open interval $(0,\infty)$ with compact support. Making use of the spectral decompositions

$$h = \int_0^\infty \lambda dp(\lambda) \quad \text{and} \quad k = \int_0^\infty \lambda dq(\lambda)$$

of h and k respectively, we define $f(h)$ and $f(k)$, $f \in K$, to be

$$f(h) = \int_0^\infty f(\lambda)dp(\lambda) \quad \text{and} \quad f(k) = \int_0^\infty f(\lambda)dq(\lambda) \quad . \tag{3.1}$$

Since each $f \in K$ is bounded, $f(h)$ and $f(k)$ both fall in M. Furthermore, since each $f \in K$ is approximated uniformly on its support by polynomials with constant term zero, we have

$$uf(h)u^* = f(k) \quad \text{and} \quad u^*f(k)u = f(h) \quad . \tag{3.2}$$

Lemma 3.1.

(a) $f(k)\xi$ *falls in* A, $f \in K$;

(b) $\langle (f(k)^*f(k)\xi)^{\#} | \xi^{\#}\rangle \geq 0$, $f \in K$;

(c) ξ *belongs to* A *if and only if there exists a constant* $\gamma > 0$ *such that*

$$\langle (f(k)^*f(k)\xi)^{\#} | \xi^{\#}\rangle \leq \gamma^2 \|f(k)\xi_0\|^2, \ f \in K \quad ; \tag{3.3}$$

if this is the case, then $\|a\| \leq \gamma$.

Proof. For each $f \in K$, we have

$$f(k)\xi = f(k)a\xi_0 = f(k)ku\xi_0 = kf(k)u\xi_0 \quad .$$

Since the function: $\lambda \in (0,\infty) \to \lambda f(\lambda) \in \mathbb{C}$ is in K for each $f \in K$, $kf(k)$ is bounded and falls in M; hence $f(k)\xi = kf(k)u\xi_0$ belongs to A.

Now, we compute that

$$(f(k)\xi)^{\#} = u^*k\bar{f}(k)\xi_0 = h\bar{f}(h)u^*\xi_0 = \bar{f}(h)a^*\xi_0 = \bar{f}(h)\xi^{\#} \quad ;$$

therefore

$$(f(k)\xi)^{\#} = \bar{f}(h)\xi^{\#}, \ f \in K \quad . \tag{3.4}$$

Hence we have

$$\langle (f(k)^*f(k)\xi)^{\#} | \xi^{\#} \rangle = \langle \bar{f}(h)f(h)\xi^{\#} | \xi^{\#} \rangle = \|f(h)\xi^{\#}\|^2 \geq 0 \quad .$$

Suppose there is a constant $\gamma > 0$ satisfying inequality (3.3). Then we have, for each $f \in K$,

$$\|kf(k)\xi_0\|^2 = \|f(k)k\xi_0\|^2 = \|f(k)ua^*\xi_0\|^2 = \|uf(h)\xi^{\#}\|^2 = \|f(h)\xi^{\#}\|^2 \leq \gamma^2 \|f(k)\xi_0\|^2 \quad .$$

Therefore, we have

$$\int_0^\infty |\lambda f(\lambda)|^2 d\|q(\lambda)\xi_0\|^2 \leq \gamma^2 \int_0^\infty |f(\lambda)|^2 d\|q(\lambda)\xi_0\|^2, \ f \in K \quad ,$$

which means that the measure $d\|q(\lambda)\xi_0\|^2$ is supported by the closed interval $[0,\gamma]$; hence $f(k)\xi_0 = 0$ if $\operatorname{supp} f \subset (\gamma,\infty)$; so $f(k) = 0$ for such an f. Thus the spectrum of h is contained in $[0,\gamma]$; hence $\|k\| \leq \gamma$. Therefore, a is bounded and $\|a\| \leq \gamma$; so $\xi = a\xi_0$ is in A.

The converse assertion is almost clear now. Since we won't use this in the sequel, we leave the proof to the reader. Q.E.D.

For a fixed $\eta \in \mathcal{D}^@$, we define $a_0'(x\xi_0) = x\eta$ and $b_0'(x\xi_0) = x\eta^@$, $x \in M$. Then $a_0'^* \supset b_0'$ and $a_0' \subset b_0'^*$, and a_0' commutes with every unitary operator in M. Hence the closure $a' = a_0'^{**}$ of a_0' is affiliated with M'. Let $a' = vh' = k'v$ be the left and right polar decompositions of a' with $h' = (a'^*a')^{\frac{1}{2}}$ and $k' = (a'a'^*)^{\frac{1}{2}}$. We define $f(h')$ and $f(k')$ for each $f \in K$ by making use of the spectral decompositions as before. Then we get the following:

<u>Lemma 3.2.</u>

(a) $f(k')\eta$ *falls in* A', $f \in K$;

(b) $\langle (f(k')^*f(k')\eta)^@ | \eta^@ \rangle \geq 0$, $f \in K$;

(c) η *falls in* A' *if and only if there exists a constant* $\gamma > 0$ *such that*

$$\langle (f(k')^*f(k')\eta)^@ | \eta^@ \rangle \leq \gamma^2 \|f(k')\xi_0\|^2, \ f \in K \quad ; \tag{3.5}$$

if this is the case, then $\|a'\| \leq \gamma$.

4. THE RESOLVENT OF THE MODULAR OPERATOR

Keep the assumptions and the notations in the previous section. Let $\tilde{\mathbb{C}}$ denote the Riemann sphere $\mathbb{C} \cup \{\infty\}$ and $[0,\infty]$ denote the extended positive half line $\{z \in \tilde{\mathbb{C}}: 0 \leq z \leq \infty\}$. Let $A[0,\infty]$ denote the space of all functions f holomorphic in a neighborhood of $[0,\infty]$ and vanishing at infinity. We shall investigate $f(\Delta)$, for each $f \in A[0,\infty]$. For each $\omega \in \tilde{\mathbb{C}} \backslash [0,\infty]$, put

$$\gamma(\omega) = (2|\omega| - \omega - \bar{\omega})^{-\frac{1}{2}} \quad ; \tag{4.1}$$

$$R(\omega) = (\omega - \Delta)^{-1} \quad . \tag{4.2}$$

Then by Lemma 2.2, we have

$$R(\omega)^T = JR(\omega)^*J = (\omega - \Delta^{-1})^{-1} \quad .$$

Theorem 4.1.

For each $\omega \in \tilde{\mathbb{C}}\backslash[0,\infty]$, *the following statements hold:*

(a) $R(\omega)A' \subset A$;

(b) $\|\pi(R(\omega)\xi')\| \leq \gamma(\omega)\|\pi'(\xi')\|$, $\xi' \in A'$;

(c) $R(\omega)^T A \subset A'$;

(d) $\|\pi'(R(\omega)^T\xi)\| \leq \gamma(\omega)\|\pi(\xi)\|$, $\xi \in A$

Proof. By symmetry, we have only to prove (a) and (b). Take and fix an arbitrary $\xi' \in A'$. Put

$$\xi = R(\omega)\xi' = (\omega - \Delta)^{-1}\xi' \quad .$$

Then ξ is in $\mathcal{D}(\Delta)$, the domain of Δ; so is in $\mathcal{D}(\Delta^{\frac{1}{2}}) = \mathcal{D}^{\#}$. We apply the arguments in Section 3 to ξ. Namely, making use of Lemma 3.1, we show that the operator a defined by ξ in Section 3 is bounded and $\|a\| \leq \gamma(\omega)\|\pi'(\xi')\|$. For each $f \in K$, we have

$$\begin{aligned}
(2|\omega| - \omega - \bar{\omega})\langle (f(k)^*f(k)\xi)^{\#}|\xi^{\#}\rangle &= (2|\omega| - \omega - \bar{\omega})\langle Sf(k)^*f(k)\xi|S\xi\rangle \\
&= (2|\omega| - \omega - \bar{\omega})\langle \Delta\xi|f(k)^*f(k)\xi\rangle = (2|\omega| - \omega - \bar{\omega})\langle f(k)\Delta\xi|f(k)\xi\rangle \\
&\leq 2|\omega|\|f(k)\Delta\xi\|\|f(k)\xi\| - 2\mathrm{Re}\,\omega\langle f(k)\Delta\xi|f(k)\xi\rangle \\
&\leq \|f(k)\Delta\xi\|^2 + |\omega|^2\|f(k)\xi\|^2 - 2\mathrm{Re}\,\omega\langle f(k)\Delta\xi|f(k)\xi\rangle \\
&= \|f(k)(\omega - \Delta)\xi\|^2 = \|f(k)\xi'\|^2 = \|f(k)\pi'(\xi')\xi_0\|^2 \\
&= \|\pi'(\xi')f(k)\xi_0\|^2 \leq \|\pi'(\xi')\|^2\|f(k)\xi_0\|^2 \quad .
\end{aligned}$$

Thus we get

$$\langle (f(k)^*f(k)\xi)^{\#}|\xi^{\#}\rangle \leq \gamma(\omega)^2\|\pi'(\xi')\|^2\|f(k)\xi_0\|^2, \; f \in K.$$

Hence ξ falls in A by Lemma 3.1 and

$$\|\pi(\xi)\| \leq \gamma(\omega)\|\pi'(\xi')\| \quad . \qquad \text{Q.E.D.}$$

Now let,

$$\Delta = \int_0^\infty \lambda dE(\lambda) \tag{4.4}$$

be the spectral decomposition of Δ. For each bounded Borel function f on $[0,\infty]$, $f(\Delta)$ is given by

$$f(\Delta) = \int_0^\infty f(\lambda)dE(\lambda)$$

and is bounded. Take a function f in $A[0,\infty]$. Since the complement U_f^c of the domain U_f of f does not meet the non-negative real half line, there exist constants $R > 1$ and $0 < \Theta < \pi$ such that $U_f^c \subset \{z \in \mathbb{C}: 1/R < |z| < R$ and $\Theta < |\arg z| < 2\pi - \Theta\}$. Let R and Θ be any such numbers and Γ_f be the contour shown in the diagram below and ℓ_f = the length of Γ_f:

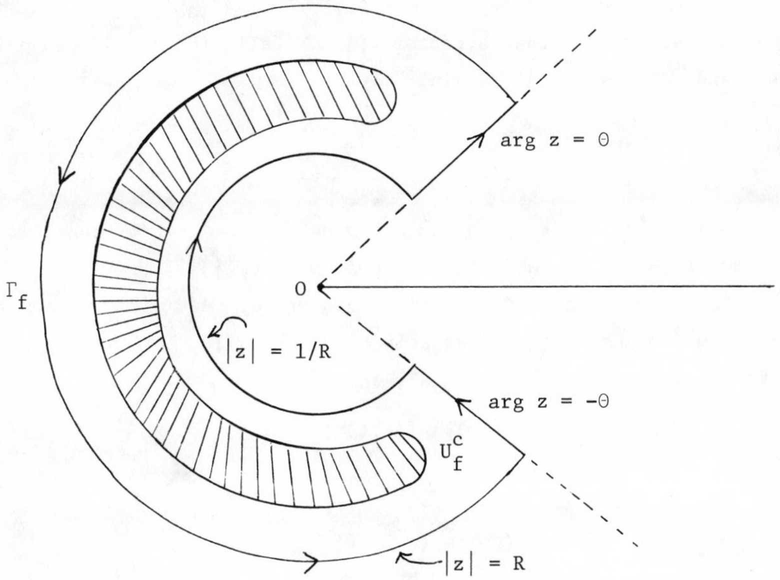

For any complex number z outside the contour Γ_f, $f(z)$ is represented in the form:

$$f(z) = \frac{1}{2\pi i}\oint_{\Gamma_f} \frac{f(\omega)}{\omega - z}\,d\omega \quad ;$$

hence we have

$$f(\Delta) = \frac{1}{2\pi i}\oint_{\Gamma_f} f(\omega)(\omega - \Delta)^{-1}d\omega \quad . \tag{4.5}$$

Then the following is an immediate consequence of Theorem 4.1.

<u>Corollary 4.2</u>

If f *is in* $A[0,\infty]$, *then*

(a) $f(\Delta)A' \subset A$, *and*

$$\|\pi(f(\Delta)\xi')\| \le \frac{1}{2\pi}\,\ell_f\|\pi'(\xi')\| \sup_{\omega\in\Gamma_f}(\gamma(\omega)|f(\omega)|),\ \xi' \in A' \quad ; \tag{4.6}$$

(b) $f(\Delta^{-1})A \subset A'$, *and*

$$\|\pi'(f(\Delta)\xi)\| \le \frac{1}{2\pi} \ell_f \|\pi(\xi)\| \sup_{\omega\in\Gamma_f} (\gamma(\omega)|f(\omega)|), \quad \xi \in A \quad . \tag{4.7}$$

Definition 4.3. We set

$$A^{\#} = \{\xi \in A' \cap \mathcal{D}(\Delta): \Delta\xi \in A'\} \quad ;$$

$$A^{@} = \{\xi \in A \cap \mathcal{D}(\Delta^{-1}): \Delta^{-1}\xi \in A\} \quad .$$

It is easily seen that a vector $\xi \in \mathcal{H}$ falls in $A^{\#}$ if and only if ξ is in $\mathcal{D}^{\#} \cap A'$ and $\xi^{\#}$ is in A' again.

Lemma 4.4

(a) *The set* $A^{\#}$ *is a self-adjoint subalgebra of* A *with respect to the #-involution and dense in the Hilbert space* $\mathcal{D}^{\#}$;

(b) *The set* $A^{@}$ *is a self-adjoint subalgebra of* A' *with respect to the @-involution and dense in the Hilbert space* $\mathcal{D}^{@}$;

(c) *If* ξ *and* η *are both in* $A^{\#}$, *then*

$$\Delta(\xi\eta) = (\Delta\xi)(\Delta\eta) \quad ; \tag{4.8}$$

(d) *If* ξ *and* η *are both in* $A^{@}$, *then*

$$\Delta^{-1}(\xi\eta) = (\Delta^{-1}\xi)(\Delta^{-1}\eta) \quad . \tag{4.9}$$

Proof. If ξ is in $A^{\#}$, then $\eta = (1 + \Delta)\xi$ is in A', so that $\xi = (1 + \Delta)^{-1}\eta$ falls in A by Theorem 4.1. Since

$$\eta^{@} = (\xi + \Delta\xi)^{@} = \xi^{@} + \xi^{\#} = (1 + \Delta)\xi^{\#} \quad ,$$

$\xi^{\#} = (\eta - \xi)^{@}$ is in A'; so $\Delta\xi^{\#} = \eta^{@} - \xi^{\#}$ falls in A'; thus $\xi^{\#}$ is in $A^{\#}$ by definition. Hence $A^{\#}$ is invariant under the involution: $\xi \mapsto \xi^{\#}$.

Now, suppose ξ and η are both in $A^{\#}$. Then we have

$$(\Delta\xi)(\Delta\eta) = \xi^{\#@}\eta^{\#@} = (\eta^{\#}\xi^{\#})^{@} = (\xi\eta)^{\#@} = \Delta(\xi\eta) \quad .$$

Hence $\xi\eta$ and $\Delta(\xi\eta)$ are both in A', so that $\xi\eta$ belongs to $A^{\#}$. Therefore, $A^{\#}$ is a subalgebra of A.

We are now going to show that $A^{\#}$ is dense in $\mathcal{D}^{\#}$ with respect to the #-norm. To do this, it suffices to show that $(1 + \Delta^{\frac{1}{2}})A^{\#}$ is dense in $\mathcal{H}$. Since the functions $z/(1 + z^2)$ and $1/(1 + z^2)$ are both in $A[0,\infty]$, we have, by Corollary 4.2,

$$\Delta^{-1}(1 + \Delta^{-2})^{-1}A \subset A' \quad \text{and} \quad (1 + \Delta^{-2})^{-1}A \subset A' \quad ,$$

which means by definition that $A^{\#}$ contains $\Delta^{-1}(1 + \Delta^{-2})^{-1}A$. Hence we get

$$(1 + \Delta^{\frac{1}{2}})A^{\#} \supset (1 + \Delta^{\frac{1}{2}})\Delta^{-1}(1 + \Delta^{-2})^{-1}A \quad .$$

But $(1 + \Delta^{\frac{1}{2}})\Delta^{-1}(1 + \Delta^{-2})^{-1} = \Delta(1 + \Delta^{\frac{1}{2}})(1 + \Delta^2)^{-1}$ is a bounded operator with dense range in $\mathcal{H}$, and A is dense in $\mathcal{H}$, so that $(1 + \Delta^{\frac{1}{2}})\Delta^{-1}(1 + \Delta^{-2})^{-1}A$ is dense in $\mathcal{H}$; so is $(1 + \Delta^{\frac{1}{2}})A^{\#}$ in $\mathcal{H}$. The assertions for $A^{@}$ follows by symmetry. Q.E.D.

Lemma 4.5

(a) $R(\omega)A^{\#} \subset A^{\#}$, $\omega \in \widetilde{\mathbb{C}}\backslash[0,\infty 1]$;

(b) *If complex numbers* ω_1, ω_2 *and* $\omega_1\omega_2$ *are all in* $\widetilde{\mathbb{C}}\backslash[0,\infty]$, *and if either* ξ_1 *or* ξ_2 *falls in* $A^{\#}$, *then the following formula holds:*

$$(R(\omega_1)\xi_1)(R(\omega_2)\xi_2) = R(\omega_1\omega_2)[\omega_1(R(\omega_1)\xi_1)\xi_2 + \xi_1(\Delta R(\omega_2)\xi_2)] \quad . \tag{4.10}$$

Proof. It is obvious that

$$\begin{cases} \Delta R(\omega) = \Delta(\omega - \Delta)^{-1} = -\omega^{-1}(\omega^{-1} - \Delta^{-1})^{-1} \quad ; \\ R(\omega) = \omega^{-1}(1 + \Delta R(\omega)), \quad \omega \in \mathbb{C}\backslash[0,\infty] \quad . \end{cases} \tag{4.11}$$

Since $A^{\#} \subset A$, we have, by Theorem 4.1,

$$\Delta R(\omega)A^{\#} = -\frac{1}{\omega} R(\omega^{-1})^{T}A^{\#} \subset A' \quad ,$$

so that

$$R(\omega)A^{\#} = \frac{1}{\omega}(1 + \Delta R(\omega))A^{\#} \subset A' \quad .$$

Therefore, $R(\omega)A^{\#}$ is contained in $A^{\#}$, so that assertion (a) follows. Thus, if either ξ_1 or ξ_2 is in $A^{\#}$, then the both sides of (4.10) are continuous functions of the other vector. Hence we may assume both ξ_1 and ξ_2 are in $A^{\#}$. Then we have

$$\Delta(R(\omega_1)\xi_1)(R(\omega_2)\xi_2) = (\Delta R(\omega_1)\xi_1)(\Delta R(\omega_2)\xi_2) = (\omega_1 R(\omega_1)\xi_1 - \xi_1)(\omega_2 R(\omega_2) - \xi_2) \quad ,$$

so that

$$(\omega_1\omega_2 - \Delta)[(R(\omega_1)\xi_1)(R(\omega_2)\xi_2)] = \omega_1(R(\omega_1)\xi_1)\xi_2 + \omega_2\xi_1(R(\omega_2)\xi_2) - \xi_1\xi_2$$

$$= \omega_1(R(\omega_1)\xi_1)\xi_2 + \xi_1(\omega_2 R(\omega_2)\xi_2 - \xi_2)$$

$$= \omega_1(R(\omega_1)\xi_1)\xi_2 + \xi_1(\Delta R(\omega_2)\xi_2) \quad . \qquad \text{Q.E.D.}$$

for $\gamma > 1$, we set

$$E_{\gamma} = \int_{1/\gamma}^{\gamma} dE(\lambda) \quad . \tag{4.12}$$

By Lemma 2.2 and the definition of the anti-automorphism $x \mapsto x^{T}$, we have

$$E_\gamma^T = E_\gamma, \quad \|\Delta E_\gamma\| \le \gamma, \quad \|\Delta^{-1} E_\gamma\| \le \gamma \quad .$$

Let f be a function holomorphic in a neighborhood U_f of $[0,\infty]$. Let $R > 1$ and $0 < \Theta < \pi$ be such that U_f^c is contained in the set $\{z \in \mathbb{C}: 1/R < |z| < R$ and $\Theta < |\arg z| < 2\pi - \Theta\}$. Choose $r > \gamma$ and $0 < \varphi < \Theta$. Consider the contours Γ_f and Γ as shown in the diagram:

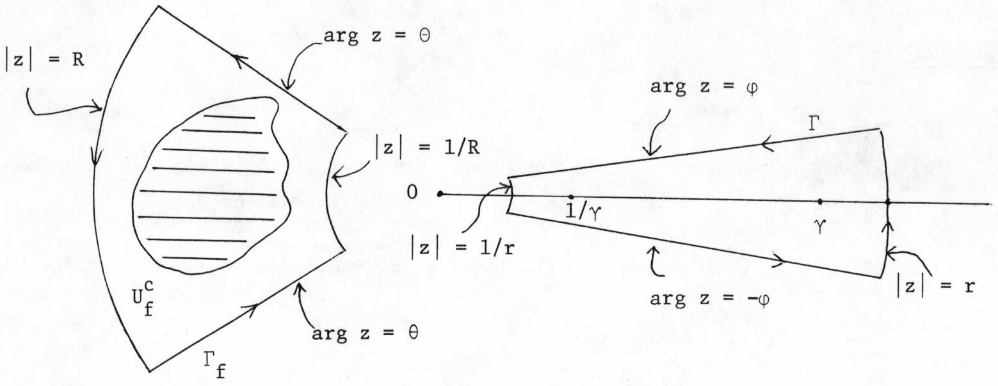

The point is that $\omega_1\omega_2$ is never a positive real number as long as ω_1 stays inside the curve Γ and ω_2 stays inside the curve Γ_f.

Lemma 4.6

(a) *For a function* f *holomorphic in a neighborhood* U_f *of* $[0,\infty]$, $f(\Delta)A^\# \subset A'$;

(b) *Choosing the contour* Γ *as above, we have, for each* $\xi \in A^\#$ *and* $\eta \in \mathcal{H}$,

$$(E_\gamma\eta)(f(\Delta)\xi) = \frac{1}{2\pi i}\oint_\Gamma f(\omega^{-1}\Delta)[(R(\omega)E_\gamma\eta)\xi]d\omega \quad . \tag{4.13}$$

Proof. The equality

$$E_\gamma = \frac{1}{2\pi i}\int_\Gamma R(\omega)E_\gamma d\omega$$

implies formula (4.13) for a constant function f. Considering the decomposition $f(z) = f(z) - f(\infty) + f(\infty)$, we may assume that f is in $A[0,\infty]$. Suppose ξ is in $A^\#$. Then $R(\omega)\xi$ is in $A^\#$ by Lemma 4.5. Furthermore, we have

$$\|\pi'(R(\omega)\xi)\| = \|\frac{1}{\omega}\pi'(\xi - \frac{1}{\omega}R(\omega^{-1})^T\xi\| \le \frac{1}{|\omega|}\|\pi'(\xi)\| + \frac{1}{|\omega|^2}\gamma(\omega^{-1})\|\pi(\xi)\| \quad ,$$

so that $\|\pi'(R(\omega)\xi)\|$ is bounded along the curve Γ_f; hence $f(\Delta)\xi$ falls in A'. Thus assertion (a) follows.

Now, we take an arbitrary $\omega_2 \in \Gamma_f$. Then, by Lemma 4.5,

$$(R(\omega_1)E_\gamma\eta)(R(\omega_2)\xi) = R(\omega_1\omega_2)[\omega_1(R(\omega_1)E_\gamma\eta)\xi + (E_\gamma\eta)(\Delta R(\omega_2)\xi)]$$

for every $\omega_1 \in \Gamma$. Multiplying by $(1/2\pi i)$ and integrating along the curve Γ with respect to ω_1, we get

$$(E_\gamma\eta)(R(\omega_2)\xi) = \frac{1}{2\pi i}\oint_\Gamma (R(\omega_1)E_\gamma\eta)(R(\omega_2)\xi)d\omega_1 \quad .$$

Since the function: $\omega_1 \mapsto R(\omega_1\omega_2)$ is holomorphic inside of the curve Γ,

$$\frac{1}{2\pi i}\oint_\Gamma R(\omega_1\omega_2)[(E_\gamma\eta)(\Delta R(\omega_2)\xi)]d\omega_1 = 0.$$

Hence

$$(E_\gamma\eta)(R(\omega_2)\xi) = \frac{1}{2\pi i}\int_\Gamma R(\omega_1\omega_2)[\omega_1(R(\omega_1)E_\gamma\eta)\xi]d\omega_1 \quad . \tag{4.14}$$

Then multiplying by $\frac{1}{2\pi i} f(\omega_2)$ and integrating with respect to ω_2 along the curve Γ_f, we obtain

$$(E_\gamma\eta)(f(\Delta)\xi) = \frac{-1}{(2\pi)^2}\int_{\Gamma_f}\int_\Gamma f(\omega_2)\omega_1 R(\omega_1\omega_2)[(R(\omega_1)E_\gamma\eta)\xi]d\omega_1 d\omega_2 \quad .$$

On the other hand, we have, for each $\omega_1 \in \Gamma$ and $t \in [0,\infty)$,

$$f(\omega_1^{-1}t) = \frac{1}{2\pi i}\oint_{\Gamma_f} f(\omega_2)\omega_1(\omega_1\omega_2 - t)^{-1}d\omega_2 \quad ,$$

so that

$$f(\omega_1^{-1}\Delta) = \frac{1}{2\pi i}\oint_{\Gamma_f} f(\omega_2)\omega_1 R(\omega_1\omega_2)d\omega_2 \quad .$$

Thus we get

$$(E_\gamma\eta)(f(\Delta)\xi) = \frac{1}{2\pi i}\oint_\Gamma f(\omega^{-1}\Delta)[(R(\omega)E_\gamma\eta)\xi]d\omega \quad . \qquad \text{Q.E.D.}$$

5. THE ONE-PARAMETER AUTOMORPHISM GROUP DEFINED BY THE MODULAR OPERATOR

We keep use of the same terminology and notation as in the previous section.

For each complex number α, the function: $z \mapsto z^\alpha = \exp(\alpha \log|z| + i\alpha \arg z)$ is defined and holomorphic except the negative half real axis $[-\infty,0]$. We set

$$\Delta^\alpha = \int_0^\infty \lambda^\alpha dE(\lambda), \quad \alpha \in \mathbb{C} \quad . \tag{5.1}$$

Theorem 5.1

The one parameter unitary group Δ^{it}, $t \in \mathbb{R}$, *forms an automorphism group of* A *and of* A' *respectively, and*

$$\pi(\Delta^{it}\xi) = \Delta^{it}\pi(\xi)\Delta^{-it}, \quad \xi \in A \quad ;$$
$$\pi'(\Delta^{it}\eta) = \Delta^{it}\pi'(\eta)\Delta^{-it}, \quad \eta \in A' \quad . \tag{5.2}$$

Proof. Consider the function

$$f_\delta(z) = \left(\frac{z+\delta}{1+\delta z}\right)^{it}, \quad 0 \le \delta < 1 \quad .$$

Then, f_δ with $\delta > 0$ is holomorphic except the segment $[-\frac{1}{\delta}, -\delta]$. We apply Lemma 4.6 to the function f_δ with $\delta > 0$.

Take η_1 and η_2 in $A^{\#}$ and ξ in A. By Lemma 4.6, we have

$$\begin{aligned}
\langle f_\delta(\Delta)\eta_1 | (E_\gamma\xi)^{\#}\eta_2\rangle &= \langle (f_\delta(\Delta)\eta_1)\eta_2^{@} | (E_\gamma\xi)^{\#}\rangle = \langle E_\gamma\xi | [(f_\delta(\Delta)\eta_1)\eta_2^{@}]^{@}\rangle \\
&= \langle E_\gamma\xi | \eta_2(f_\delta(\Delta)\eta_1)^{@}\rangle \quad \text{since } f_\delta(\Delta)\eta_1 \in A' \quad , \\
&= \langle (E_\gamma\xi)(f_\delta(\Delta)\eta_1) | \eta_2\rangle \\
&= \frac{1}{2\pi i}\oint_\Gamma \langle f_\delta(\omega^{-1}\Delta)[(R(\omega)E_\gamma\xi)\eta_1] | \eta_2\rangle d\omega \\
&= \frac{1}{2\pi i}\oint_\Gamma \langle (R(\omega)E_\gamma\xi)\eta_1 | (f_\delta(\omega^{-1}\Delta)^*\eta_2\rangle d\omega \\
&= \frac{1}{2\pi i}\oint_\Gamma \langle R(\omega)E_\gamma\xi | (f_\delta(\omega^{-1}\Delta)^*\eta_2)\eta_1^{@}\rangle d\omega \quad .
\end{aligned}$$

Since

$$f_\delta(\omega^{-1}\Delta)^* = \{(\omega^{-1}\Delta + \delta)^{it}(1 + \omega^{-1}\delta\Delta)^{-it}\}^* = (\bar{\omega}^{-1}\Delta + \delta)^{-it}(1 + \bar{\omega}^{-1}\delta\Delta)^{it} \quad ,$$

Lemma 4.6 assures that $f_\delta(\omega^{-1}\Delta)^*\eta_2$ is in A'; hence the above integral equals

$$\begin{aligned}
&\frac{1}{2\pi i}\oint_\Gamma \langle R(\omega)E_\gamma\xi | [\eta_1(f_\delta(\omega^{-1}\Delta)^*\eta_2)^{@}]^{@}\rangle d\omega \\
&= \frac{1}{2\pi i}\oint_\Gamma \langle \eta_1(f_\delta(\omega^{-1}\Delta)^*\eta_2)^{@} | (R(\omega)E_\gamma\xi)^{\#}\rangle d\omega \\
&= \frac{1}{2\pi i}\oint_\Gamma \langle \eta_1 | (R(\omega)E_\gamma\xi)^{\#}(f_\delta(\omega^{-1}\Delta)^*\eta_2)\rangle d\omega \quad .
\end{aligned}$$

Thus we get, for each $\xi \in A$ and $\eta_1, \eta_2 \in A^{\#}$,

$$\langle f_\delta(\Delta)\eta_1 | (E_\gamma\xi)^{\#}\eta_2\rangle = \frac{1}{2\pi i}\oint_\Gamma \langle \eta_1 | (R(\omega)E_\gamma\xi)^{\#}(f_\delta(\omega^{-1}\Delta)^*\eta_2)\rangle d\omega \tag{5.3}$$

Since we have

$$f_\delta(\omega^{-1}\Delta)^* = (1 + \bar{\omega}\delta\Delta^{-1})^{-it}(\delta + \bar{\omega}\Delta^{-1})^{it} \quad ,$$

if we define a function g_δ^ω by

$$g_\delta^\omega(z) = \left(\frac{\delta + \bar{\omega} z}{1 + \bar{\omega}\delta z}\right)^{it} - \delta^{-it} \quad ,$$

then

$$f_\delta(\omega^{-1}\Delta)^* = g_\delta^\omega(\Delta^{-1}) + \delta^{-it}$$

and g_δ^ω belongs to $A[0,\infty]$. Hence by Corollary 4.2, we have

$$\|\pi'(f_\delta(\omega^{-1}\Delta)^*\eta_2)\| \leq \|\pi'(\eta_2)\| + \frac{1}{2\pi}\, \ell \|\pi(\eta_2)\| \sup_{z \in \Gamma_{f_\delta}} (\gamma(z)\, |g_\delta^\omega(z)|) \quad ,$$

where ℓ is the length of the contour Γ_{f_δ}. Hence there exists a constant γ_0 not depending on ω such that

$$\|\pi'(f_\delta(\omega^{-1}\Delta)^*\eta_2)\| \leq \gamma_0 \quad .$$

The operator: $\xi \mapsto (E_\gamma \xi)^\#$ is bounded, so that the function: $\omega \mapsto (R(\omega)E_\gamma\xi)^\#(f_\delta(\omega^{-1}\Delta)^*\eta_2)$ is bounded. Thus both sides of equality (5.3) are continuous functions of η_1, so that equality (5.3) holds for every $\eta_1 \in \mathcal{H}$. Therefore, we get, for each $\eta_1 \in A'$, $\eta_2 \in A^\#$ and $\xi \in A$,

$$\langle f_\delta(\Delta)\eta_1 | (E_\gamma\xi)^\#\eta_2\rangle = \frac{1}{2\pi i}\oint_\Gamma \langle (R(\omega)E_\gamma\xi)\eta_1 | f_\delta(\omega^{-1}\Delta)^*\eta_2\rangle d\omega \quad . \tag{5.4}$$

For each $\omega \in \Gamma$ and $\lambda \in [0,\infty]$, we have

$$|f_\delta(\omega^{-1}\lambda)| = \left|\left(\frac{\lambda + \omega\delta}{\omega + \delta\lambda}\right)^{it}\right| = \exp\left[-t \arg\left(\frac{\lambda + \omega\delta}{\omega + \delta\lambda}\right)\right]$$

$$= \exp[t(\arg(\omega + \delta\lambda) - \arg(\lambda + \delta\omega))] \quad .$$

Recalling the definition of the contour Γ, we get

$$|\arg(\omega + \delta\lambda) - \arg(\lambda + \delta\omega)| \leq |\arg\omega| \leq \varphi \quad .$$

Hence we get the estimate

$$|f_\delta(\omega^{-1}\lambda)| \leq e^{\varphi|t|} \quad ;$$

so $f_\delta(\omega^{-1}\lambda)$ is uniformly bounded on the contour Γ and converges to $f_0(\omega^{-1}\lambda)$ as δ tends to zero. Therefore, the Lebesgue convergence theorem implies that

$$\langle f_0(\Delta)\eta_1 | (E_\gamma\xi)^\#\eta_2\rangle = \frac{1}{2\pi i}\oint_\Gamma \langle (R(\omega)E_\gamma\xi)\eta_1 | f_0(\omega^{-1}\Delta)^*\eta_2\rangle d\omega$$

for every $\eta_1 \in A'$, $\eta_2 \in A^\#$ and $\xi \in A$. But $f_0(\Delta) = \Delta^{it}$ and $f_0(\omega^{-1}\Delta)^* = \bar{\omega}^{-it}\Delta^{-it}$; hence we get

$$\langle \Delta^{it}\eta_1 | (E_\gamma\xi)^\#\eta_2\rangle = \frac{1}{2\pi i}\oint_\Gamma \langle (R(\omega)E_\gamma\xi)\eta_1 | \bar{\omega}^{-it}\Delta^{-it}\eta_2\rangle d\omega =$$

$$= \frac{1}{2\pi i}\oint_\Gamma \langle (\omega^{-it}R(\omega)E_\gamma\xi)\eta_1|\Delta^{-it}\eta_2\rangle d\omega = \langle (\Delta^{-it}E_\gamma\xi)\eta_1|\Delta^{-it}\eta_2\rangle \quad .$$

Thus we get

$$\langle \Delta^{it}\eta_1|(E_\gamma\xi)^{\#}\eta_2\rangle = \langle (\Delta^{-it}E_\gamma\xi)\eta_1|\Delta^{-it}\eta_2\rangle \quad .$$

Since we have

$$\lim_{\gamma\to\infty} (E_\gamma\xi)^{\#} = \lim_{\gamma\to\infty} SE_\gamma\xi = \lim_{\gamma\to\infty} J\Delta^{\frac{1}{2}}E_\gamma\xi = \lim_{\gamma\to\infty} E_\gamma J\Delta^{\frac{1}{2}}\xi = \lim_{\gamma\to\infty} E_\gamma\xi^{\#} = \xi^{\#} \quad ,$$

we get

$$\begin{aligned}\langle \pi(\xi)\Delta^{it}\eta_1|\eta_2\rangle &= \langle \Delta^{it}\eta_1|\xi^{\#}\eta_2\rangle = \lim_{\gamma\to\infty} \langle \Delta^{it}\eta_1|\pi'(\eta_2)(E_\gamma\xi)^{\#}\rangle \\ &= \lim_{\gamma\to\infty} \langle (\Delta^{-it}E_\gamma\xi)\eta_1|\Delta^{-it}\eta_2\rangle = \lim_{\gamma\to\infty} \langle \pi'(\eta_1)(\Delta^{-it}E_\gamma\xi)|\Delta^{-it}\eta_2\rangle \\ &= \langle \pi'(\eta_1)\Delta^{-it}\xi|\Delta^{-it}\eta_2\rangle = \langle \Delta^{it}\pi'(\eta_1)\Delta^{-it}\xi|\eta_2\rangle \quad ;\end{aligned}$$

hence we obtain, for every $\xi \in A$ and $\eta \in A'$,

$$\pi(\xi)\Delta^{it}\eta = \Delta^{it}\pi'(\eta)\Delta^{-it}\xi \quad ,$$

equivalently

$$\Delta^{-it}\pi(\xi)\Delta^{it}\eta = \pi'(\eta)\Delta^{-it}\xi \quad .$$

Thus Δ^{it} leave A and A' both invariant, and

$$\pi(\Delta^{it}\xi) = \Delta^{it}\pi(\xi)\Delta^{-it}, \quad \xi \in A \quad ;$$

$$\pi'(\Delta^{it}\eta) = \Delta^{it}\pi'(\eta)\Delta^{-it}, \quad \eta \in A' \quad .$$

Hence, for every pair ξ_1 and ξ_2 in A,

$$\Delta^{it}(\xi_1\xi_2) = \Delta^{it}\pi(\xi)\xi_2 = \Delta^{it}\pi(\xi_1)\Delta^{-it}\Delta^{it}\xi_2 = \pi(\Delta^{it}\xi_1)\Delta^{it}\xi_2 = (\Delta^{it}\xi_1)(\Delta^{it}\xi_2) \quad .$$

Thus the one parameter unitary group Δ^{it} is an automorphism group of A. Similarly we conclude that Δ^{it} is an automorphism group of A'. Furthermore, we have, for each $\xi \in A$ and $\eta \in A'$,

$$\Delta^{it}\xi^{\#} = \Delta^{it}J\Delta^{\frac{1}{2}}\xi = J\Delta^{it}\Delta^{\frac{1}{2}}\xi = J\Delta^{\frac{1}{2}}\Delta^{it}\xi = (\Delta^{it}\xi)^{\#} \quad ;$$

$$\Delta^{it}\eta^{@} = \Delta^{it}J\Delta^{-\frac{1}{2}}\eta = J\Delta^{it}\Delta^{-\frac{1}{2}}\eta = J\Delta^{-\frac{1}{2}}\Delta^{it}\eta = (\Delta^{it}\eta)^{@} \quad .$$

Q.E.D.

Lemma 5.2

Let $\gamma > 0$. *If* ξ *is in the domain* $\mathcal{D}(\Delta^\gamma)$, *then*

(a) ξ *is in the domain* $\mathcal{D}(\Delta^\alpha)$ *for* $0 \le \mathrm{Re}\,\alpha \le \gamma$;

(b) *the function* $F(\alpha) = \langle \Delta^\alpha\xi|\eta\rangle$ *for any* $\eta \in \mathcal{H}$ *is bounded, holomorphic in and continuous on the strip,* $0 \le \mathrm{Re}\,\alpha \le \gamma$.

Proof. Let $\alpha = s + it$ with $0 \le s \le \gamma$ and $t \in \mathbb{R}$. Let $f(\lambda) = 1$ for $0 \le \lambda \le 1$ and $f(\lambda) = \lambda^{2\gamma}$ for $1 < \lambda$. Then $|\lambda^\alpha|^2 \le f(\lambda)$ for every $\lambda \ge 0$. By assumption, $f(\lambda)$ is integrable with respect to the measure $d\langle E(\lambda)\xi|\eta\rangle$. Therefore, the function F is bounded, and continuous by the Lebesgue dominated convergence theorem. Consider any simply closed smooth curve C contained in the strip, $0 \le \mathrm{Im}\,\alpha \le \gamma$. Then we have, by the Fubini Theorem,

$$\oint_C F(\alpha)d\alpha = \oint_C \int_0^\infty \lambda^\alpha d\langle E(\lambda)\xi|\eta\rangle = \int_0^\infty \left(\oint_C \lambda^\alpha d\alpha\right) d\langle E(\lambda)\xi|\eta\rangle = 0 \quad .$$

Thus Morera's theorem assures that $F(\alpha)$ is holomorphic in the strip. Q.E.D.

Definition 5.3. We set

$$A_0 = \{\xi \in A : \xi \in \bigcap_{\alpha\in\mathbb{C}} \mathcal{D}(\Delta^\alpha) \text{ and } \Delta^\alpha\xi \in A \text{ for every } \alpha \in \mathbb{C}\} \quad .$$

Lemma 5.4

(a) $\Delta^\alpha A_0 = A_0$, $\alpha \in \mathbb{C}$;

(b) $A_0 \subset A \cap A'$;

(c) $SA_0 = A_0$, $FA_0 = A_0$ *and* $JA_0 = A_0$;

(d) A_0 *is an algebra;*

(e) $J(\xi\eta) = (J\eta)(J\xi)$, $\xi,\eta \in A_0$.

Proof. Assertion (a) follows directly from the definition of A_0. If ξ is in A_0, then $\Delta^{-1}\xi$ is in A again, so that ξ belongs to $A^@$; hence to A'. This proves assertions (b).

Since $J\mathcal{D}(\Delta^\alpha) = \mathcal{D}(\Delta^{-\bar{\alpha}})$ for any $\alpha \in \mathbb{C}$, we have $J \bigcap_{\alpha\in\mathbb{C}} \mathcal{D}(\Delta^\alpha) = \bigcap_{\alpha\in\mathbb{C}} \mathcal{D}(\Delta^\alpha)$. If ξ is in A_0, then we have, for any $\alpha \in \mathbb{C}$,

$$\Delta^\alpha S\xi = \Delta^\alpha J\Delta^{\frac{1}{2}}\xi = J\Delta^{-\bar{\alpha}}\Delta^{\frac{1}{2}}\xi = J\Delta^{\frac{1}{2}}\Delta^{-\bar{\alpha}}\xi = S\Delta^{-\bar{\alpha}}\xi \in A \quad .$$

Therefore, $S\xi$ belongs to A_0.

Suppose ξ and η are in A_0. Then $(\Delta^\alpha\xi)(\Delta^\alpha\eta)$ is in A for any $\alpha \in \mathbb{C}$ and the function: $\alpha \in \mathbb{C} \to (\Delta^\alpha\xi)(\Delta^\alpha\eta) \in \mathcal{H}$ is entire by Lemma 5.2. But we know by Theorem 5.1 that $(\Delta^{it}\xi)(\Delta^{it}\eta) = \Delta^{it}(\xi\eta)$ for every $t \in \mathbb{R}$. Hence the function: $t \in \mathbb{R} \to \Delta^{it}(\xi\eta)$ is extended holomorphically to $(\Delta^{i\alpha}\xi)(\Delta^{i\alpha}\eta)$, which means that $\xi\eta$ belongs to $\mathcal{D}(\Delta^{i\alpha})$ for any $\alpha \in \mathbb{C}$ and $\Delta^{i\alpha}(\xi\eta) = (\Delta^{i\alpha}\xi)(\Delta^{i\alpha}\eta) \in A$. Thus $\xi\eta$ falls in A_0, and Δ^α, $\alpha \in \mathbb{C}$, is a complex one parameter group of automorphisms of A_0.

The last assertion (e) is now verified as follows: $J(\xi\eta) = \Delta^{\frac{1}{2}}S(\xi\eta)$ $= \Delta^{\frac{1}{2}}(S\eta)(S\xi) = (\Delta^{\frac{1}{2}}S\eta)(\Delta^{\frac{1}{2}}S\xi) = (J\eta)(J\xi)$. Q.E.D.

Lemma 5.5

If a function f *on* $\mathbb{R}$ *is the Fourier-Stieljes transform of a finite measure* μ, *then*

(a) $f(\log \Delta)A \subset A$, *and*

$$[f(\log \Delta)\xi]^{\#} = \overline{f}(-\log \Delta)\xi^{\#}, \quad \xi \in A; \tag{5.5}$$

(b) $f(\log \Delta)A' \subset A'$, *and*

$$[f(\log \Delta)\eta]^{@} = \overline{f}(-\log \Delta)\eta^{@}, \quad \eta \in A' \quad . \tag{5.6}$$

Proof. Since we have

$$f(t) = \int_{-\infty}^{\infty} e^{ist}\, d\mu(s), \quad t \in \mathbb{R} \ ,$$

we get

$$f(\log \Delta) = \int_{-\infty}^{\infty} e^{is \log\Delta}\, d\mu(s) = \int_{-\infty}^{\infty} \Delta^{is}\, d\mu(s) \quad .$$

For each $\xi \in A$ and $\eta \in A'$, we have

$$\pi'(\eta)f(\log\Delta)\xi = \int_{-\infty}^{\infty} \pi'(\eta)\Delta^{is}\xi\, d\mu(s) = \int_{-\infty}^{\infty} \pi(\Delta^{is}\xi)\eta\, d\mu(s) = \int_{-\infty}^{\infty} \Delta^{is}\pi(\xi)\Delta^{-is}\eta\, d\mu(s) \quad ,$$

so that

$$\|\pi'(\eta)f(\log\Delta)\xi\| \le \int_{-\infty}^{\infty} \|\Delta^{is}\pi(\xi)\Delta^{-is}\eta\| d|\mu|(s) \le (\int_{-\infty}^{\infty} d|\mu|(s))\ \|\pi(\xi)\|\|\eta\| \quad ,$$

where $|\mu|$ denotes the total variation of the measure μ. Hence $f(\log \Delta)\xi$ falls in A. Similarly, $f(\log \Delta)\eta$ falls in A'. Equalities (5.5) and (5.6) follow from the calculation:

$$\begin{aligned} J\Delta^{\frac{1}{2}}f(\log \Delta) &= Jf(\log \Delta)\Delta^{\frac{1}{2}} \\ &= \overline{f}(J(\log \Delta)J)J\Delta^{\frac{1}{2}} \\ &= \overline{f}(\log \Delta^{-1})J\Delta^{\frac{1}{2}} = \overline{f}(-\log \Delta)J\Delta^{\frac{1}{2}} \quad . \end{aligned}$$

Q.E.D.

Lemma 5.6

If f *and* g *are continuous functions on* $\mathbb{R}$ *with compact support, then*

$$(f * g)(\log \Delta)A \subset A_0 \quad \textit{and} \quad (f * g)(\log \Delta)A' \subset A_0$$

where

$$f * g(t) = \int_{-\infty}^{\infty} f(s)g(t-s)ds \quad .$$

Proof. Since $f * g$ is the Fourier transform of an integrable function on $\mathbb{R}$, $(f * g)(\log\Delta)A \subset A$. Since the support of $f * g$ is compact, the range of $(f * g)(\log\Delta)$ is contained in $\bigcap_{\alpha\in\mathbb{C}} \mathcal{D}(\Delta^{\alpha})$. Furthermore, putting $e_{\alpha}(t) = e^{\alpha t}$, $\alpha \in \mathbb{C}$, we have

$$e_{\alpha}(f * g) = (e_{\alpha}f) * (e_{\alpha}g) \ ,$$

$$e_{\alpha}(f * g)(\log \Delta) = \Delta^{\alpha}(f * g)(\log \Delta) \quad .$$

Hence Δ^{α} maps $(f * g)(\log \Delta)A$ into A. Therefore we conclude that $(f * g)(\log\Delta) A \subset A_0$. By symmetry, the assertion for A' follows. Q.E.D.

Lemma 5.7

The algebra A_0 *is dense in the Hilbert space* $\mathcal{D}^{\#}$ *with respect to the* $\#$*-norm and also dense in* $\mathcal{D}^{@}$ *with respect to the* $@$*-norm.*

Proof. Let $\mathcal{E}$ denote the linear space spanned by $f * g$ with continuous functions f and g on $\mathbb{R}$ of compact support. By Lemma 5.6, $f(\log\Delta)A \subset A_0$. Let $\{f_n\}$ be a bounded sequence in $\mathcal{E}$ such that $\lim_{n\to\infty} f_n(t) = 1$. For each $\xi \in A \subset \mathcal{D}(\Delta^{\frac{1}{2}})$, we have, as $n \to \infty$,

$$\|f_n(\log\Delta)\xi - \xi\|^2 = \int_0^\infty |f_n(\log\lambda) - 1|^2 d\|E(\lambda)\xi\|^2 \to 0 \quad ;$$

$$\|Sf_n(\log\Delta)\xi - S\xi\|^2 = \|J\Delta^{\frac{1}{2}}f_n(\log\Delta)\xi - J\Delta^{\frac{1}{2}}\xi\|^2$$
$$= \|\Delta^{\frac{1}{2}}f_n(\log\Delta)\xi - \Delta^{\frac{1}{2}}\xi\|^2$$
$$= \int_0^\infty |f_n(\log\lambda) - 1|^2 |\lambda| d\|E(\lambda)\xi\|^2 \to 0 \quad .$$

Q.E.D.

Now, we are in the position to see the following:

Theorem 5.8

The von Neumann algebra $\{M,\mathcal{H}\}$ *with a separating and cyclic vector* ξ_0 *is standard with respect to the unitary involution* J *defined in* Section 2. *That is,*

$$JMJ = M', \quad JM'J = M \quad ;$$
$$JaJ = a^* \quad \text{for} \quad a \in M \cap M' \quad .$$

Proof. By Lemma 5.4 (e), we have

$$\pi'(J\xi) = J\pi(\xi)J, \quad \xi \in A_0 \quad ;$$

hence $J\pi(A_0)J = \pi'(A_0)$. Therefore, $J\pi(A_0)''J = \pi'(A_0)''$. Let $\eta = y\xi_0$ for an arbitrary $y \in \pi(A_0)'$. Then for each $\xi \in A_0$, we have

$$\|\pi(\xi)\eta\| = \|\pi(\xi)y\xi_0\| = \|y\pi(\xi)\xi_0\| = \|y\xi\|$$
$$\leq \|y\|\|\xi\| \quad .$$

If ξ is an element of A, then there exists, by Lemma 5.7, a sequence $\{\xi_n\}$ in A_0 such that

$$\lim_{n\to\infty}\|\xi_n - \xi\| = 0 \quad \text{and} \quad \lim_{n\to\infty}\|\xi_n^\# - \xi^\#\| = 0 \quad .$$

For every $\zeta \in A'$, we have

$$\langle y\xi|\zeta\rangle = \lim_{n\to\infty}\langle y\xi_n|\zeta\rangle = \lim_{n\to\infty}\langle\pi(\xi_n)\eta|\zeta\rangle = \lim_{n\to\infty}\langle\eta|\pi(\xi_n)^*\zeta\rangle = \lim_{n\to\infty}\langle\eta|\pi'(\zeta)\xi_n^\#\rangle$$

$$= \langle\eta|\pi'(\zeta)\xi^\#\rangle = \langle\eta|\pi(\xi)^*\zeta\rangle = \langle\pi(\xi)\eta|\zeta\rangle \quad ,$$

which means that $y\xi = \pi(\xi)\eta$. Hence η belongs to A' and $y = \pi'(\eta) \in M'$. Therefore, $\pi(A_0)' \subset M'$, which implies that $\pi(A_0)'' \supset M'' = M$. Thus $\pi(A_0)'' = M$. Similarly, we get $\pi'(A_0)'' = M'$. Thus $JMJ = M'$.

If a is in the center $M \cap M'$, then

$$Sa\xi_0 = a^*\xi_0 = Fa\xi_0 \quad ,$$

so that $FSa\xi_0 = a\xi_0$, that is, $\Delta a\xi_0 = a\xi_0$. Hence $\Delta^{\frac{1}{2}}a\xi_0 = a\xi_0$ and $a^*\xi_0 = J\Delta^{\frac{1}{2}}a\xi_0 = Ja\xi_0$; so $JaJ = a^*$. Q.E.D.

6. THE KUBO-MARTIN-SCHWINGER BOUNDARY CONDITION

Suppose φ is a faithful normal positive linear functional on a von Neumann algebra M. Let $\{\pi_\varphi, \mathcal{H}_\varphi, \xi_\varphi\}$ be the cyclic representation of M induced by φ. Since φ is faithful, π_φ is faithful too and the normality of φ assures that $\pi_\varphi(M)'' = \pi_\varphi(M)$. Hence π_φ is an isomorphism of M onto $\pi_\varphi(M)$, so that we can identify M itself with $\pi_\varphi(M)$. Then the functional φ of M is given by $\varphi(x) = (x\xi_\varphi|\xi_\varphi)$, $x \in M$. If e denotes the projection to $[M'\xi_\varphi]$, then e belongs to M and $\varphi(1 - e) = 0$; hence $e = 1$ by the faithfulness of φ. Thus ξ_φ is a separating vector for M. Therefore, $\{M, \mathcal{H}_\varphi, \xi_\varphi\}$ may be regarded as the triplet $\{M, \mathcal{H}, \xi_0\}$ discussed before. Therefore, we use the same notations as before. But we write

$$\sigma_t^\varphi(x) = \Delta^{it}x\Delta^{-it}, \quad x \in M \quad , \tag{6.1}$$

because the modular operator Δ depends on the choice of a cyclic and separating vector ξ_0, hence on φ.

Theorem 6.1

In the above situation, the faithful, normal positive functional φ *of* M *satisfies the* KMS-*condition at* $\beta = 1$ *for the one parameter automorphism group* σ_t^φ *of* M.

Proof. For an arbitrary pair x, y in M, let $\xi = x\xi_0$ and $\eta = y\xi_0$. Then we have

$$f(t) = \varphi(\sigma_t(x)y) = \langle \Delta^{it}x\Delta^{-it}y\xi_0|\xi_0\rangle = \langle \Delta^{-it}y\xi_0|x^*\xi_0\rangle = \langle \Delta^{-it}\eta|\xi^{\#}\rangle \quad ;$$

$$g(t) = \varphi(y\sigma_t(x)) = \langle y\Delta^{it}x\Delta^{-it}\xi_0|\xi_0\rangle = \langle \Delta^{it}x\xi_0|y^*\xi_0\rangle = \langle \Delta^{it}\xi|\eta^{\#}\rangle \quad .$$

Since η and ξ are in $\mathcal{D}(\Delta^{1/2})$, Lemma 5.2 tells us that $f(t)$ and $g(t)$ are extended holomorphically to functions $F(\alpha)$ defined on the strip, $0 \le \operatorname{Im} \alpha \le 1/2$ and $G(\alpha)$ defined on the strip, $-1/2 \le \operatorname{Im} \alpha \le 0$, respectively. These functions F and G have the properties:

$$\begin{aligned} F(t + \tfrac{1}{2}i) &= \langle \Delta^{-i(t+\frac{1}{2}i)}\eta|\xi^{\#}\rangle = \langle \Delta^{\frac{1}{2}}\Delta^{-it}\eta|\xi^{\#}\rangle = \langle \Delta^{-it}\eta|\Delta^{\frac{1}{2}}\xi^{\#}\rangle = \langle \Delta^{-it}\eta|J\xi\rangle \\ &= \langle \xi|J\Delta^{-it}\eta\rangle = \langle \xi|\Delta^{-it}J\eta\rangle = \langle \Delta^{it}\xi|J\eta\rangle \quad ; \\ G(t - \tfrac{1}{2}i) &= \langle \Delta^{i(t-\frac{1}{2}i)}\xi|\eta^{\#}\rangle = \langle \Delta^{\frac{1}{2}}\Delta^{it}\xi|\eta^{\#}\rangle = \langle \Delta^{it}\xi|\Delta^{\frac{1}{2}}\eta^{\#}\rangle = \langle \Delta^{it}\xi|J\eta\rangle \quad . \end{aligned}$$

Therefore, we have $F(t + \frac{1}{2}i) = G(t - \frac{1}{2}i)$, $t \in \mathbb{R}$. Hence, the functions F and G define a bounded function H holomorphic in and continuous on the strip, $0 \le \operatorname{Im} \alpha \le 1$ such that $H(\alpha) = F(\alpha)$ if $0 \le \operatorname{Im} \alpha \le 1/2$ and $H(\alpha) = G(\alpha - i)$ if $1/2 \le \operatorname{Im} \alpha \le 1$. This function H assures the KMS-condition for φ. Q.E.D.

Theorem 6.2

The automorphism group σ_t^φ *of* M *is the unique one parameter automorphism group for which* φ *satisfies the* KMS-*condition at* $\beta = 1$.

Proof. Suppose α_t is another one-parameter automorphism group of M for which φ satisfies the KMS-condition at $\beta = 1$. Let x (resp. y) be such an element of M that the function: $t \mapsto \sigma_t^\varphi(x)$ (resp. $t \mapsto \alpha_t(y)$) is extended to an M-valued entire function, whose value at $\omega \in \mathbb{C}$ will be denoted by $\sigma_\omega^\varphi(x)$ (resp. $\alpha_\omega(y)$). Note that the concept of holomorphy does not depend on the operator topologies in M, see [11; page 92]. Let $F(\omega,\zeta) = \varphi(\sigma_\omega^\varphi(x)\alpha_\zeta(y))$. Then the KMS-condition for σ_t^φ implies that

$$F(t + i,\zeta) = \varphi(\alpha_\zeta(y)\sigma_t^\varphi(x)), \quad t \in \mathbb{R} \quad ;$$

the KMS-condition for α_t implies that

$$F(t + i,s + i) = \varphi(\sigma_t^\varphi(x)\alpha_s(y)), \quad s \in \mathbb{R} \quad .$$

It is obvious that $F(\omega) = F(\omega,\omega) = \varphi(\sigma_\omega(x)\alpha_\omega(y))$ is a bounded holomorphic function of ω. The periodicity, $F(t) = F(t + i)$, implies by the Sturm-Liouville Theorem that F is constant. Hence we have

$$\varphi(x\sigma_{-t}^\varphi \cdot \alpha_t(y)) = \varphi(\sigma_t^\varphi(x)\alpha_t(y)) = \varphi(xy) \quad ,$$

so that

$$\langle \sigma_{-t}^\varphi \cdot \alpha_t(y)\xi_0 | x^*\xi_0 \rangle = \langle y\xi_0 | x^*\xi_0 \rangle \quad .$$

Since the set of such x is strongly dense in M, we get $\sigma_{-t}^\varphi \cdot \alpha_t(y) = y$. Again the density of the set of such y implies that $\sigma_{-t}^\varphi \cdot \alpha_t$ is the identity automorphism of M, so that α_t and σ_t^φ must coincide. Q.E.D.

Definition 6.3. This unique one parameter automorphism group σ_t^φ of M is called the *modular automorphism group of* M *associated with* φ

7. THE CONDITIONAL EXPECTATIONS

In the usual probability theory, the conditional probability and the conditional expectation play important roles. Since the theory of operator algebras, especially the theory of states, is regarded as the non-commutative extension of probability theory, it is natural to ask if one can generalize the usual conditional expectation to the non-commutative situation. If we interpret the usual conditional expectation in terms of operator algebras, then we easily find the following: Let A be an abelian von Neumann algebra and $\mathcal{B}$ is a von Neumann subalgebra. Suppose φ is a faithful normal state of A. Then the conditional expectation in this situation is the map ε of A onto $\mathcal{B}$ such that

$$\varphi(xy) = \varphi(\varepsilon(x)y), \quad x \in A, \quad y \in \mathcal{B} \quad . \tag{7.1}$$

It is easy to see that ε has the properties:

$$\varepsilon(x) = x, \quad x \in \mathcal{B} \quad ; \tag{7.2}$$

$$\|\varepsilon(x)\| \le \|x\|, \quad x \in A \quad ; \tag{7.3}$$

$$\varepsilon(x^*) = \varepsilon(x)^*, \quad x \in A \quad ; \tag{7.4}$$

$$\varepsilon(x^*x) \ge 0, \quad x \in A \quad ; \tag{7.5}$$

$$\varepsilon(x^*x) = 0 \text{ implies } x = 0, \ x \in A \quad ; \tag{7.6}$$

$$\varepsilon(ax) = a\varepsilon(x), \quad a \in \mathcal{B}, \ x \in A \tag{7.7}$$

$$\varepsilon(x)^*\varepsilon(x) \le \varepsilon(x^*x), \quad x \in A \tag{7.8}$$

$$\sup \varepsilon(x_i) = \varepsilon(\sup x_i) \tag{7.9}$$

for every bounded increasing net $\{x_i\}$ of positive elements in A. It is known, by Tomiyama [31], that a linear map ε of an arbitrary C^*-algebra A onto a C^*-subalgebra $\mathcal{B}$ satisfying conditions (7.2) and (7.3) enjoys properties (7.4), (7.5), (7.7) and (7.8). Such a map ε is called a *projection of norm one of* A *onto* $\mathcal{B}$.

Now, we are at the position to see the situation for non-commutative conditional expectations:

Theorem 7.1

Let M *be a von Neumann algebra and* φ *a faithful normal state of* M *with the associated modular automorphism group* σ_t^φ. *For a von Neumann subalgebra* N *of* M, *the following two statements are equivalent:*

(a) N *is invariant under* σ_t^φ, *that is,* $\sigma_t^\varphi(N) = N$ *for every* $t \in \mathbb{R}$;

(b) *There exists a mapping* ε *of* M *onto* N *satisfying* (7.1)-(7.9).

Proof. (a) $\Rightarrow$ (b): As in the previous section, we may assume that M acts on a Hilbert space $\mathcal{H}$ with a cyclic and separating vector ξ_0 such that $\varphi(x) = (x\xi_0|\xi_0)$, $x \in M$. Let $A = M\xi_0$ and $B = N\xi_0$. Let K denote the closure of B. Then ξ_0 is a cyclic and separating vector of K for the von Neumann algebra N_K obtained by the restriction of N to K. Let Δ denote the modular operator for A. Then we have $\Delta^{it}x\xi_0 = \Delta^{it}x\Delta^{-it}\xi_0 = \sigma_t^\varphi(x)\xi_0$ for every $x \in M$ and $t \in \mathbb{R}$. Hence $\Delta^{it}B = B$, so that Δ^{it} leaves the subspace K invariant. Therefore, the restriction Δ_K of Δ to K is a non-singular positive self-adjoint operator in K. Since Δ_K^{it} induces the one parameter automorphism group σ_t^φ of N for which the restriction φ_N of φ to N satisfies the KMS-condition, Δ_K must be the modular operator for B. Therefore, the domain $\mathcal{D}^\#(B)$ of the #-operator in K constructed from B is simply the intersection of K and $\mathcal{D}^\#(A)$, where $\mathcal{D}^\#(A)$ means, of course, the domain of the #-operator S defined by A. Namely, we have

$$\mathcal{D}^{\#}(\mathcal{B}) = \mathcal{D}(\Delta_K^{\frac{1}{2}}) = \mathcal{D}(\Delta^{\frac{1}{2}}) \cap K = \mathcal{D}^{\#}(\mathcal{A}) \cap K \quad ;$$

$$\mathcal{D}^{@}(\mathcal{B}) = \mathcal{D}(\Delta_K^{-\frac{1}{2}}) = \mathcal{D}(\Delta^{-\frac{1}{2}}) \cap K = \mathcal{D}^{@}(\mathcal{A}) \cap K \quad ,$$

where $\mathcal{D}^{@}(\mathcal{B})$ and $\mathcal{D}^{@}(\mathcal{A})$ should be naturally understood. Let E denote the projection of $\mathcal{H}$ onto K. Since K is invariant under N, E belongs to N', that is,

$$Ex = xE, \quad x \in N \quad . \tag{7.10}$$

Since $Sx\xi_0 = x^*\xi_0$ belongs to $\mathcal{B}$ if $x \in N$, S leaves $\mathcal{B}$ invariant, that is, $J\Delta^{\frac{1}{2}}\mathcal{B} = \mathcal{B}$. Since $\Delta^{\frac{1}{2}}\mathcal{B}$ is dense in K, J leaves K invariant; therefore we get

$$JE = EJ, \quad \text{so} \quad SE = ES \quad .$$

We claim that $Ex\xi_0 \in \mathcal{B}$ for each $x \in M$. Let $x\xi_0 = \xi$. Then $J \in \mathcal{A}'$ and for every $\eta \in \mathcal{B}$, we have

$$\|\eta EJ\xi\| = \|\pi(\eta)EJ\xi\| = \|E\pi(\eta)J\xi\| \leq \|\pi(\eta)J\xi\| = \|\pi'(J\xi)\eta\| \leq \|\pi'(J\xi)\|\|\eta\| \quad .$$

Therefore, $J\xi$ belongs to $\mathcal{B}'$, where $\mathcal{B}'$ is considered in K as the one corresponding to $\mathcal{B}$. Thus $E\xi = JEJ\xi$ belongs to $J\mathcal{B}' = \mathcal{B}$. Therefore, there exists a unique element $\varepsilon(x)$ in N such that $Ex\xi_0 = \varepsilon(x)\xi_0$. Thus we get, for every $y \in N$,

$$\begin{aligned}\varphi(xy) = \langle xy\xi_0|\xi_0\rangle = \langle y\xi_0|x^*\xi_0\rangle = \langle y\xi_0|Ex^*\xi_0\rangle = \langle y\xi_0|(Ex\xi_0)^{\#}\rangle \\ = \langle y\xi_0|\varepsilon(x)^*\xi_0\rangle = \varphi(\varepsilon(x)y) \quad .\end{aligned}$$

The linearity and the other properties follows easily from the unicity of $\varepsilon(x)$.

(b) ⇒ (a): Let K denote the closure of $N\xi_0$ as before and E the projection of $\mathcal{H}$ onto K. For each $x \in M$ and $y \in N$, we have

$$\langle Ex\xi_0|y\xi_0\rangle = \langle x\xi_0|y\xi_0\rangle = \varphi(y^*x) = \varphi(y^*\varepsilon(x)) = \langle y^*\varepsilon(x)\xi_0|\xi_0\rangle = \langle \varepsilon(x)\xi_0|y\xi_0\rangle \quad ,$$

so that

$$Ex\xi_0 = \varepsilon(x)\xi_0, \quad x \in M \quad .$$

Furthermore, we have, for every $x \in M$,

$$ES_0x\xi_0 = Ex^*\xi_0 = \varepsilon(x^*)\xi_0 = \varepsilon(x)^*\xi_0 = S_0\varepsilon(x)\xi_0 = S_0Ex\xi \quad ,$$

where S_0 denotes the pre-closed operator: $x\xi_0 \in M\xi_0 \mapsto x^*\xi_0$. Hence we have $(1 - 2E)M\xi_0 \subset M\xi_0$; hence $(1 - 2E)M\xi_0 = M\xi_0$ and

$$S_0 = (1 - 2E)S_0(1 - 2E) \quad .$$

Therefore, we get, since $(1 - 2E)$ is unitary,

$$S = (1 - 2E)S(1 - 2E) \quad ;$$

so

$$F = (1 - 2E)F(1 - 2E) \quad ;$$

$$\Delta = (1 - 2E)\Delta(1 - 2E) \quad ;$$

$$J = (1 - 2E)J(1 - 2E) \quad .$$

Hence we get

$$\Delta^{it} = (1 - 2E)\Delta^{it}(1 - 2E) \quad .$$

Thus Δ^{it} leaves K invariant. Since $EM\xi_0 = \varepsilon(M)\xi_0 = N\xi_0$, we have

$$\Delta^{it}N\xi_0 = \Delta^{it}EM\xi_0 = E\Delta^{it}M\xi_0 = EM\xi_0 = N\xi_0 \quad .$$

Hence for any $x \in N$, there exists $x(t)$ in N with $\Delta^{it}x\xi_0 = x(t)\xi_0$. Then we have

$$\sigma_t^{\varphi}(x)\xi_0 = \Delta^{it}x\Delta^{-it}\xi_0 = \Delta^{it}x\xi_0 = x(t)\xi_0 \quad ,$$

so that $\sigma_t^{\varphi}(x) = x(t)$. Thus σ_t^{φ} leaves N invariant. Q.E.D.

Definition 7.2. The mapping ε is called the *conditional expectation* of M onto N with respect to φ.

8. THE RADON-NIKODYM THEOREMS

As is well known, the representation theory of operator algebras is highly linked with the analysis of positive linear functionals; and the latter is quite naturally regarded as a non-commutative analogue of the usual integration theory on locally compact spaces. In the usual integration theory, the Radon-Nikodym theorem plays an important role. In this section, we will investigate how one can generalize this theorem from the abelian case to the non-commutative case.

We begin with discussion for semi-finite von Neumann algebras. Let M be a semi-finite von Neumann algebra with a faithful, semi-finite and normal trace τ. Let φ be a faithful normal state on M. Then it is known, see [21], that there exists uniquely a positive non-singular self-adjoint closed operator h affiliated with M such that

$$\varphi(x) = \tau(xh) = \int_0^{\infty} \lambda d\tau(xe(\lambda)), \quad x \in M \quad , \qquad (8.1)$$

where $h = \int_0^{\infty} \lambda de(\lambda)$ is the spectral decomposition of h.

Theorem 8.1

If a faithful normal state φ *is given by* (8.1), *then the associated modular automorphism group* σ_t^{φ} *of* M *is obtained by*

$$\sigma_t^{\varphi}(x) = h^{it}xh^{-it}, \quad x \in M, \quad t \in \mathbb{R} \quad . \qquad (8.2)$$

Proof. Let $e_n = \int_{1/n}^{n} de(\lambda)$, $n = 1, 2, \ldots$, and $M_0 = \bigcup_{n=1}^{\infty} e_n M e_n$. Then M_0 is a σ-weakly dense $*$-subalgebra of M. For each $x \in M_0$, the map: $t \in \mathbb{R} \mapsto h^{it}xh^{-it} \in M$ is extended to an M-valued entire function, whose value at $\alpha \in \mathbb{C}$ is denoted symbolically by $h^{i\alpha}xh^{-i\alpha}$. For any $y \in M$, the function $F_{x,y}(\alpha) = \varphi(h^{i\alpha}xh^{-i\alpha}y)$

is entire and $F_{x,y}(t) = \varphi(h^{it}xh^{-it}y)$. Furthermore, we have

$$F_{x,y}(t + i) = \varphi(h^{i(t+i)}xh^{-i(t+i)}y) = \tau(h^{i(t+i)}xh^{-i(t+i)}yh)$$
$$= \tau(yhh^{i(t+i)}xh^{-(it+i)}) = \tau(yh^{it}xh^{-it}h) = \varphi(yh^{it}xh^{-it}) \quad .$$

Therefore, the function $F_{x,y}$ assures the KMS-conditions for the pair x, y with respect to the one parameter automorphism group $h^{it}xh^{-it}$. The same arguments as in the proof of Proposition 1.3 shows that φ satisfies the KMS-condition for this one parameter automorphism group. Therefore, the modular automorphism group σ_t^φ is implemented by the one parameter unitary group h^{it} lying in M. Q.E.D.

As the converse, we have the following:

<u>Theorem 8.2</u>

Suppose φ *is a faithful normal state of a von Neumann algebra* M. *If the modular automorphism group* σ_t^φ *is inner in the sense that there exist a one parameter unitary group* $\Gamma(t)$ *lying in* M *such that*

$$\sigma_t^\varphi(x) = \Gamma(t)x\Gamma(t)^{-1}, \quad x \in M \quad , \tag{8.2a}$$

then M *is semi-finite.*

<u>Proof</u>. By assumption, there exists a positive, non-singular self-adjoint operator h affiliated with M such that $\Gamma(t) = h^{it}$. Let $h = \int_0^\infty \lambda de(\lambda)$ be the spectral decomposition of h. Set $e_n = \int_{1/n}^n de(\lambda)$, $n = 1, 2, \ldots$. Then the non-singularity of h implies that e_n converges strongly to the identity 1; hence the algebra $M_0 = \bigcup_{n=1}^\infty e_nMe_n$ is σ-weakly dense in M. Noticing that if x is in M_0, then $h^{-1}x$ is bounded and falls in M_0, we define the linear functional τ on M_0 by

$$\tau(x) = \varphi(h^{-1}x), \quad x \in M_0 \quad . \tag{8.3}$$

Let x be an element in M_0. Then $x = e_nxe_n$ for some n. Let s be a real number. Let $F(\alpha)$ be the bounded function holomorphic in and continuous on the strip, $0 \leq \operatorname{Im} \alpha \leq 1$, such that

$$F(t) = \varphi(\sigma_t^\varphi(h^se_n)x) \quad ;$$

$$F(t + i) = \varphi(x\sigma_t^\varphi(h^se_n)) \quad .$$

Since h^se_n is fixed by the inner automorphism $\sigma_t^\varphi = h^{it} \cdot h^{-it}$, $F(t)$ is constant; so is $F(\alpha)$; hence $F(t) = F(t + i)$, which means that

$$\varphi(h^se_nx) = \varphi(xh^se_n), \quad s \in \mathbb{R} \quad . \tag{8.4}$$

Therefore, we get, putting $s = -1/2$,

$$\tau(x) = \varphi(h^{-1}e_nx) = \varphi(h^{-\frac12}e_nh^{-\frac12}e_nx) = \varphi(h^{-\frac12}e_nxh^{-\frac12}e_n) \quad .$$

Therefore, τ is positive and $\tau(x^*x) = 0$ implies $x = 0$.

Let x and y be any pair in M_0. Then there exists the bounded function $F(\alpha)$ holomorphic in and continuous on the strip with

$$F(t) = \varphi(\sigma_t^{\varphi}(h^{-1}x)y) \quad ;$$

$$F(t + i) = \varphi(y\sigma_t^{\varphi}(h^{-1}x)) \quad .$$

Since $h^{-1}x$ is in M_0, the function: $t \mapsto \sigma_t^{\varphi}(h^{-1}x) = h^{it}h^{-1}xh^{-it}$ is extended holomorphically to the function: $\alpha \in \mathbb{C} \mapsto h^{i(\alpha+i)}xh^{-i\alpha}e_n$, where n is chosen so that $e_nxe_n = x$. Therefore, we get, from the equality for $F(t + i)$,

$$F(t) = \varphi(yh^{it}xh^{-i(t-i)}e_n) = \varphi(yh^{it}xh^{-it}h^{-1}e_n) = \varphi(h^{-1}yh^{it}xh^{-it}) \text{ by (8.4)} \quad .$$

Thus, we get

$$\tau(\sigma_t^{\varphi}(x)y) = \varphi(h^{-1}\sigma_t^{\varphi}(x)y) = \varphi(\sigma_t^{\varphi}(h^{-1}x)y) = \varphi(h^{-1}yh^{it}xh^{-it}) = \tau(y\sigma_t^{\varphi}(x)) \quad .$$

In particular, we have

$$\tau(xy) = \tau(yx), \quad x,y \in M_0 \quad . \tag{8.5}$$

Suppose p is a non-zero projection in M_0 and u is a partial isometry in M such that $u^*u = p$ and $uu^* = q \le p$. Then u falls in M_0. Hence equality (8.6) says that $\tau(p) = \tau(u^*u) = \tau(uu^*) = \tau(q)$. Hence $\tau(q - p) = 0$. Therefore $p = q$. Therefore, all projections in M_0 are finite, which means that M is semi-finite because M_0 contains the increasing sequence $\{e_n\}$ of finite projections converging strongly to the identity. Q.E.D.

From Theorems 8.1 and 8.2, we can say that the modular automorphism group σ_t^{φ} associated with φ is the shadow of the Radon-Nikodym derivative of φ with respect to the non-existent trace. Keeping this fact in mind, we will further investigate the relation between faithful normal states and the associated modular automorphism group.

Let M be a von Neumann algebra with faithful normal state φ. Considering the cyclic representation of M induced by φ, we assume that M acts on a Hilbert space $\mathcal{H}$ with the cyclic and separating vector ξ_0 such that $\varphi(x) = \langle x\xi_0|\xi_0\rangle$, $x \in M$. Let M_* denote the predual of M, the Banach space of all σ-weakly continuous linear functionals on M. In the predual M_*, the *-operation: $\omega \in M_* \to \omega^* \in M_*$ is defined by $\omega^*(x) = \overline{\omega(x^*)}$, $x \in M$. For each $\xi \in \mathcal{H}$, we define two functionals: one is an element φ_ξ in M_* and the other is an element φ'_ξ in M'_* as follows:

$$\begin{cases} \varphi_\xi(x) = \langle x\xi|\xi_0\rangle, & x \in M \quad ; \\ \varphi'_\xi(x) = \langle x\xi|\xi_0\rangle, & x \in M' \quad . \end{cases} \tag{8.6}$$

Let V (resp. V') denote the set of all φ_ξ (resp. φ'_ξ), $\xi \in \mathcal{H}$. Then we can easily verify the following based on Lemma 2.1:

Lemma 8.3

The following statements (a) *and* (b) (resp. (a′) *and* (b′)) *are equivalent:*

(a) ξ *belongs to* $\mathcal{D}^{\#}$, (a′) ξ *belongs to* $\mathcal{D}^{@}$;

(b) $\varphi_\xi'^*$ *belongs to* V′, (b′) φ_ξ^* *belongs to* V.

If ξ *is in* $\mathcal{D}^{\#}$ (resp. $\mathcal{D}^{@}$), *then*

$$\varphi_\xi'^* = \varphi'_{\xi^{\#}} \quad (\text{resp. } \varphi_\xi^* = \varphi_{\xi^{@}}) \quad . \tag{8.7}$$

Let $P^{\#}$ (resp. $P^{@}$) denote the set of all $\xi \in \mathcal{D}^{\#}$ (resp. $\xi \in \mathcal{D}^{@}$) such that $\varphi_\xi' \geq 0$ (resp. $\varphi_\xi \geq 0$). Then $P^{\#}$ and $P^{@}$ are both convex cones in $\mathcal{H}$.

Proposition 8.4

The linear spaces $\mathcal{D}^{\#}$ *and* $\mathcal{D}^{@}$ *are both algebraically spanned by* $P^{\#}$ *and* $P^{@}$ *respectively. Furthermore,* $P^{\#}$ *and* $P^{@}$ *are the dual cones of each other in the following sense:*

(a) *A vector* $\xi \in \mathcal{H}$ *falls in* $P^{\#}$ *if and only if* $\langle \xi | \eta \rangle \geq 0$ *for every* $\eta \in P^{@}$;

(b) *A vector* $\eta \in \mathcal{H}$ *falls in* $P^{@}$ *if and only if* $\langle \xi | \eta \rangle \geq 0$ *for every* $\xi \in P^{\#}$.

Therefore, $P^{\#}$ *and* $P^{@}$ *are both closed.*

Proof. By symmetry, we have only to prove the assertion for $\mathcal{D}^{\#}$ and $P^{\#}$. Since $\mathcal{D}^{\#}$ is spanned by its self-adjoint part, we shall prove that every self-adjoint $\xi \in \mathcal{D}^{\#}$, i.e. $\xi = \xi^{\#}$, is of the form $\xi_1 - \xi_2$ for some ξ_1 and ξ_2 in $P^{\#}$. Define two actions of each $a \in M$ (resp. $a \in M'$) to $\omega \in M_*$ (resp. M_*') by

$$\langle x, a\omega \rangle = \langle xa, \omega \rangle \quad \text{and} \quad \langle x, \omega a \rangle = \langle ax, \omega \rangle$$

for every $x \in M$ (resp. $x \in M'$) and $\omega \in M_*$ (resp. $\omega \in M_*'$), where $\langle x, \omega \rangle$ means the value of ω at x. Then we have

$$a\varphi_\xi = \varphi_{a\xi}, \quad a'\varphi_\xi' = \varphi'_{a'\xi}$$

for each $a \in M$ (resp. $a' \in M'$). If φ_ξ' is self-adjoint, then there exists a projection e' in M' such that $e'\varphi_\xi' \geq 0$ and $(1 - e')\varphi_\xi' \leq 0$. Hence putting $\xi_1 = e'\xi$ and $\xi_2 = -(1 - e')\xi$, we get the desired expression: $\xi = \xi_1 - \xi_2$, since ξ_1 and ξ_2 are both in $P^{\#}$.

Let ξ be a vector in $\mathcal{H}$. Suppose $\langle \xi | \eta \rangle \geq 0$ for every $\eta \in P^{@}$. Let M_+ (resp. M_+') denote the set of all positive elements in M (resp. M'). Then we have, for every $h \in M_+$ and $k \in M_+'$,

$$\langle h\xi_0 | k\xi_0 \rangle = \langle h^{\frac12}h^{\frac12}\xi_0 | k^{\frac12}k^{\frac12}\xi_0 \rangle = \langle h^{\frac12}k^{\frac12}\xi_0 | k^{\frac12}h^{\frac12}\xi_0 \rangle = \|h^{\frac12}k^{\frac12}\xi_0\|^2 \geq 0 \quad .$$

Hence $M_+'\xi_0 \subset P^{@}$. Thus we get

$$0 \le \langle \xi | k\xi_0 \rangle = \varphi'_\xi(k), \quad k \in M'_+ \quad ,$$

so that ξ belongs to $P^\#$.

Suppose ξ is an element of $P^\#$. We shall prove that $\langle \xi|\eta\rangle \ge 0$ for every $\eta \in P^@$. Define an operator h_0 on $A' = M'\xi_0$ by

$$h_0 x\xi_0 = x\xi, \; x \in M' \quad .$$

Then, as have already been seen enough times, h_0 is a densely defined operator commuting with every unitary operator in M'. The positivity condition for φ'_ξ implies that h_0 is positive and hence symmetric. Let h denote the Friedrichs self-adjoint extension of h_0. Then h is affiliated with M. Let $h = \int_0^\infty \lambda de(\lambda)$ be the spectral decomposition of h. Put $h_n = \int_0^n \lambda de(\lambda)$. Then h_n belongs to M_+, and $h_n\xi_0$ converges to $h\xi_0$. Thus we get, for every $\eta \in P^@$,

$$\langle \xi|\eta\rangle = \lim_{n\to\infty} \langle h_n\xi_0|\eta\rangle = \lim_{n\to\infty} \langle \xi_0|h_n\eta\rangle = \lim_{n\to\infty} \overline{\langle h_n\eta|\xi_0\rangle} \ge 0 \quad ,$$

since $\varphi_\eta \ge 0$. Q.E.D.

Remark. The above proof shows that $P^\#$ (resp. $P^@$) is the closure of $M_+\xi_0$ (resp. $M'_+\xi_0$).

For each pair ξ, η in $\mathcal{H}$, define a normal linear functional $\omega_{\xi,\eta}$ (resp. $\omega'_{\xi,\eta}$) of M (resp. M') by

$$\begin{cases} \langle x, \omega_{\xi,\eta}\rangle = \langle x\xi|\eta\rangle, & x \in M \quad ; \\ \langle x, \omega'_{\xi,\eta}\rangle = \langle x\xi|\eta\rangle, & x \in M' \quad . \end{cases} \tag{8.8}$$

It is clear that $\varphi_\xi = \omega_{\xi,\xi_0}$ and $\varphi'_\xi = \omega'_{\xi,\xi_0}$. Furthermore,

$$\omega^*_{\xi,\eta} = \omega_{\eta,\xi}; \; (\omega'_{\xi,\eta})^* = \omega'_{\eta,\xi} \quad .$$

Theorem 8.5

(a) *Any normal positive linear functional* ψ *of* M *is uniquely represented in the form* $\psi = \omega_{\xi,\xi}$, $\xi \in P^\#$. *Therefore, there exists a positive self-adjoint closed operator* h *affiliated with* M *such that*

$$\psi(x) = \langle xh\xi_0|h\xi_0\rangle, \quad x \in M \quad . \tag{8.9}$$

(b) *If* $\psi \le \varphi$, *then the above* h *is unique and* $0 \le h \le 1$.

Proof. It is known, (see[4]; Théorème 4, p.222), that there exists a vector $\eta \in \mathcal{H}$ with $\psi = \omega_{\eta,\eta}$. We consider the polar decomposition of φ'_η in M'_*:

$$\varphi'_\eta = v\psi', \quad \psi' = v^*\varphi_\eta \quad ,$$

where ψ is a normal positive functional of M' and v is a partial isometry in

M', see [4; Théorème 4, p. 61]. Put $\xi = v^*\eta$. Then $\psi' = v^*\varphi'_\eta = \varphi'_{v^*\eta} = \varphi'_\xi$, so that ξ is in $P^\#$. Furthermore, we have $\varphi'_\eta = v\psi' = v\varphi'_\xi = \varphi'_{v\xi}$. Since the map: $\zeta \in \mathcal{H} \to \varphi'_\zeta \in M'_*$ is injective, we have $\eta = v\xi$. Therefore, we get, for every $x \in M$,

$$\langle x,\psi\rangle = \langle x\eta|\eta\rangle = \langle xv\xi|\eta\rangle = \langle vx\xi|\eta\rangle = \langle x\xi|v^*\eta\rangle = \langle x\xi|\xi\rangle = \langle x,\omega_{\xi,\xi}\rangle \quad .$$

Hence ψ is of the desired form $\psi = \omega_{\xi,\xi}$ with $\xi \in P^\#$. Suppose there is another vector $\xi_1 \in P^\#$ with $\psi = \omega_{\xi_1,\xi_1}$. Then we have, for every $x \in M$,

$$\|x\xi\|^2 = \langle x\xi|x\xi\rangle = \langle x^*x\xi|\xi\rangle = \langle x^*x, \rangle = \langle x^*x\xi_1|\xi_1\rangle = \langle x\xi_1|x\xi_1\rangle = \|x\xi_1\|^2 \quad .$$

Define an operator u_0 by $u_0x\xi = x\xi_1$, $x \in M$. Then u_0 is an isometry of $M\xi$ onto $M\xi_1$, so that it is extended to a partial isometry u which vanishes on $[M\xi]^\perp$. It is easy to check that u commutes with every element in M, so that u falls in M'. Thus we get the partial isometry $u \in M'$ with $u\xi = \xi_1$. Hence we have $\varphi'_{\xi_1} = \varphi'_{u\xi} = u\varphi'_\xi$. Since φ'_{ξ_1} and φ'_ξ are both positive, φ'_{ξ_1} and φ'_ξ must coincide due to the unicity of the polar decomposition in M'_*. Thus ξ and ξ_1 are identical. This shows the unicity of the above ξ.

Suppose $\psi \le \varphi$. This assumption is equivalent to the fact that ξ falls in A' and $\|\pi'(\xi)\| \le 1$. We have to show that ξ falls in A and $\|\pi(\xi)\| \le 1$. If this is the case, then the positivity of $\pi(\xi)$ automatically follows from ξ being in $P^\#$. The fact known about ξ is that $\xi \in A' \cap P^\#$ and $\|\pi'(\xi)\| \le 1$. Then we have, since $\xi = \xi^\#$,

$$\xi^@ = \xi^{\#@} = \Delta\xi \in A' \quad .$$

Hence ξ is in $A^\#$. Furthermore, putting $\eta = \xi^@ + \xi$, we have $\xi = (1+\Delta)^{-1}\eta$. Hence, by Theorem 4.1, we have

$$\|\pi(\xi)\| \le \gamma(-1)\|\pi'(\eta)\| = \tfrac{1}{2}\|\pi'(\eta)\| = \tfrac{1}{2}(\|\pi'(\xi)\| + \|\pi'(\xi)^*\|) \le 1 \quad .$$

Therefore, $h = \pi(\xi)$ is the desired one. The unicity of h follows from the unicity of ξ and the boundedness of h. Q.E.D.

Remark. Theorem 8.5 tells us that if a positive operator a on a Hilbert space is majorized by a positive nuclear operator b, then there exists *uniquely* a *positive* operator h of norm ≤ 1 with $a = hbh$. The author does not know if this is true for general pair a and b with $0 \le a \le b$.

Now, we shall consider how one can describe the relation between φ and ψ in terms of the associated modular automorphism groups.

Definition 8.6. For each normal positive functional φ of a von Neumann algebra, the set $M_\varphi = \{x \in M: x\varphi = \varphi x\}$ is called the *centralizer* of φ. Of course, the centralizer M_φ is a von Neumann subalgebra of M.

If M is semi-finite, and if φ is of the form $\varphi(x) = \tau(xh)$, $x \in M$, with a faithful, semi-finite trace τ and a positive self-adjoint operator h affiliated with M, then the centralizer M_φ of φ is nothing but the relative commutant of h in M. In general, the centralizer M_φ of a faithful normal state φ is precisely the fixed point subalgebra of M under the associated modular automorphism group σ_t^φ as seen in the following:

Proposition 8.7

If φ is a faithful normal positive linear functional of a von Neumann algebra M, then the centralizer M_φ of φ is precisely the set of all fixed points of the associated modular automorphism group σ_t^φ.

Proof. Suppose $a \in M$ is a fixed point of σ_t^φ. By the KMS-condition, for any $x \in M$, there exists a bounded function $F(\alpha)$ holomorphic in and continuous on the strip, $0 \le \text{Im}\,\alpha \le 1$, such that

$$F(t) = \varphi(\sigma_t^\varphi(a)x) \quad ;$$

$$F(t+i) = \varphi(x\sigma_t^\varphi(a)) \quad .$$

But $\sigma_t^\varphi(a) = a$ for every $t \in \mathbf{R}$, so that $F(t)$ and $F(t+i)$ are constant functions of t; hence $F(\alpha)$ must be constant. Thus, we have $F(t) = F(t+i)$, which means that $\varphi(ax) = \varphi(xa)$. Since x is arbitrary, we get $\varphi a = a\varphi$; so a is in the centeralizer M_φ of φ.

Conversely, suppose a is in the centralizer M_φ. Since φ is σ_t^φ-invariant, it is obvious that $\sigma_t^\varphi(M_\varphi) = M_\varphi$. Hence we have

$$\varphi\sigma_t^\varphi(a) = \sigma_t^\varphi(a)\varphi, \quad t \in \mathbf{R} \quad . \tag{8.11}$$

Let x be an arbitrary element of M. Then there exists a bounded function $F(\alpha)$ on the strip which assures the KMS-condition for the pair a and x. But equality (8.11) mentions that $F(t) = F(t+i)$, $t \in \mathbf{R}$. Therefore, by the Sturm-Liouville Theorem, the periodic bounded holomorphic function F must be constant. Thus $\varphi(\sigma_t(a)x) = \varphi(ax)$, $t \in \mathbf{R}$ and $x \in M$. Hence we have

$$\langle x\xi_0 | a^*\xi_0\rangle = \langle ax\xi_0|\xi_0\rangle = \varphi(ax) = \varphi(\sigma_t^\varphi(a)x) = \langle\sigma_t^\varphi(a)x\xi_0|\xi_0\rangle = \langle x\xi_0|\sigma_t^\varphi(a^*)\xi_0\rangle \quad .$$

Therefore, $\sigma_t^\varphi(a^*) = a^*$, equivalently $\sigma_t^\varphi(a) = a$. Thus a is a fixed point of σ_t^φ.

Q.E.D.

Theorem 8.8

In the same situation as before, the following statements for a normal positive linear functional ψ of M are equivalent:

(a) *ψ is σ_t^φ-invariant;*

(b) ψ *is of the form* $\psi = \omega_{\xi,\xi}$ *with* $\xi \in P^{\#}$ *and* $\Delta^{it}\xi = \xi$, $t \in \mathbf{R}$;

(c) ψ *is of the form* $\psi = \omega_{h\xi_0, h\xi_0}$ *with* h *a positive self-adjoint operator affiliated with the centralizer* M_φ *of* φ.

Under the assumption that ψ *is faithful, the following statements are also equivalent to the above:*

(d) σ_t^φ *and* σ_s^ψ *commute,* $s,t \in \mathbf{R}$;

(e) φ *is* σ_t^ψ*-invariant.*

If a faithful ψ satisfies any one of the above conditions, then the modular automorphism group σ_t^ψ associated with ψ is given by

$$\sigma_t^\psi(x) = \sigma_t^\varphi(h^{2it}xh^{-2it}), \quad x \in M \quad ,$$

where h is the one described in statement (c)

<u>Proof</u>. The implications (c) $\Rightarrow$ (b) $\Rightarrow$ (a) are almost trivial.

(a) $\Rightarrow$ (b) Suppose ψ is σ_t^φ-invariant. By Theorem 8.5, ψ is uniquely of the form $\psi = \omega_{\xi,\xi}$ with $\xi \in P^{\#}$. Then we have, for each $x \in M$,

$$\omega_{\Delta^{it}\xi,\Delta^{it}\xi}(x) = \langle x\Delta^{it}\xi | \Delta^{it}\xi\rangle = \langle \Delta^{-it}x\Delta^{it}\xi|\xi\rangle = \psi \cdot \sigma_{-t}^\varphi(x) = \psi(x) \quad ,$$

so that ψ $\omega_{\Delta^{it}\xi,\Delta^{it}\xi}$. But it is obvious that $\Delta^{it}P^{\#} = P^{\#}$. Hence the unicity of ξ implies that $\xi = \Delta^{it}\xi$, $t \in \mathbf{R}$; thus (b) follows.

(b) $\Rightarrow$ (c) Suppose $\Delta^{it}\xi = \xi$, $t \in \mathbf{R}$. Recall the construction of h in the proof of Proposition 8.4. Then the operator h_0 constructed there commutes with Δ^{it}, so that the Friedrichs extension h of h_0 and Δ^{it} commute. Hence all the spectral projections of h commute with Δ^{it}, which means that σ_t^φ leaves all such projections fixed; hence they belong to M_φ by Proposition 8.7. Thus h is affiliated with M_φ.

The equivalence of (d) and (a) follows from the next more general result, Theorem 8.11, and that of (d) and (e) follows by symmetry.

Now, we shall prove the last statement for σ_t^ψ. Since ψ is faithful, h is non-singular. Let $h = \int_0^\infty \lambda de(\lambda)$ be the spectral decomposition and let $e_n = \int_{1/n}^n de(\lambda)$, n = 1, 2, ... Then $\{e_n\}$ converges strongly to the identity 1. Let A_0 be the subset of $A = M\xi_0$ considered in Definition 5.3. Since e_n and Δ^{it} commute, $\Delta^{it}e_n\xi_0 = e_n\xi_0 \in A$, so that $e_n\xi_0$ falls in A_0. Since $\pi(e_n\xi_0) = e_n$, $e_n\pi(A_0)e_n$ is contained in $\pi(A_0)$. Let $M_0 = \bigcup_{n=1}^\infty e_n\pi(A_0)e_n$. Then M_0 is a σ-weakly dense $*$-subalgebra of M and for each $x \in M_0$, the maps: $t \mapsto \sigma_t^\varphi(x)$ and $t \mapsto h^{it}xh^{-it}$ are both extended to M_0-valued entire functions, whose values at $\alpha \in \mathbb{C}$ are formally denoted by $\sigma_\alpha^\varphi(x)$ and $h^{i\alpha}xh^{-i\alpha}$ respectively. Let x and y be arbitrary elements in M_0. Then we have $x = e_nxe_n$ and $y = e_nye_n$ for some n. Let $h_n = he_n \in M$. Define an entire function $F(\alpha)$ by

$$F(\alpha) = \langle h^{2i\alpha}xh^{-2i\alpha}\sigma_{-\alpha}^\varphi(y)h\xi_0|h\xi_0\rangle \quad .$$

For each $t \in \mathbb{R}$, we have

$$F(t) = \langle h^{2it}xh^{-2it}\sigma^{\varphi}_{-t}(y)h\xi_0 | h\xi_0\rangle = \psi(h^{2it}xh^{-2it}\sigma^{\varphi}_{-t}(y))$$

$$= \psi(\sigma^{\varphi}_t(h^{2it}xh^{-2it})y) \quad \text{by the } \sigma^{\varphi}_t\text{-invariance of } \psi ;$$

$$F(t+i) = \langle h^{2i(t+i)}xh^{-2i(t+i)}\sigma^{\varphi}_{-(t+i)}(y)h\xi_0 | h\xi_0\rangle$$

$$= \langle h^{2i(t+i)}e_nxe_nh^{-2i(t+i)}\sigma^{\varphi}_{-(t+i)}(e_nye_n)h\xi_0 | h\xi_0\rangle$$

$$= \langle \Delta^{-i(t+i)}e_nye_nh\xi_0 | h^{2i(t-i)}e_nx^*e_nh^{-2i(t-i)}h\xi_0\rangle$$

$$= \langle Jh^{2(it+1)}e_nx^*e_nh^{-2(it+1)}h\xi_0 | J\Delta^{-it+1}e_nye_nh\xi_0\rangle$$

$$= \langle \Delta^{\frac{1}{2}}Sh^{2(it+1)}e^nx^*e_nh^{-(2it+1)}h\xi_0 | \Delta^{-it-1}\Delta^{\frac{1}{2}}Se_nye_nh\xi_0\rangle$$

$$= \langle \Delta^{\frac{1}{2}}hh^{2(it-1)}e_nxe_nh^{-2(it-1)}\xi_0 | \Delta^{-it-\frac{1}{2}}h_ny^*\xi_0\rangle$$

$$= \langle h_n^2h^{2(it-1)}e_nxe_nh^{-2it}h_n^2\xi_0 | \Delta^{-it}y^*\xi_0\rangle$$

$$= \langle h^{2it}e_nxe_nh^{-2it}h_n^2\xi_0 | \Delta^{-it}y^*\xi_0\rangle$$

$$= \langle y\Delta^{it}h^{2it}xh^{-2it}h_n^2\xi_0 | \xi_0\rangle$$

$$= \varphi(y\sigma^{\varphi}_t(h^{2it}xh^{-2it})h_n^2)$$

$$= \varphi(h_ny\sigma^{\varphi}_t(h^{2it}xh^{-2it})h_n) \quad \text{by Proposition 8.7,}$$

$$= \langle h_ny\Delta^{it}h^{2it}e_nxe_nh^{-2it}h_n\xi_0 | \xi_0\rangle$$

$$= \langle \Delta^{it}h^{2it}xh^{-2it}h\xi_0 | y^*h_n\xi_0\rangle$$

$$= \langle \Delta^{it}h^{2it}xh^{-2it}h\xi_0 | y^*h\xi_0\rangle$$

$$= \langle y\Delta^{it}h^{2it}xh^{-2it}h\xi_0 | h\xi_0\rangle$$

$$= \psi(y\sigma^{\varphi}_t(h^{2it}xh^{-2it})) .$$

Thus, this function $F(\alpha)$ satisfies the condition:

$$F(t) = \psi(\sigma^{\varphi}_t(h^{2it}xh^{-2it})y) ;$$

$$F(t+i) = \psi(y\sigma^{\varphi}_t(h^{2it}xh^{-2it})) .$$

Suppose x and y are an arbitrary pair in M. Then there exist sequences $\{x_n\}$ and $\{y_n\}$ in M_0 such that

$$x\xi_0 = \lim_{n\to\infty} x_n\xi_0, \quad x^*\xi_0 = \lim_{n\to\infty} x^*_n\xi_0 ,$$

$$xh\xi_0 = \lim_{n\to\infty} x_nh\xi_0, \quad x^*h\xi_0 = \lim_{n\to\infty} x^*_nh\xi_0 ;$$

$$y\xi_0 = \lim_{n\to\infty} y_n\xi_0 , \quad y^*\xi_0 = \lim_{n\to\infty} y^*_n\xi_0 ,$$

$$yh\xi_0 = \lim_{n\to\infty} y_nh\xi_0, \quad y^*h\xi_0 = \lim_{n\to\infty} y^*_nh\xi_0 .$$

Let $\{F_n\}$ be the sequence of entire functions such that

$$F_n(t) = \psi(\sigma^{\varphi}_t(h^{2it}x_nh^{-2it})y_n) ;$$

$$F_n(t + i) = \psi(y_n \sigma_t^\varphi(h^{2it} x_n h^{-2it})) \quad .$$

Let $f(t) = \psi(\sigma_t^\varphi(h^{2it} x h^{-2it})y)$ and $g(t) = \psi(y\sigma_t^\varphi(h^{2it} x h^{-2it}))$. Then we have

$$\begin{aligned}
|f(t) - F_n(t)| &= |\psi(\sigma_t^\varphi(h^{2it} x h^{-2it})y) - \psi(\sigma_t^\varphi(h^{2it} x_n h^{-2it})y_n)| \\
&\le |\psi(\sigma_t^\varphi(h^{2it}(x - x_n)h^{-2it})y)| + |\psi(\sigma_t^\varphi(h^{2it} x_n h^{-2it})(y - y_n))| \\
&\le \psi(\sigma_t^\varphi(h^{2it}(x - x_n)(x - x_n)^* h^{-2it}))^{\frac{1}{2}} \psi(y^* y)^{\frac{1}{2}} \\
&\quad + \psi(\sigma_t^\varphi(h^{2it} x_n x_n^* h^{-2it}))^{\frac{1}{2}} \psi((y - y_n)^*(y - y_n))^{\frac{1}{2}} \\
&= \psi(x - x_n)(x - x_n)^*)^{\frac{1}{2}} \psi(y^* y)^{\frac{1}{2}} + \psi(x_n x_n^*)^{\frac{1}{2}} \psi(y - y_n)^*(y - y_n))^{\frac{1}{2}} \\
&= \|(x^* - x_n^*)h\xi_0\| \|yh\xi_0\| + \|x_n \xi_0\| \|(y - y_n)h\xi_0\| \quad ,
\end{aligned}$$

so that $F_n(t)$ converges uniformly to $f(t)$. Similarly, $F_n(t + i)$ converges uniformly to $g(t)$. Thus, by the Phragmen-Lindelöf Theorem, $F_n(\alpha)$ converges uniformly to the bounded function $F(\alpha)$ holomorphic in and continuous on the strip, $0 \le \text{Im}\,\alpha \le 1$, such that $F(t) = f(t)$ and $F(t + i) = g(t)$. Thus ψ satisfies the KMS-condition for the one parameter automorphism group: $x \mapsto \sigma_t^\varphi(h^{2it} x h^{-2it})$, $t \in \mathbb{R}$. Therefore, we get $\sigma_t^\psi(x) = \sigma_t^\varphi(h^{2it} x h^{-2it})$, $x \in M$, $t \in \mathbb{R}$. Q.E.D.

Corollary 8.9

In the same situation as before, if a normal positive linear functional ψ *of* M *satisfies the* KMS-*condition for* σ_t^φ, *then* ψ *is of the form* $\psi = \omega_{h\xi_0, h\xi_0}$ *with* h *a positive self-adjoint operator affiliated with the center of* M.

Proof. Since ψ is invariant under σ_t^φ, ψ satisfies the conditions in Theorem 8.8. Hence ψ is of the form $\psi = \omega_{h\xi_0, h\xi_0}$ with h a positive self-adjoint operator affiliated with M_φ. We claim that the support projection of ψ, which is the range projection of h, is central. In fact, if $\psi(x^* x) = 0$, then the function $\psi(\sigma_t^\varphi(x^*)x)$ vanishes identically by the Schwarz inequality; hence $\psi(x\sigma_t^\varphi(x^*))$ vanishes identically too by the KMS-condition. In particular, $\psi(xx^*) = 0$. Thus the left kernel of ψ is actually a two-sided ideal. Therefore, the support projection z of ψ is central. Considering Mz only, we may assume that ψ is faithful. Then again by Theorem 8.8, $\sigma_t^\psi(x) = \sigma_t^\varphi(h^{2it} x h^{-2it})$ for every $x \in M$ and $t \in \mathbb{R}$. But by assumption, $\sigma_t^\psi = \sigma_t^\varphi$. Therefore, we get

$$\sigma_t^\varphi(x) = \sigma_t^\varphi(h^{2it} x h^{-2it}) \quad ;$$

hence $x = h^{2it} x h^{-2it}$, which means that h is affiliated with the center. Q.E.D.

Corollary 8.10

Suppose M *is of type* III. *Then the modular automorphism group* σ_t^φ *of*

M *associated with a faithful normal state* φ *does not admit a normal positive functional satisfying the KMS-condition at* $\beta \neq 1$.

Proof. Suppose ψ is a normal positive linear functional satisfying the the KMS-condition at β for σ_t^φ. Then, as in the proof of Corollary 8.9, we may assume that ψ is faithful. Then we get $\sigma_t^\psi = \sigma_{\beta t}^\varphi$ by assumption. On the other hand, Theorem 8.8 says that $\sigma_t^\psi(x) = \sigma_t^\varphi(h^{2it}xh^{-2it})$, $x \in M$, with h a positive self-adjoint operator affiliated with M_φ. Therefore, we get

$$\sigma_{\beta t}^\varphi(x) = \sigma_t^\varphi(h^{2it}xh^{-2it}),\ x \in M \quad ,$$

so that

$$\sigma_{t(\beta-1)}^\varphi(x) = h^{2it}xh^{-2it}, \quad x \in M \quad .$$

Hence if $\beta \neq 1$, then

$$\sigma_t^\varphi(x) = h^{2it/(\beta-1)}xh^{-2it/(\beta-1)}, \quad x \in M,$$

so that the modular automorphism group σ_t^φ is inner. Then Theorem 8.2 says that M is semi-finite. This contradicts the assumption. Q.E.D.

Theorem 8.11

Let σ *be an automorphism of* M.

(a) *If* σ *leaves* φ *invariant, then* σ *commutes with the modular automorphism group* σ_t^φ.

(b) *Conversely, if* σ *commutes with* σ_t^φ *and if* σ *leaves the center of* M *elementwise fixed, then* σ *leaves* φ *invariant.*

Proof. Let x and y be elements in M. Let $F_{x,y}(\alpha)$ be the bounded function holomorphic in and continuous on the strip, $0 \leq \operatorname{Im} \alpha \leq 1$, such that

$$F_{x,y}(t) = \varphi(\sigma_t^\varphi(x)y); \quad F(t+i) = \varphi(y\sigma_t^\varphi(x)) \quad .$$

Then we have

$$F_{\sigma(x),\sigma(y)}(t) = \varphi(\sigma_t^\varphi(\sigma(x))\sigma(y)) = \varphi \cdot \sigma(\sigma^{-1}\sigma_t^\varphi\sigma(x)y) \quad ;$$

similarly

$$F_{\sigma(x),\sigma(y)}(t+i) = \varphi \cdot \sigma(y\sigma^{-1}\sigma_t^\varphi\sigma(x)) \quad .$$

Therefore, we get

$$\sigma_t^{\varphi\cdot\sigma} = \sigma^{-1}\sigma_t^\varphi\sigma \quad .$$

Hence if $\varphi = \varphi \cdot \sigma$, then σ and σ_t^φ must commute.

Suppose σ leaves the center of M elementwise fixed and commutes with σ_t^φ. Let $\psi = \varphi \cdot \sigma$. Then we have shown that $\sigma_t^\psi = \sigma^{-1}\sigma_t^\varphi\sigma = \sigma_t^\varphi$. Hence ψ satisfies the KMS-condition for σ_t^φ, so that $\psi(x) = (xh\xi_0|h\xi_0)$, $x \in M$, with h a positive self-adjoint operator affiliated with the center Z of M.

Suppose $h \neq 1$. Then there exists non-zero positive element k in Z such that hk is bounded and $h^2k^2 > k^2$ or $h^2k^2 < k^2$. Then we have

$$\|hk\xi_0\|^2 = \langle h^2k^2\xi_0|\xi_0\rangle \neq \langle k^2\xi_0|\xi_0\rangle = \|k\xi_0\|^2 \quad ;$$

but

$$\begin{aligned}\|hk\xi_0\|^2 &= \langle kh\xi_0|kh\xi_0\rangle = \langle k^2h\xi_0|h\xi_0\rangle \\ &= \psi(k^2) = \varphi\cdot\sigma(k^2) = \varphi(k^2) \\ &= \langle k^2\xi_0|\xi_0\rangle = \|k\xi_0\|^2 \quad .\end{aligned}$$

This is impossible. Hence $h = 1$, equivalently $\varphi = \psi$. Q.E.D.

SECTION 9. NOTES

Much of the theory presented here was devoted to showing the existence of the symmetry between a von Neumann algebra M and its commutant M'. The existence of a unitary involution J which sets up the symmetry between M and M' has long been predicted. We can trace its history back to Murray-von Neumann's work in 1930's. Actually, they showed in 1937 that if a finite factor M on $\mathcal{H}$ admits a cyclic trace vector ξ_0 (necessarily separating), then the involution: $x\xi_0 \in M\xi_0 \mapsto x^*\xi_0 \in M\xi_0$ is extended to a unitary involution J with $JMJ = M'$, [15]. This result was generalized by H.A. Dye in 1952, [8], to finite von Neumann algebras with a cyclic and separating trace vector. Motivated by Ambrose's work [1], H. Nakano introduced Hilbert algebras in 1950, [17]. These algebras received further investigation by several authors in 1951-55, notably, J. Dixmier [5], R. Godement [9], R. Pallu de la Barriere [18], L. Pukanszky [21], I. E. Segal [22], O. Takenouchi [24] and so on.

Extending Dye's work on the non-commutative Radon-Nikodym theorem to non-commutative integration theory, I. E. Segal showed in 1953, [22], that (a) the left von Neumann algebra $L(A)$ and the right one $R(A)$ of a Hilbert algebra are commutants of each other; (b) the involution in A is extended to a unitary involution J with $JL(A)J = R(A)$; (c) $L(A)$, hence $R(A)$, is semi-finite; (d) every semi-finite von Neumann algebra is isomorphic to the left von Neumann algebra of some Hilbert algebra; (e) the predual M_* of a semi-finite von Neumann algebra is represented as the Banach space $L^1(M,\tau)$ of all integrable operators with respect to a faithful, semi-finite, normal trace τ on M. R. Godement also proved (a), (c) and (d) shortly after Segal, [8]. J. Dixmier verified (e) in an abstract fashion, developing the L^p-theory, [6].

Generalizing the notion of Hilbert algebra, J. Dixmier introduced quasi-Hilbert algebras and proved in 1952, [5], statement (a) for quasi-Hilbert algebras together with a criterion for the semi-finiteness of the left von Neumann algebra of a quasi-Hilbert algebra under some technical assumptions. The standardness of $L(A)$, i.e., statement (b), was not mentioned (though it is trivial to deduce standardness from his results). Furthermore, he proved in the same paper that the von Neumann algebra obtained by the so-called group-measure space construction is the left von Neumann algebra of a quasi-Hilbert algebra. The technical assumptions for the semi-finiteness criterion were later removed by L. Pukanszky in 1955, [20].

In the work of both Dixmier and Pukanszky, the modular operator Δ, or more precisely $\Delta^{\frac{1}{2}}$ and $\Delta^{-\frac{1}{2}}$, did not receive much attention. They regarded it as merely an adjusting term of the involution rather than as the vitally important object in the theory. It was Tomita who considered $\Delta^{\frac{1}{2}}$ as the absolute value of the involution of M in the Hilbert space structure. He discovered this fact as early as in 1959, [29], and raised various important questions. But no one followed up his very suggestive idea (due partly to the notational complexity of his paper). Finally, Tomita in 1967 succeeded in proving our Theorem 5.8 together with the commutation theorem for the tensor product of von Neumann algebras, [30,31]. The theory developed in sections 1-5 is exclusively due to him.

At the same time as Tomita's work appeared, very important input from quantum statistical mechanics was brought into the theory of standard representations of operator algebras by R. Haag, N.M. Hugenholtz and M. Winnink [10]. They showed that, given a C^*-algebra equipped with a one parameter automorphism group, the cyclic representation of that algebra induced by a state satisfying the Kubo-Martin-Schwinger condition is standard. Comparing Tomita's work and the results of Haag, Hugenholtz and Winnink, M. Takesaki found that the KMS-condition is the essence of standard representations, [25]. A technical background of Theorems 8.1 and 8.2 is found in the works of Dixmier [5] and Pukanszky [20].

Theorem 6.2 was proved independently by Takesaki [25] and Winnink [36]. The proof of it presented in this note was given to the author during the conference by Hugenholtz.

Conditional expectations in operator algebras were first discussed by M. Nakamura and T. Turumaru in 1954, [15], who abstracted the basic properties of the usual conditional expectations in probability theory. See properties (7.1) - (7.8) above. A similar object was discussed by Dixmier somewhat earlier, [6], Umegaki took over this subject in his succeeding papers [33] and discussed in detail conditional expectations for a von Neumann algebra with a finite trace. Tomiyama also has been working on this subject from a slightly different point of view [32,33].

The results in section 8 are mostly found in Takesaki's work, [25,26,27]. A part of Theorem 8.5, namely the existence of such an h in (b), was conjectured by J. Dixmier in his first edition of [4] p. 63 and proved later by S. Sakai, [21]. F. Perdrizet gave an example [19] in which the positive self-adjoint operator h in (8.5) of Theorem 8.5 is not unique while the ξ is unique. Theorem 8.11 is due to R. Herman and M. Takesaki, [11].

Of course, the theory presented here is not completed. There are various things that should be considered: for example, what is the dependence of the modular automorphism group on the faithful normal state determining it? Is the map: $\varphi \mapsto \sigma_t^{\varphi}$ continuous in any sense? Is there any relation between the modular automorphism groups and the algebraic type of von Neumann algebras? Etc, Etc, ... The author hopes these notes will help in the further development of the theory.

REFERENCES

[1] Ambrose, W., "The L^2-system of a Unimodular Group I", *Trans. Amer. Math. Soc.*, 65, 27-48, (1949).

[2] Combes, F., "Poids Associé á une Algébre Hilbertienne á Gauche", *Compositio Math.*, 23, 49-77, (1971).

[3] _______, "Poids et Espérances Conditionnelles dans les Algébres de von Neumann," *Bull. Soc. Math. France*, 99, 73-112, (1971).

[4] Dixmier, J., *Les Algébres d'Opérateurs dans l'Espace Hilbertien*, Gauthier-Villars, Paris 2é edition, (1969).

[5] _______, "Algébres Quasi-unitaires," *Comm. Math. Helv.*, 26, 275-322, (1952); see also its abstract, C.R. Acad. Sci., Paris, 233, 837-839, (1951).

[6] _______, "Formes Linéaires sur un Anneau d'Opérateurs," *Bull. Soc. Math. France*, 81, 9-39, (1953).

[7] Dunford, N., and Schwartz, J. T., *Linear Operators II*, Interscience Publications, New York, (1963).

[8] Dye, H. A., "The Radon-Nikodym Theorem for Finite Rings of Operators," *Trans. Amer. Math. Soc.*, 72, 243-280, (1952).

[9] Godement, R., "Théorie des Caractéres, I. Algébres unitaires," *Ann. Math.*, 59, 47-62, (1954).

[10] Haag, R., Hugenholtz, N., and Winnink, M., "On the Equilibrium States in Quantum Statistical Mechanics", *Comm. Math. Phys.*, 5, 215-236, (1967).

[11] Herman, R., and Takesaki, M., "States and Automorphism Groups of Operator Algebras", *Comm. Math. Phys.*, 19 142-160, (1970).

[12] Hille, E., and Phillios, R., "Functional Analysis and Semi-groups", *Amer. Math. Colloquium Publication*, 31, (1957).

[13] Hugenholtz, N., and Wieringa, J., "On Locally Normal States in Quantum Statistical Mechanics", *Comm. Math. Phys.*, 11, 183-197, (1969).

[14] Kastler, D., Pool, J., and Thue Poulsen, E., "Quasi-unitary Algebras attached to Temperature States in Statistical Mechanics - a comment on the work of Haag, Hugenholtz and Winnink," *Comm. Math. Phys.*, 12, 175-192, (1969).

[15] Murray, F. J., and von Neumann, J., "On Rings of Operators", *Ann. Math.*, 37, 116-229, (1936), II, *Trans. Amer. Math. Soc.*, 41, 208-248, (1937).

[16] Nakamura, M., and Turumaru, T., "Expectations in an Operator Algebra", *Tôhoku Math. J.*, 6, 182-188, (1954).

[17] Nakano, H., "Hilbert Algebras", *Tôhoku Math. J.*, 2, 4-23, (1950).

[18] de la Barriére, Pallu., "Algébres Unitaires et Espaces de Ambrose", *C.R. Acad. Sci.*, 233, 997-999, (1951).

[19] Perdrizet, F., "Elements Positif Relativement à une Algebre Hilbertienne à Gauche", to appear.

[20] Pukanszky, L., "On the Theory of Quasi-unitary Algebras", *Math. Szeged*, 16, 103-121, (1955).

[21] Sakai, S., "A Radon-Nikodym Theorem in W*-algebras", *Bull. Amer. Math. Soc.*, 71, 149-151, (1965).

[22] Segal, I.E., "A Non-commutative Extension of Abstract Integration", *Ann, Math.*, 57, 401-457, (1953).

[23] Sirugue,M., and Winnink, M., "Constraints Imposed Upon a State of a System that Satisfies the K.M.S. Boundary Condition", to appear.

[24] Takenouchi, O., "On the Maximal Hilbert Algebras", *Tôhoku Math. J.*, 3, 123-131, (1951).

[25] Takesaki, M., *Tomita's Theory of Modular Hilbert Algebras and its Appliactions*, (Lecture Notes), Springer, 128, (1970).

[26] _______, "Disjointness of the KMS-States of Different Temperatures:, *Comm. Math. Phys.*, 17, 33-41, (1970).

[27] _______, "Conditional Expectations in von Neumann Algebras,"to appear in *J. Functional Analysis*.

[28] _______, *The Theory of Operator Algebras*, (Lecture Notes), UCLA, (1969/1970).

[29] Tomita, M., "Spectral Theory of Operator Algebras I", *Math. J. Okayama Univ.*, 9, 63-98, (1959).

[30] _______, "Quasi-standard von Neumann Algebras", (Mimeographed note). (1967).

[31] _______, "Standard Forms of von Neumann Algebras", *The Vth Functional Analysis Symposium of the Math. Soc. of Japan, Sendai.*, (1967).

[32] Tomiyama, J., "On the Projection of Norm One in W*-algebras", *Proc. Japan Acad.*, 33, 608-612, (1957).

[33] _______, *Tensor Products and Projections of Norm one in von Neumann Algebras*, (Lecture Notes), University of Copenhagen, (1970/71).

[34] Umegaki, H., "Conditional Expectation in Operator Algebras I", *Tôhoku Math. J.*, 6, 177-181, (1954); II, *Tôhoku Math. J.*, 8, 86-100, (1956); III, *Kōdai Math. Sem. Rep.*, 11, 51-64, (1959); IV, *Kōdai Math. Sem. Rep.*, 14, 59-85, (1962).

[35] Winnink, M., "An Application of C*-algebras to Quantum Statistical Mechanics of Systems in Equilibrium", *Thesis*, University of Groningen, (1968).

[36] Winnink, M., "Algebraic Aspects of the Kubo-Matin-Schwinger Condition", *Cargèse Lecture in Phys.*, Volume 4, 235-255, Gordon and Beach Publishers, New York, (1969).

ATTENDEES
1971 RENCONTRES

Battelle Seattle Research Center
Seattle, Washington

Dr. Alain Connes
Dr. Bernard H. Duane
Dr. Jean-Peirre Eckmann
Dr. William Faris
Dr. Giovanni Gallavotti
Dr. Alan G. Gibbs
Prof. Jean Ginibre
Dr. M. Lawrence Glasser
Dr. William Greenber
Prof. Robert B. Griffiths
Dr. Christian Gruber
Dr. F.A. Grünbaum
Dr. Rudolf Haag
Dr. Ole Heilmann
Dr. Richard Holley
Prof. N. M. Hugenholtz
Prof. Richard. V. Kadison
Dr. Daniel A. Kastler
Dr. Oscar E. Lanford, III
Prof. Joel Lebowitz
Prof. Andrew Lenard
Prof. Elliot Leib
Dr. Anders Martin-Löf
Dr. Frederick J. Milford
Dr. Salvador Miracle-Solé
Prof. David P. Ruelle
Dr. M. Beth Ruskai
Dr. Seymour Sherman
Prof. Masamichi Takesaki
Dr. Martin Walter
Dr. Wilbert Wils

Lecture Notes in Physics

Bisher erschienen / Already published

Vol. 1: J. C. Erdmann, Wärmeleitung in Kristallen, theoretische Grundlagen und fortgeschrittene experimentelle Methoden. 1969. DM 20,–

Vol. 2: K. Hepp, Théorie de la renormalisation. 1969. DM 18,–

Vol. 3: A. Martin, Scattering Theory: Unitarity, Analyticity and Crossing. 1969. DM 16,–

Vol. 4: G. Ludwig, Deutung des Begriffs physikalische Theorie und axiomatische Grundlegung der Hilbertraumstruktur der Quantenmechanik durch Hauptsätze des Messens. 1970. DM 28,–

Vol. 5: M. Schaaf, The Reduction of the Product of Two Irreducible Unitary Representations of the Proper Orthochronous Quantummechanical Poincaré Group. 1970. DM 16,–

Vol. 6: Group Representations in Mathematics and Physics. Edited by V. Bargmann. 1970. DM 24,–

Vol. 7: R. Balescu, J. L. Lebowitz, I. Prigogine, P. Résibois, Z. W. Salsburg, Lectures in Statistical Physics. 1971. DM 18,–

Vol. 8: Proceedings of the Second International Conference on Numerical Methods in Fluid Dynamics. Edited by M. Holt. 1971. DM 28,–

Vol. 9: D. W. Robinson, The Thermodynamic Pressure in Quantum Statistical Mechanics. 1971. DM 16,–

Vol. 10: J. M. Stewart, Non-Equilibrium Relativistic Kinetic Theory. 1971. DM 16,–

Vol. 11: O. Steinmann, Perturbation Expansions in Axiomatic Field Theory. 1971. DM 16,–

Vol. 12: Statistical Models and Turbulence. Edited by M. Rosenblatt and C. Van Atta. 1972. DM 28,–

Vol. 13: M. Ryan, Hamiltonian Cosmology. 1972. DM 18,–

Vol. 14: Methods of Local and Global Differential Geometry in General Relativity. Edited by D. Farnsworth, J. Fink, J. Porter and A. Thompson. 1972. DM 18,–

Vol. 15: M. Fierz, Vorlesungen zur Entwicklungsgeschichte der Mechanik. 1972. DM 16,–

Vol. 16: H.-O. Georgii, Phasenübergang 1. Art bei Gittergasmodellen. 1972. DM 18,–

Vol. 17: Strong Interaction Physics. Edited by W. Rühl and A. Vancura. 1973. DM 28,–

Vol. 18: Proceedings of the Third International Conference on Numerical Methods in Fluid Mechanics, Vol. I. Edited by H. Cabannes and R. Temam. 1973. DM 18,–

Vol. 19: Proceedings of the Third International Conference on Numerical Methods in Fluid Mechanics, Vol. II. Edited by H. Cabannes and R. Temam. 1973. DM 26,–

Vol. 20: Statistical Mechanics and Mathematical Problems. Edited by A. Lenard. 1973. DM 22,–

Vol. 21: Optimization and Stability Problems in Continuum Mechanics. Edited by P. K. C. Wang. 1973. DM 16,–